Animal Space Use, Second Edition

Animal Space Use, Second Edition:

Memory Effects, Scaling Complexity and Biophysical Model Coherence

By

Arild O. Gautestad

Cambridge Scholars Publishing

Animal Space Use, Second Edition:
Memory Effects, Scaling Complexity and Biophysical Model Coherence

By Arild O. Gautestad

This book first published 2021

Cambridge Scholars Publishing

Lady Stephenson Library, Newcastle upon Tyne, NE6 2PA, UK

British Library Cataloguing in Publication Data
A catalogue record for this book is available from the British Library

ISBN (10): 1-5275-7206-4
ISBN (13): 978-1-5275-7206-5

TABLE OF CONTENTS

PREFACE

A preface normally presents a brief background for a book's content. The present outline is more extensive. Given the theoretical challenge with respect to making sense of animal space use in a realistic and coherent manner, I have chosen to enlarge this section in two directions. First, some basic memory- and scale-related concepts are described. Second, some main points from each chapter are outlined and given some introductory comments. Hopefully, these enlargements of the preface will provide a valuable theoretical preparation to the main body of the book. It should in fact be read as an introduction to the introduction.

Animal space use may be conceptualized and modelled at two levels, individual and population. At the individual level movement, ecology is a field in rapid theoretical progress (Nathan *et al.* 2008, Smouse *et al.* 2010, Viswanathan *et al.* 2011, Lewis *et al.* 2013, Humphries and Sims 2014, Méndez *et al.* 2014). Population ecological theory has also matured; with emphasis on spatially explicit model designs, age-structured dynamics, and multi-species interactions (Hiebeler 1997, Keeling 1999, Turchin 2003, Hanski and Gaggiotti 2004, Ranta *et al.* 2006). However, the main theme for the present book is not ecological inference, but rather a critical evaluation of the biophysical framework implicitly assumed when making ecological inference. "Biophysical" does in this context specify a meso-scale observational level of movement and space use.

For example, when studying individual space use, this level is reflected in a sample of GPS fixes where behavioural modes and movement-influencing events are hidden at finer temporal resolutions than the sampled path. The temporal scale interval from the fine-resolved movement path to the sampled path (leading to a set of relocation dots on the map, rather than a continuous line) is referred to as "the hidden layer" in this book. At the population level the hidden layer is best reflected by the spatial resolution of the study. This resolution determines local population density; number of individuals per spatial unit at this resolution; and at a chosen temporal resolution (a day, week, or year, depending on context). Again, the actual biological events and interactions like individual searching, feeding, courting, resting, and a myriad of other aspects are spatially and temporally fine-grained processes being executed by the population's constituents at micro-scale below the resolution for the study; i.e., below the hidden layer.

Whether we consider local density of GPS fixes from a given individual as accumulated over a sampling period or local population density representing a snapshot of individuals' whereabouts, we have changed focus away from direct observation of biological processes at micro-scales. Instead we perform indirect observation of these processes at a coarser resolution; the biophysical system representation. The hidden layer, which has both a temporal, a spatial, and a scale range dimension, is a core concept under the extended framework for analysis of animal space use to be presented in this book. Observe that I define scale range as an independent aspect of time and space. I define this dimension as a scale range, not a specific scale. This dimension reflects the depth of the hidden layer. In process terms the deeper the layer the clearer the difference between the fine-grained biological scales and the coarse-grained biophysical scales.

Such system coarse-graining is common in ecological research both at the individual and the population level, but the transition from a biological to a biophysical system abstraction is only rarely explicitly acknowledged. This is unfortunate, for three main reasons. First, the biophysical approach opens the door towards a huge theoretical framework, "statistical mechanics," which few biologists are trained in or even consider relevant. Second, statistical mechanics in its present state is—in my view—unfortunately not sufficiently extensive to cover the kind of biophysics generally applicable to animal space use. Third, this discrepancy has led to implicit application in ecological theory of principles from classical statistical mechanics—as a consequence of invoking the hidden layer for system modelling—despite the intrinsic issue that the current framework is based on assumptions that are frequently violated by animal space use behaviour. This book circulates among these three aspects and in particular the third one. Hence, proposals for extensions of the traditional biophysical framework percolate the chapters, for the sake of—hopefully—stepping towards a more realistic biophysical representation of animal space use.

Scrutinizing the foundation of ecological models based on the hidden layer is not a walk in the park. It implies that movement ecology's standard toolbox (containing model designs and statistical protocols, see Box 1 towards the end of the preface) is rattled with respect to applicability of very general assumptions (Box 2). A critical evaluation of the foundation of individual movement models by necessity percolates to the population level, where I put a corresponding set of core assumptions under scrutiny. At respective levels—individual space use behaviour and population dynamics—the specific assumptions I refer to are in fact so deeply buried under layers of historic consensus that they are rarely explicitly mentioned in ecological analyses that depend on their support (Gautestad and Mysterud 2005, Gautestad 2011, 2013b). For some, this consensus may simply reflect "case solved"; others are not so confident. Here are examples of opposing views:

> *It seems to me to be beyond doubt that population ecology has a mature theory.*
> Turchin (2003), p. 392

> *It is time to acknowledge the need for a scathing indictment of the poor practices that have accumulated in species distribution modelling over the decades.*
> O'Connor (2002), p. 33

Does the book's focus on modelling theory and accompanying assumptions make it peripheral to a general audience of animal ecologists? Definitely not! Ecology regards the branch of biology that deals with the relations of organisms to one another and to their physical surroundings. What regards animal space use and population dynamics, the basic level of ecological inference regards exploration of (a) why intensity of habitat use is stronger in some places than other places (and/or other times), (b) how this insight may produce predictions about space use intensity at other locations, and (c) what kind of statistical consistency may be expected when changing observational scale (spatial or temporal resolution; changing the depth of the hidden layer). From this broad perspective the present book drills to the core of animal ecology by proposing both revisions and enlargements of the generally applied biophysical space-use framework, which currently provides conceptual and predictive models for ecological inference.

Any ecologist with field experience will agree that animal space use is more challenging than appears in textbooks. From a trivial perspective this may be because of insufficient data or too simplistic models. Even larger samples of GPS relocations, improved GIS data, and more fine-grained population monitoring should then suffice to resolve the challenges at hand. The model might also gain from additional terms to cover the system in more detail. Such a quest for improved system details may also need to be supplemented by considering the given model foundation at a deeper theoretical level. I advocate that an explicit application of the biophysics of space use that is most relevant for the given system should determine the most realistic set of models for ecological inference. In this book I present eight such conditions, called biophysical universality classes, for animal movement and space use. In statistical mechanics, universality is the observation that there are properties for a large class of systems that are independent of the dynamical details of the system. Unfortunately, at present only one of these classes provides the foundation for most of ecological theory, and two others are only now slowly making a broader entry. The other five classes are probably totally unfamiliar to you. In my view, the broadly applied class of movement is stretched to the extent that the model assumptions become unrealistic for many scenarios. Old challenges and even paradoxes that haunt the present arena may thus find better solutions in the extended system containing a broader range of universality classes. Hopefully you will agree after reading the following chapter that the eight classes present great opportunities for a more diversified arena for realistic model representation of animal space use both at the individual and the population level.

In a concrete example of the potential shortcoming of the dominating class (yes, I will define it in more detail shortly), consider space-use intensity. What is space-use intensity? According to the dominating system class, it may be represented by local variability of *density*, either the spatial dispersion of GPS relocations (individual level) or the dispersion of individuals themselves (population level). However, an individual or a population may express higher space-use intensity in some parts of available space even when the habitat characteristics seem similar to less utilized areas. There may be many reasons for this apparently paradoxical pattern, but in this book I'm focusing on some intrinsic drivers. In particular, by diversifying between eight biophysical university classes for movement and space use, I bring you into the arena of process complexity. Complexity is qualitatively different from complicatedness (for example, for adding more terms to a model). Basically, space-use complexity may boil down to explicitly considering the "scaring" properties (from a system modeller's perspective) of site fidelity and philopatric behaviour on the individual level and intra-specific population cohesion from conspecific attraction at the population level. In other words, it regards the behavioural tendency to constrain space use despite living in an open environment.

The common denominator of complexity seems to be influence from temporal and spatial memory influence on space use. Why are memory effects so scaring from a theoretical point of view? Because the structuring force that emerges from this important biological capacity involves positive (self-reinforcing) feedback loops, which is mathematically difficult to handle and formulate in models under the dominant biophysical universality class. The "balance of nature" idea, which basically rests on negative rather than positive feedback loops, becomes further undermined in the context of animal space use, lending support to critics of this dogma (Andrewartha and Birch 1954, Nicholson 1954, Pimm 1991, den Boer and Reddingius 1996) but from another and perhaps surprising theoretical angle. Consideration of spatial memory effects as an important intrinsic driver for space-use complexity opens for a new direction of system conceptualization and process modelling. In other words, a wider range of biophysical universality classes is needed to make complexity less scaring and, in fact, both theoretically and *empirically* fascinating! The broader system offers great opportunities for ecological inference.

A system abstraction is represented by models, whether these regard equations for static relationships (like statistical correlations) or dynamical simulation algorithms. If some core assumptions for the actual model are violated by the dynamics of a specific individual or population it is meant to represent, one may easily run into a dead end: even more data will not produce substantially improved results. On the contrary, it is common in animal space-use research to find that more data tend to blur results that emerged from shorter series. From such frustrating experience one should be open-minded to explore alternative paths—a wider range of universality classes. If such an extended toolbox provides theory for better compliance with the actual organism at the given place and time, one should retest the actual hypothesis under more context-realistic assumptions. One may then expect improved clarity with respect to ecological inference from a given data set. Consequently, the study of the relations of organisms to one another and to their physical surroundings has two main directions for improvement: extended data and extended theoretical foundation with respect to universality classes. Acknowledging this trivial duality will bring animal ecology closer to hard science, like we envy physicists who achieved this level of theoretical maturity a long time ago.

This book presents a general process approach (Box 3) towards understanding and modelling individual movement and population dynamics, in quite cotton-less wrapping in some sections, I must admit (a benefit that presents itself from writing a book rather than a journal paper). The reason to drill to the core of the biophysical aspect of animal ecology is a pile of quite generally verified empirical patterns which seem to defy classical system representations. Despite a growing menagerie of sophisticated statistical analyses under what I will define as the classical toolbox compartment (under the dominating universality class), interpretation and dynamic modelling of some of these aspects seem to ignite both confusion and controversies, in particular in movement ecology, where studies involving additional universality classes are now emerging (Benhamou 2007, 2008, Reynolds 2008, Edwards *et al.* 2012, Gautestad 2012, Humphries *et al.* 2012, Sims and Humphries 2012, Boyer and Solis-Salas 2014). The distance to consensus still seems to lie well beyond the horizon. In fact, while the horizon for some looked close a few years ago, others may—from the ongoing theoretical discussions in movement ecology—advocate that a general agreement on the sticking issues is sliding further away owing to the multitude of complex aspects that continue to expand (Fagan and Calabrese 2014, Fleming *et al.* 2014). In this book I explore whether an explicit definition and exploration of eight universality classes—in tandem with an explicit consideration of the hidden layer—may contribute to resolving some paradoxes. Exciting times, indeed!

I'm confident that the present book provides food for thought, whether you are a theoretician, an empiricist in a field of applied ecology, or somewhere in between. I suspect that if you are working in wildlife management but are also keeping an eye on developments in theoretical ecology, you may get the largest upshot. Personal experience with the study objects may already have led you towards a similarly critical attitude with respect to some aspects of mainstream theoretical models. Perhaps you will be surprised by some of the statistical model assumptions under present scrutiny, and you may even immediately agree on the critique and/or ideas for an extended model framework that is outlined. Alternatively, you may find the ideas provocative but sufficiently interesting to pilot test some of them on your own data. Whether you find empirical support for the alternative system or not, your results will be important for further clarification of the proposed directions. In order to determine success or failure, the present ideas and directions do of course require the same process of scientific scrutiny from peers as I'm performing here on the standard model designs. This book thus provides a combination of paradigm critique, presentation of alternative approaches, and respective outlines for how these extensions may be tested empirically.

Many of the nagging issues in focus here are not new. Here are some quotes from twenty- to thirty-year-old publications:

... when they are nectar-feeding, the flight patterns of butterflies cannot be simply described as correlated random walks. We do not know what complications might be involved in nectar-feeding that violate the simplifying assumptions ...
Kareiva and Shigesada (1983)

It has proven extremely difficult to model how animals interact with the habitat dynamics through processes such as habitat selection, foraging site choice, and juvenile dispersal. The basic problem is that animals interact with the habitat at several spatial and temporal scales, using behavioral cues and experience (memory maps) in complex and poorly understood ways.
Walters (1992)

... real-life movement distributions may have complex forms that are not amenable to explanations using simple models.
Porter and Dooley (1993)

Exactly how the translation from individual movements to population distribution and interactions should be accomplished is one of the most vexing problems confronting a metapopulation-landscape synthesis.
Wiens (1997)

Where do these old challenges stand today? The complexity revealed by analysis of individual space-use data—lately mainly based on GPS-based relocation series—may be summarized by two properties; on one hand the quite broad verification of multi-scaled movement (Viswanathan *et al.* 2011, Humphries *et al.* 2012, Sims and Humphries 2012), and on the other hand a general recognition that individual memory map utilization by vertebrates and perhaps some branches of invertebrates is the rule rather than the exception wolf(Frank 1964, Morris 1981, Collett *et al.* 1986, Garber 1989, Rodriguez *et al.* 1994, Bailey 1995, Ostfeld and Manson 1996, Sherry 1996, Mendl *et al.* 1997, Beekman and Ratnieks 2000, Burt de Perera 2004, Boyer *et al.* 2006, Merkle and Wehner 2009, Wolf *et al.* 2009, Levett and Walls 2011, Pokorny *et al.* 2014). Both aspects, memory and scaling, may turn out to represent two sides of the same coin, as will be explained later in this book. The concepts of single-scaled, multi-scaled, scale-free, and memory-influenced movement are defined in Box 4. In fact, this scaling/memory duality hypothesis is, to my knowledge, brand-new and is presented here in explicit detail for the first time.

To account for the observed process complexity from scaling and memory, various directions of research have emerged. The dominating approach in movement ecology is to apply mechanistic model designs with increased level of mathematical and computational sophistication (van Moorter *et al.* 2009, Song *et al.* 2010, Giuggioli and Bartumeus 2011, Moorcroft 2012, Reynolds 2012, Spencer 2012, Fronhofer *et al.* 2013, Lewis *et al.* 2013, Nabe-Nielsen *et al.* 2013, Benhamou 2014, Humphries and Sims 2014, Méndez *et al.* 2014, Merkle *et al.* 2014). These approaches are generally constructed under the umbrella of classic model assumptions, referred to as the standard group of universality classes in the biophysical framework. They have contributed not only to greatly improved model realism by explicitly considering memory and scaling effects, but they have also led to a broader realisation of the system complexity—and theoretical challenges—at hand. Consequently, an extremely dynamic melting pot for biologists, physicists, and statisticians has now emerged.

However, this dynamic field of research is still for the most part contained within the walls of the theoretical pool provided by classical biophysics. In this book I explore an alternative direction to "stretching the string" under mechanistic modelling principles when approaching the complexity of memory effects. I describe concepts and models where I invoke elements from an extended framework, based on *non*-mechanistic system assumptions for some of the universality classes. In this respect, the term "non-mechanistic" in the context of system dynamics is theoretically challenging (and provocative!) and explicitly elaborated on here for the first time: I argue for the need to distinguish between two types of dynamics: mechanistic and non-mechanistic. Non-mechanistic dynamics? Strange, indeed! However—as will be described in this book—a potentially problem-solving concept with respect to animal space use. Non-mechanistic dynamics require—according to the arguments and accompanying support—model formulations based on statistical-mechanical principles (Box 5), which in this system representation (and opposed to classical statistical mechanics) explicitly rests on premises of scale-free and memory-dependent space use.

The term "statistical-mechanical" may sound strange and perhaps even outright scary from a biologist's perspective, invoking concepts from thermodynamics like mathematical- and probability-theoretical heavy concepts. However, rest assured that the present book is relaxingly simple in this respect. I apply some core principles from statistical mechanics to design the extended theoretical framework, but the equations are in the present context very simple. The challenge is primarily at the conceptual and cognitive level: how to imagine the difference between the mechanistic and the non-mechanistic modelling approach, how to define and formulate the difference between micro- and meso-scale process properties in this respect, and so on. This is where the hidden layer, already referred to, becomes a key concept.

I argue that observationally coarse-graining the process under scrutiny and thus invoking the hidden layer may provide a necessary operation to make sense of some central parts of a complex system's dynamic and statistical properties, in particular with respect to its emergent properties (to be defined in other chapters). In other words, where spatial memory and scaling are involved, the hidden layer approach (and thus a statistical-mechanical system representation) may be necessary for improved predictive power of ecological models of animal space use. The following chapters basically regard exploration of this obviously controversial working hypothesis from many angles, both from the individual and the population level of system abstraction. With this biophysical foundation one should be better equipped to perform studies involving ecological inference.

While mechanistic model designs regard micro-scale behaviour like detailed movement rules and local rates of change at fine temporal resolutions, a statistical-mechanical description describes the system states and dynamics at meso- and macro-scale resolutions (coarse-graining; see Box 4 and Box 5). In a nutshell, the distinction between the stretching-the-string approach (enhanced mechanistic modelling) and the alternative path outlined here regards the toolbox-extension referred to above, where the traditional compartment embeds the mechanistic framework and its complementary classical statistical-mechanical representation. The other compartment carries the novel label "non-mechanistic," and contains the (extended) statistical-mechanical system representation only. This implies that—at least at this stage—I have not been able to formulate a logically and mathematically coherent dynamic representation of micro-scale non-mechanistic animal space use that fills the role of mechanistic models *below* the hidden layer in the classical framework. My present working hypothesis is that it does not exist. In other words, non-mechanistic dynamics may actually *require* conceptualization as a meso-scale (statistical-mechanical) process, as will be described from many angles in this book.

"Non-mechanistic" does not—in this context—represent "static." Non-mechanistic and mechanistic models represent different subsets of universality classes for dynamic processes. As you will learn in various chapters of the book, mechanistic models will appear as the limiting case where non-mechanistic dynamics are squeezed towards a mechanistic system, conceptually similar to a non-linear equation tuned towards a linear one. Thus, non-linearity is the general case, while linearity represents the exception (albeit more frequently applied, owing to simpler mathematics). These concepts will become clearer over the following chapters. The goal for both toolbox compartments—mechanistic and non-mechanistic—is of course improved theoretical coherence between movement theory and real data, for the sake of improved protocols for ecological studies involving both animal movement and population dynamics. However, owing to a broadened toolbox for system representations and ecological analyses, new pathways for analysis of complex systems become available. An additional string to stretch becomes available.

I describe how the alternative model premises may be conceptually and mathematically formulated to allow the respective hypotheses to be statistically tested against the traditional model premises at a very basic level of system design. In other words, does the contrarian set of ideas and hypotheses presented here make sense? Do they provide advantages when confronted with real data and applied in ecological studies? Some of the proposed tests are summaries of already published methods; others are described here for the first time. A potential for actual applications of these ideas and model constructions in follow-up ecological research is outlined. I also highlight some consequences of ignoring a closer scrutiny of the actual premises in contemporary theory on movement and population ecology. In particular, a statistical and/or dynamic modelling framework that depends on compliance with a memory-less process—while the actual dynamics do not comply with this premise—will both inflate the error term in the analyses, leave it open for misinterpretation, and even create paradoxical results.

All these aspects are exemplified in many contexts throughout the book. For example, in the last chapter; *11 Implications and applications*, I argue that the empirically verified knowledge that animals are able to relate and respond to environmental conditions over a range of spatio-temporal scales (the concept of multi-scaled movement) seems to have become theoretically derailed. The mechanistic class of individual-based models and accompanying mathematical formulations that dominate this field depend on the study object actually ignoring its memory capacity to execute goals at several spatio-temporal scales in parallel—perhaps the strongest source of information-sorting the animal has when relating to its environment beyond its current perceptual field!

Attacking the memory and scaling aspects from the present book's perspective is a formidable theoretical and perceptual challenge, despite relatively simple mathematical and statistical-mechanical formulations at this stage. This volume is only a toe-dipping into what may emerge as a deep and wide ocean for future research—or conversely, the approaches may in due course turn out to be a futile attempt to penetrate the complex issues at hand. One should always seek to balance optimism and confidence with a self-critical attitude. However, given the fact that I finally dare to synthesize the extended individual/population level theory into a book format reflects that a certain level of confidence has been reached. Each chapter could indeed be the theme for separate books. I hope that the collection of ideas, concepts, mathematical formulations, simulations, and empirical tests should at this stage appear sufficiently mature and empirically supported by pilot tests to be of interest for a broad audience. However, be warned that some of the elements of the proposed framework extensions are still highly speculative, and others are at this stage only supported by a limited range of empirical pilot tests. Thus, in these sections I'm rattling a wasp nest with limited protective gear.

The presentations should not be considered reviews of the respective aspects; the topics are too diverse and wide in scope to allow for this. As a background for the evolution of general movement models towards improved realism, the recent review by Smouse et al. (2010) is a good choice, since it covers—in an introductory manner—both the memory and the scaling challenges. Additionally, as an introduction to the physical aspects of scale-free movement, I recommend the review by Zaburdaev et al. (2014).

The multi-scale space use aspect of animal ecology is of particular importance in the present context. The recent critics of basic models in this field have been raised by and large in two disparate theoretical environments. On one hand a growing group of scientists are studying the property of scale-free movement (Box 4), which recently has been empirically verified in large GPS data sets from a broad range of species (Sims *et al.* 2008, Humphries *et al.* 2010, Humphries *et al.* 2012, Sims *et al.* 2012). On the other hand, a consensus is now emerging where it is acknowledged that spatial memory effect on movement and habitat choice has received insufficient attention by ecologists (Frank 1964, Börger *et al.* 2008, van Moorter *et al.* 2009, Levett and Walls 2011, Nabe-Nielsen *et al.* 2013, Pokorny *et al.* 2014). The first group generally ignores aspects related to spatial memory (site fidelity from cognitive map utilization). The other group generally ignores the scale-free aspect of movement. An integrative approach is thus clearly needed, as already indicated in Box 5. Such an integrative approach would fit nicely into classic ecological concepts like top-down and bottom-up space-use strategies [for a recent overview and model formulation, see Vergara *et al.* (2015)].

To summarize, in the present book I support the view that the classical toolbox for models in the respective fields of scale-free movement and memory-utilizing space use may be insufficient to mimic many animals' behaviour (Gautestad and Mysterud 2005, 2010b, Song *et al.* 2010, Boyer *et al.* 2012). Consequently, one should—from this premise—build an extended theory brick by brick, starting by defining testable hypotheses related to universal or generic system descriptions. In this book I advocate that even a redesign of the biophysical frame of reference may be needed to understand systems showing intrinsic process complexity from the combined effects from scaling and memory. This regards what I called non-mechanistic system dynamics above, which—without this redesign and framework extension—will present itself as a contradiction in terms! By defining eight universality classes of animal space use and their respective transitions, this statistical-mechanical foundation hopefully provides a constructive starting point for deeper theoretical spin-offs. The ultimate goal is—from my personal background—a more coherent modelling toolbox for ecological inference both at the individual and the population level of analysis. However, by challenging the mechanistic paradigm in the context of modelling memory-influenced space use—and additionally proposing an alternative kind of dynamical framework for this condition—I'm definitely rattling the wasp nest.

In Chapter 0, *Introduction: the broad picture*, I elaborate on the general themes raised above, with weight on the individual/population duality from the perspective of complex system dynamics. The chapters following this introduction will then start out with a detailed practical illustration of studies on individual space use and the underlying system assumptions, continuing in the direction of stronger reference to population level dynamics in general and between the two levels in particular.

Chapter 1, *Large mystery in a little detail*, gives a brief overview of how the individual-level topics of this book got their initial kick-start with respect to biophysics of space use. As in all research, locking in on a new direction often starts with a frustration. In the present case it was the contradiction that observed space use by some animals on one hand seemed to be spatially stationary while on the other hand expanded non-trivially with observation intensity (sample of spatial relocations, N).

Observed space use not only expanded non-asymptotically and thus violated expectation, but in accordance to a power law and with a power exponent that was consistent between data sets from various species. In Chapter 1, I provide a walk-through of the stepwise developments towards a solution of the paradox after raising and falsifying the most obvious explanations based on classical system assumptions.

It should take more than twenty years before an explicit adjustment for sample size effect according to the "home range ghost" formula (presented in Chapter 1) is applied by others as a premise for ecological analysis of relocation data (Morellet *et al.* 2013, Campos *et al.* 2014). Finally, other researchers in the field of complex movement are also beginning to publish what seems to support home-range ghost-compliant space-use patterns, with accompanying model proposals that resemble some of the ideas summarized and extended in this book (Song *et al.* 2010, Boyer *et al.* 2012, Boyer and Solis-Salas 2014, Fleming *et al.* 2014).

More explicitly, in a recent paper Merkle et al. (2014) tested a set of contemporary hypotheses on memory-influenced movement in free-ranging bison, *Bison bison*, and found support for a central premise of the present theory (the MRW model) in the summer ranges of bison movement. Multi-scaled random walk (MRW) is our approach towards statistical-mechanical simulation of the combination of scale-free and memory-influenced space use. Returns of bison to familiar patches were quite independent of physical distance to these patches and time since last visit, implying that even relatively distant resource experiences were appreciated despite energy costs to reach these locations. Hence, spatial memory utilization seemed to be an important component of this species' optimal foraging strategy and connection to familiar sites-influenced movement. In another study, foraging by Japanese Black cows *Bos taurus*, the animals discriminated preferred patches from others before they visited the patches to eat (Hirata *et al.* 2014), in compliance with previous findings (Gross *et al.* 1995) but with more explicit revelation of patch utilization as a two-step process; choice of where to move and decision when to leave. In other words, they utilized spatial memory to return non-randomly to previously rewarding patches. These results also showed that time since last visit was subordinate, given that the actual patches had been sufficiently rewarding in the past.

I find the results by Merkle et al. and Hirata et al. both encouraging and inspirational since they lend support to a basic premise of the generic MRW model (ignorance of time and distance with respect to return targets, but both properties may be adjusted slider-like), in addition to supporting memory map utilization. However, it remains to be seen if these aspects of grazing behaviour are generally applicable beyond ungulate species. Theoretical developments that resemble the MRW approach are now also taking place from a mathematical angle; with recent papers from Boyer et al. (Boyer and Romo-Cruz 2014, Boyer and Solis-Salas 2014) as prime examples.

In Chapter 1, I suggest a feasible reason—backed up by recent third party research—why the home-range ghost paradox (Gautestad and Mysterud 1995) still has not been more generally recognized among wildlife ecologists, more than twenty years since its presentation and subsequent empirical support. The most widely acknowledged method to demarcate space use from a set of relocations—the Kernel Density Estimate (KDE) (Worton 1989)—has an Achilles' heel, as verified by relatively recent simulations and application of KDE on real data. KDE produces isopleths; contour lines representing expected equal-sized intensity of space use given the actual sample of relocations (fixes), N. If the dynamics behind the pattern do not comply with standard statistical assumptions, the demarcated area from KDE will overcompensate for the effect from change in N (Belant and Follmann 2002, Barg *et al.* 2005, Fieberg 2007, Fieberg and Börger 2012, Schuler *et al.* 2014). In other words, the home-range ghost paradox from non-asymptotic space-use demarcation with increasing N (based on a non-parametric protocol for area demarcations) may be camouflaged by violation of a specific statistical assumption under the KDE protocol. Crucially, this hypothesis is supported by applying simple, non-parametric demarcation methods described in Chapter 1.

In Chapter 2, *Basics of movement physics: the standard models*, I turn more explicitly towards the biophysics of animal movement. This is important background for what follows. Starting with very basic principles and illustrations (and a few simple equations), I describe the current paradigm for movement dynamics and the transitions between (a) deterministic-mechanistic models, (b) stochastic-mechanistic models, and (c) a mixture of the two as representation of animal space use. In particular, I describe how the common denominator—the mechanistic approach—depends on a shared assumption with respect to the dynamics: Markov compliance. In statistical-mechanical terms a Markovian process reflects a system close to local equilibrium. An animal makes a move based on a superposition of local movement-influencing factors at that point in time. These factors embed both environmental influence (for example, attraction and repellence) and internal state (for example, hungry or frightened). At successive moments previous factors are already reflected in the animal's current location, and new influences are leading to new moves. In absence of further change in these conditions the animal is expected either to stay put or to unidirectionally continue to move forward with a constant speed.

In simple systems, like a bouncing billiard ball, a deterministic-mechanistic model reflects the actual process well; principles from differential and integral calculus may be applied to describe it realistically. In other systems like a moving individual where motion is influenced by a myriad of factors from one point in time to the next, the deterministic-mechanistic approach is typically replaced by a stochastic-mechanistic model. Creating a large equation with many terms for the respective influences (in order to keep the model deterministic-mechanistic) will in this case be counterproductive. The archetypical stochastic-mechanistic function is a Brownian motion model. Movement satisfying a Brownian motion process leads to classical (so-called Fickian) diffusion at the population level and, consequently, provides the necessary support for application of classical calculus to model the dynamics of both individual space use and population dispersion. Again—as for the deterministic system description—a stochastic-mechanistic model may provide realism and thus high predictive power, given that the Markov premise is satisfied.

Markov compliance is a very convenient assumption both from a modeller's perspective and with respect to statistical analysis of empirical data. Accepting the Markov premise *a priori* allows for application of a mature and broad toolbox of classical and contemporary mathematics and statistical methods. This approach rests on the theory of standard statistical mechanics. Fine-grained Markov-based theory, represented by mechanistic models and "standard" statistics, provides the backbone of movement ecology in particular and idealization of animal space use in general both at the individual and the population level.

What if Markov is not satisfied? The answer seems simple, but difficult to digest: remove differential and integral calculus from population dynamical theory, and little remains. Remove Brownian motion (and variants like correlated random walk with or without environmental bias) from movement ecology and little is left there as well. These broad theoretical foundations rest on the Markov assumption as satisfied for the underlying process. Consequently, the core question raised in this book is this: may a Markovian biophysical framework be extended to account for spatial memory and multi-scaled dynamics in a realistic manner? There are currently many theoreticians in the Yes camp, where sophisticated mechanistic models are now flourishing. This book—on the other hand—represents the (non-crowded) No camp, where a conditional "No" is raised as a testable working hypothesis by scrutinizing the statistical-mechanical premises on which the Yes camp depend. The No answer is conditional, since I advocate that a distinction between mechanistic and non-mechanistic dynamics is needed. There are obviously a large number of systems and a range of observational scales where the Markovian framework not only suffices but is the correct choice. However, in my view there are also a large number of systems for which the Markovian framework should be considered unrealistic. Thus, the answer Yes should not be accepted *a priori*, but should be verified empirically as a realistic premise for the system in question.

Alternatives to the Markovian methods in general terms are the theme of Chapter 3, *Approaching memory*, and Chapter 4, *Movement physics beyond Markov compliance*. Here the Markov assumption in the classical biophysics of animal movement is scrutinized theoretically. As already mentioned, the rapid growth of the field of movement ecology has been sparked by two main directions of research; on one hand a recognition of scale-free movement and on the other hand a more sophisticated modelling of spatial memory influence. Designing dynamic models where memory-influenced space use (site fidelity) is explicitly considered are now popping up and growing in sophistication (Gautestad and Mysterud 2010b, Song *et al.* 2010, Boyer *et al.* 2012, Gautestad and Mysterud 2013, Nabe-Nielsen *et al.* 2013).

Interesting models have begun also to invoke non-Markovian mathematics in this respect: continuous time random walk (CTRW) to mimic scale-free movement and biased random walk to mimic scale-specific but memory-influenced movement (van Moorter *et al.* 2009, Bartumeus *et al.* 2010, Giuggioli and Bartumeus 2011, Spencer 2012, Nabe-Nielsen *et al.* 2013). Implementation of temporal (but not spatially explicit) memory effects on movement by applying the CTRW approach (Giuggioli *et al.* 2006, Bartumeus *et al.* 2010) may seem to contradict my claim above that a mechanistic model by necessity has to be Markov compliant. However, I argue in Chapters 3 and 4 that the respective model classes are—*in a statistical-mechanical* frame of reference—in fact Markov compliant. The apparent contradiction is resolved by considering that the mathematics of scale-free movement may be non-Markov (current behaviour is influenced by the past in a slowly decaying manner), while the statistical-mechanical application of this class of mathematics implies a Markovian interpretation. This aspect will also be elaborated on later in the book.

Similarly, mechanistic models describing spatially explicit memory-influenced movement—resulting in site fidelity under some parameter range—also describe a slowly decaying influence from past experiences on present movement behaviour, in the form of a potential drift towards locations that were visited in the past (Börger *et al.* 2008, Dalziel *et al.* 2008, van Moorter *et al.* 2009, Song *et al.* 2010, Spencer 2012, Fronhofer *et al.* 2013, Nabe-Nielsen *et al.* 2013, Boyer and Solis-Salas 2014). While this may sound non-Markovian, it is not. Biophysically, the memory influence is implemented as a "field" force. This field regards the directional bias towards historically visited locations, owing to utilization of spatial memory. At every point along a movement path, the direction, directional persistence, and movement speed are determined by a resultant vector where the influence from the memory field represents one of many components of movement. The memory-invoked vector field implies that the next displacement is recalculated on a moment-to-moment basis.

In summary, I advocate the need to make a distinction between models where (a) the memory influence is processed in a sequential and immediate manner and thus leading to micro-scale local/temporal equilibrium in statistical-mechanical terms at the model's time resolution (increments), and (b) memory influence where such micro-scale equilibrium cannot be expected. The first variant regards mechanistic and thus Markov-compliant modelling while the latter regards the extended toolbox proposal; non-Markovian and non-mechanistic system dynamics represented by parallel processing.

In Chapter 5, *Parallel processing: the concept*, I confront the third direction of contemporary space-use models at the individual level: integration of scale-free movement (direction 1) and spatial memory-influenced movement (direction 2). Applying a statistical-mechanical approach, the concept of parallel processing—as opposed to Markov-compliant processing—is outlined. Parallel processing regards system dynamics, but in a non-mechanistic sense. I suggest that a statistical-mechanical representation is not only an option but a necessity to model this kind of process. This direction of memory implementation is—according to the parallel-processing postulate—incompatible with mechanistic principles in biophysical terms. In a stepwise manner the concept of parallel processing has matured from a loosely defined graphical illustration in our publications in the 1990s to more mathematically and biophysically formulated models over the last ten years or so. In the present description additional elements of the parallel-processing framework are introduced.

I show how the application of the parallel-processing postulate has resulted in theory and space-use simulations that seem to resolve specific paradoxes. Under this conjecture the actual system properties appear paradoxical only from the perspective of classical premises. Crucially, statistical expectations from parallel processing are now in the onset of being supported by larger and better empirical datasets, where the falsified null hypothesis is based on Markovian statistics (Gautestad *et al.* 2013, Gautestad and Mysterud 2013).

Chapter 6, *Modelling parallel processing*, transforms for the first time the concept of non-mechanistic space-use dynamics at the individual level to the framework of statistical mechanics at the population level. Several new aspects and specifications are introduced. For example, the principle of ergodicity and inter-level system coherence between individual movement and population dynamics is here outlined in more detail. I propose in a biophysical (statistical-mechanical) context how kinetic energy of scale-free space use should be expected to be distributed in a self-organized manner over a scale range. Also for the first time, I describe a population-level simulation model of parallel processing, the "Zoomer model". The model allows for "fine-tuning" of the dynamics along the continuum from a scale-specific and Markovian-compliant (mechanistic) kind of population kinetics to a non-mechanistic and parallel processing-compliant kind of process. By this approach the alternative framework not only connects individual and population dynamics in a coherent manner; it also presents a transition between apparently disparate system classes by connecting non-mechanistic dynamics to its mechanistic counterpart.

Interesting statistical properties of population dispersion emerge from the simulations of the Zoomer model, which brings up what may appear to be a potential solution to an old and still unsolved controversy in ecology: Taylor's power law (Taylor 1961, 1986). This empirically derived law has over the years been subject to a large number of attempts to explain it, but a consensus has still not been obtained (Kendal and Jørgensen 2011). Taylor's power law is one of the most widely tested empirical patterns in ecology and is the subject of an estimated thousand papers (Eisler *et al.* 2008)! It describes how variance (V) of population abundance (M) in a time series or over a spatial transect tends to comply empirically with $V(M)=aM^b$, with $b>>1$ (often close to 2). However, one aspect of this law has by and large been ignored: the V(M) pattern unexpectedly emerges also when abundance is measured at different spatial resolutions within a given extent (Taylor 1986). Larger spatial units (grain size) give larger M on average, and the scaling law with $b\approx2$ even in spatially non-auto-correlated data appears even more paradoxical in the context of this property. I call this the Z-paradox, and propose (with support from simulations) that a parallel-processing kind of population kinetics may contribute to solve it. Will it succeed under closer scrutiny and when confronted with real data? Time will show, but in a later chapter I provide a pilot test.

In Chapter 7, *Parallel processing at the individual level*, I show that individual-level statistics for V(M) when generated from the parallel processing- compliant MRW model is coherent with population-level statistics for the same aspect in output from the Zoomer model. The MRW model is also specified in more detail than in previous publications, with weight on clarifying some aspects that have led to common misinterpretations. Chapter 7 also summarizes various methods—called protocols—to test for compliance with MRW, using respective standard models in respective fields as null hypotheses. The methods now span many statistical aspects of space use, like the distribution of step lengths of a path that is sampled at regular intervals (the Lagrangian aspect), the extent of space use as a function of sample size of fixes (the Eulerian aspect, also containing home-range ghost paradox), the fractal dimension of fix dispersion, and so on. In all respects, MRW is now a testable hypothesis, and I summarize already published empirical support for the model.

Chapter 8, *Movement classes at the statistical-mechanical level*, focuses on one main topic: a conceptual model I call the scaling cube, which is presented here for the first time. In a statistical-mechanical frame of reference, I show how a three-dimensional system (the cube) may diversify between Markov-compliant types of movement (represented by the x- and y-dimension) and parallel processing-compliant space-use dynamics (the z-dimension). The x-axis regards degree of spatial memory utilization, the y-axis regards degree of compliance with scale-free movement in the time dimension, and the z-axis represents hierarchical scaling. The latter expresses the degree of parallel-processing compliance by linking past experiences to execution of future goals. In contrast, a Markovian-based memory model only implements past experience as a term in the model for current movement decision in a stepwise manner (the memory field vector, as explained above). It does not allow embedment of parallel execution of future goals at different temporal resolutions. For example, an animal may decide to move towards a given target even if the direction deviates from a parallel goal to reach another target in a different direction within a wider time frame, and thus higher in the temporal hierarchy of parallel goals. Under the Markov premise only one goal may be executed at any point in time, but this goal may contain influence from past experiences. This goal is embedded in the resultant vector for the directional bias under Markovian memory execution; see above. Consequently, in the scaling cube spatial memory is represented by two qualitatively different processes: with and without hierarchical execution of goals. This aspect thus requires an independent axis, orthogonal on the Markov floor of degree of scaling and degree of memory implementation. A cube has eight corners tentatively representing eight universality classes of movement. In Chapter 8, various methods to distinguish individual space use under the respective eight movement classes that are embedded in the scaling cube are summarized.

Chapter 9, *Is parallel processing applicable to invertebrates?*, returns focus to the population level. In 1990, I sampled a local population of sycamore aphids, *Drepanosiphum platanoides*, at the campus of University of Oslo. The purpose was to test compliance with the parallel-processing concept, and how it might cast new light on Taylor's power law and scale-free (fractal-compliant) population kinetics. Owing to the challenge of the home-range ghost paradox that received the focus over the following years, the aphid material had to be put on the shelf for a later day. This day has hereby finally arrived twenty-five years later. Chapter 9 provides a synthesis of novel theoretical aspects—as outlined in the previous chapters—and what appear to be matching empirical patterns in the spatial dispersion of the aphid population. In particular, the analysis shows coherence between how the home-range ghost paradox was resolved at the individual level and how the Z-paradox gets its proposed solution at the population level. Thus, the Z-paradox may represent the population-level analogue to the individual-level home-range ghost! Both concepts turn up as respective levels' expressions of parallel processing, and the aphid data apparently support compliance with this memory-extended kind of statistical mechanics. This is obviously a speculative and controversial statement, since it implies a quite advanced memory capacity of these tiny insects with respect to spatial reorientation and non-random path crossing. I'm looking forward to learning whether others may find less contrarian (read: Markov-compliant and mechanistic) biophysical explanations for the results. With respect to the variance-mean relationship, the aphid population shows scaling compliance with Taylor's power law, and the empirical pattern is thus encouragingly similar also to the simulated data from the Zoomer model.

In Chapter 10, *Intrinsic complexity: 1/f noise and the scaling paradox*, I turn towards yet another nagging issue of statistical mechanics, 1/f noise (also called flicker noise, fractal noise, and pink noise). An alternative way of demonstrating the increase of V over spatial or temporal scales in a series is to decompose population variability into cycles using techniques like Fast Fourier Transforms (FFT) and then study how the power (squared magnitude of amplitudes) of those cycles varies with their frequency. The frequency equals the inverse of cycle length, and the power spectrum is an expression for cycle-specific variance. A larger frequency means a smaller cycle length. Cycle length is proportional with the lag in the series, which is the time interval or space distance between successive measurements we use for calculating the power. In the spectra this analysis produces, the longer cycles typically have the greatest power (larger squared amplitude).

The theoretical issue in this respect regards systems which show power increasing inversely proportional with frequency; $P \approx 1/f^{\eta}$ with $\eta \approx 1$. This statistical pattern is problematic when trying to explain the underlying process, since it apparently does not comply with classical Markovian-compliant statistical mechanics ($\eta \approx 2$ for Brownian motion and $\eta \approx 0$ for white noise; the latter satisfies increments of Brownian motion). On one hand I show in Chapter 10 how simulated space use from the population-level—the Zoomer model—seem to confirm a 1/f noise pattern. On the other hand I verify 1/f noise also empirically. Both the sycamore aphid population and a multi-year time series of leaf miners, *Leucoptera meyricki* (Bigger and Tapley), apparently comply with 1/f fluctuations. Again the parallel-processing extended kind of statistical mechanics allows for a potential solution of a paradox, in this case the issue of 1/f noise. Again I hope for the emergence of presentations and discussions of alternative model proposals that may show similar statistical coherence between simulation output and real data, both with respect to variability at a specific scale (a time series or a spatial transect) and with respect to variability over a range of scales.

In Chapter 10, I show how the Z-paradox in V(M) and the 1/f paradox are inter-related. Thus, apparently for the first time, Taylor's power law is studied in the context of 1/f noise. The coherence between inter-scale V(M) analysis and output from power spectrograms at the population level and a similarly coherent V(M) and power pattern at the individual level (from population and MRW simulations, respectively) indicates an interesting path forward for the parallel processing concept, which represents a key postulate for the present theory for scaling/memory implementation. The fact that parallel processing is a postulate rather than being verified from first principles at this stage is not necessarily a critical weakness. As I mention in Chapter 10, Boltzmann's famous definition of entropy under condition of system equilibrium in classical statistical mechanics was also based on a specific postulate. In fact, it is still just a postulate, which has proven tremendously successful since its formulation in 1875.

In summary, this book's chapters zigzag between various properties of *complex* systems, with emphasis on individual space use and population kinetics. I hope that the reader is not too overwhelmed by all the new concepts, model formulations, and occasionally provocative hypotheses. Provocations may spark creative thinking and constructive proposals, bringing some elements of the proposed framework perhaps closer to general consensus and other aspects perhaps closer to their termination from lack of empirical support. Again, time will show.

As referred to above, in Chapter 1 and elsewhere throughout the book, the road towards the present theoretical synthesis has been both steep and long-lasting. One of the largest theoretical challenges has been the continuous struggle to cognitively depress the bias from standard system principles while trying out new directions for the sake of better compliance between theory and empirical results. All researchers working on complexity theory—whether related to biophysics or other fields—are continuously experiencing similar frustrations (www.santafe.edu is a good entry point to meet contemporary complexity theory). To some extent, cognitive "un-learning" of long-lasting imprinting of model principles and concepts in classical textbooks on animal space use and population dynamics has been necessary. In particular, this regards the concepts of parallel processing and non-mechanistic dynamics from a statistical-mechanical perspective. In this exploration I do not argue that Markov-based models should be generally abandoned! I argue that they should be supplemented to allow for a broadened scope of models for a potentially more realistic description and analysis of systems where long-term memory and scaling aspects of the dynamics are verified. For the moment the non-mechanistic toolbox extensions are in their infancy, but they do already seem to offer many potentially interesting research opportunities in the time to come. Hopefully, all the rattling of the wasp nest over the following pages may contribute to spark further clarifications in this active field of research.

The theoretical development that has culminated with this book has been driven forward by a mixture of inspirational events, personal stubbornness, and dialogue with peers. However, by and large it has been a lonely journey. Part of the reason for this fate percolates Chapter 1, A *large mystery in a little detail*, which illuminates the self-invoked problem of choosing confrontation over conformity. I have repeatedly locked myself out from various collaboration projects by stirring up the group harmony. I recall one occasion in particular where I was invited to discuss a spatially explicit population model. The starting point was a proposal based on a coupled map lattice equation. However, I wouldn't continue until the hidden system assumptions for using this approach were scrutinized and accepted by the group. I argued that the constituent parts of the actual population—the individuals—did not comply with the behaviour that was assumed when transforming this kind of dynamic to the population level. At that point I unfortunately only had the critical arguments ready, not the model substitutes. The group preferred to move on rather than getting stuck in such assumptions that "everybody" seemed to accept anyway. Thus, the collaboration led nowhere. My stubbornness made me continue circling around the critical assumptions more or less on my own.

From a personality perspective one may consider such a confronting attitude from my side a weakness or strength, depending on perspective. Anyway, instead of leaving the home-range ghost and other paradoxical system behaviours alone and moving on, I decided from the onset—with never-ending acclamation from Ivar Mysterud (see 1. *A large mystery in a little detail*), who has a similar personality in this respect—that the home-range ghost was theoretically exciting and thus needed deeper exploration. In retrospect not such a wise choice on my part, from the perspective of career and network-building. However, given the rapid sequence of theoretical breakthroughs and more deep-drilling empirical tests of the present ideas in particular over the last five years or so, I'm finally confident that I made the right choice back then. However, the first ten to fifteen years of development (1990–2005) were challenging, albeit sufficiently rewarding to keep the ball rolling despite lack of much paper production.

Theory needs to show coherence with real data. The statistical-mechanical approach to study animal space use has been tested not only on Ivar's and his students' database on summer season sheep movement, but also on extensive data on black bear and red deer movement, provided by Mike Pelton and his group (University of Tennessee) and Atle Mysterud and his group (University of Oslo), respectively. I thank Ivar, Mike and Atle for their open-minded attitude with respect to allowing for projects that involve scrutinizing their data from an unconventional angle. As the years go by, some events stand out with particularly strong influence on the following path. In my case one such period was the truly inspirational two-month stay at University of Tennessee in 1994, and I thank—in particular—Mike Pelton and Stuart L. Pimm for their support and energetic impact on my research during that period (see also Chapter 6.5.2 *South state hospitality*). At present I have status as guest researcher at the Centre for Ecological and Evolutionary Synthesis University of Oslo, courtesy of Nils Chr. Stenseth. Nils and Atle offered affiliation as researcher during 2012–2013, and Ivar supported me with a doctoral fellowship at the Department of Biology 1993–1997 in addition to enabling some additional funding for research and travel during that period.

This book (first edition, 2015) has not been subject to traditional peer reviewing. Instead the manuscript has been distributed to a selection of outstanding scientists in the field of movement ecology. Most of them I know by reputation but not personally, and I invited them to make a critical evaluation of my project. The response was encouraging, and an online forum was created for chapter-specific discussions. Thus, I want to thank members of this forum for their attendance in this rather unconventional "crowd-reviewing" process. I particularly want to thank Don DeAngelis (University of Miami) and Daniel Campos Moreno (Universitat Autònoma de Barcelona) for important contributions on the forum. The initial presentation of the book project can be seen at *www.thescalingcube.com*, and updates can be found on my personal site *www.gautestad.com* and *www.animalspaceuse.net*.

Ulsteinvik, Norway, April 2021
Arild O. Gautestad

Box 1—Example from movement ecology's toolbox

A toolbox contains tools and instruments. All parts of science have their own toolbox, where much of the content is shared among the various disciplines. Other parts are more adjusted or even tailor-made to fit specific needs. In each sub-discipline like movement ecology, the toolbox contains statistical methods, sometimes referred to as protocols, as a recipe for pattern description and hypothesis testing. In this respect the toolbox contains mathematical equations (models) that formulate relationships and hypotheses in a condensed and logically precise manner. The toolbox also contains dynamical models, for example simulation algorithms, which are used to explore time-dependent progression of a specific set of relationships (referred to as the system). An algorithm describes the "if-then" rules for the dynamics.

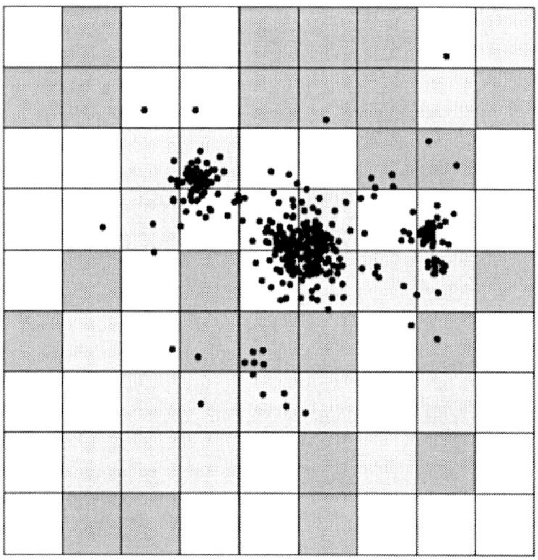

Figure Box-1

As an example of application of the statistical part of the toolbox of movement ecology, consider that you have collared an animal with a GPS transmitter, and you have collected 1,000 successive locations (called fixes) of the individual over a time span T at intervals t_{obs}=T/1000 (called lag). You superimpose the fixes onto a vegetation map, as illustrated here, where three vegetation types are classified by shades of grey. You then observe several patterns: (a) the scatter of fixes is not homogeneously dispersed over the landscape, but concentrated on a part of the map; (b) also inside the area that contains the fixes, the scatter is heterogeneous; (c) you get the general impression that the highest concentration of fixes are located in areas of some specific vegetation type (light shades in the illustration), while other types embed no or just a few fixes. To get a step further in your analysis, you apply statistics to describe the pattern more formally, and you also use statistics to test two apparently obvious hypotheses: H1, the animal moved in a home range, and H2: inside the home range the animal seemed to prefer specific vegetation types over others.

Box 2—Hypothesis testing in the context of movement classes

Statistical description is often trivial, while the follow-up steps—the analyses including hypothesis testing—are not. Under the hypothetico-deductive method, an analysis implies that a model needs to be defined and tested. A statistical model describes in a simplistic manner how a specific process (for example, a specified kind of movement) is expected to produce a pattern; for example, the spatial scatter of fixes in the illustration. If this expectation can be taken for granted (owing to many previous analyses, performed independently by many researchers, that have confirmed the model), the model is part of the theory. If not, the model itself is a hypothesis. From this model one or more hypotheses may be deducted as spin-offs. In the example in Box 1, the hypotheses were H1, the animal moved in a home range, and H2: inside the home range the animal seemed to prefer specific vegetation types over others. From these hypotheses the respective null hypotheses (also called null models) are $H0_1$: the animal did not constrain its space use, and $H0_2$: the space use inside the demarcated area was random, implying homogeneously dispersed fixes. With respect to H1 one of the assumptions is that the individual has either by behaviour or by environmental constraint on movement restricted its movement to a specific sub-section of the available environment to a larger extent than expected from random choice.

Statistical methods are then applied to verify the hypotheses H1 and H2 by falsifying the respective null models. In the present example such a test seems to be trivial. H2 requires somewhat more sophisticated statistical protocol (among the many that are available), but even this should be trivial in principle. However, it is not, and this book is dedicated to exploring and explaining why not. While it is common to consider H1 a model that is part of the theory of animal ecology, I will argue that it should be considered a hypothesis that depends on the given universality class of movement (to be described in Box 3).

Many of these methods available to test H2 are quite generally applicable to many situations beyond testing animal space use and habitat selection. In fact, they are typically developed as general statistical methods, and just applied and possibly slightly modified for ecological inference with respect to animal movement and space use.

Box 3—What is a general process approach?

In a complementary manner the statistics described for the scenario in Box 1 are always accompanied by processes behind the pattern. Pattern implies heterogeneity, and this heterogeneity has to be created by a set of processes. The model that is formulated to explain a specific aspect of the pattern depends on a set of assumptions which are taken for granted when performing the statistical test of the hypothesis. Some of the assumptions for testing hypotheses like H1 and H2 (see Box 1) regard specification of the movement process. In fact, many of these assumptions are almost universally applied, meaning that they are invoked by default by the choice of the generally accepted methods to test H1 and H2. Consequently, the assumptions are implicit to the given methods, and rarely explicitly stated or questioned. They are part of the theory of animal space use, with respect to the intrinsic movement process.

The term "general process approach" refers to a study and scrutiny of the biophysical process side of the two-sided coin where pattern and statistics are reflected on one side and the complementary processes are reflected on the other. In the example given in Box 1, the animal's space-use behaviour is one such general process, formulated and simplified by a model. On one hand the general process approach then regards analysis of how changes of model details (conditions) are expected to be reflected on changes in the statistics. On the other hand the model may also be shown to be quite insensitive—resilient—to specific conditions. This insensitivity to process details may in fact be so prominent that the movement model may be said to describe a universality class in physical terms.

In physics, universality is the observation that there are properties for a large class of systems that are independent of the dynamical details of the system. In this book eight universality classes of animal movement are described, and for the first time unified into a common biophysical framework, conceptualized as "the scaling cube" (to be explained in a later chapter of the book). An important aspect of the model simplification and generalization thus regards the distinction and formulation of main types (classes) of the intrinsic movement process in biophysical terms.

Movement classes are central parts of animal ecology, and one particular class of the eight to be presented here is dominating ecological theory. The general properties of this class, Brownian motion (a physical concept), imply that given sufficiently large lag, t_{obs} is applied when sampling the actual movement path. Under this premise the assumption of H1 (see Box

1 for definitions)—each visit to any of the virtual grid cells in the illustration is statistically independent of previous visits to this cell. Thus, the process should under this assumption reflect the general properties of an area-restricted classic random walk, i.e., Brownian motion inside a container. The assumption of area-restricted Brownian motion, which leads to independent successive revisits to a given section of the home range, is also traditionally applied when testing H2 (habitat selection within the home range based on local variation of density of fixes relative to the underlying type of habitat).

Box 4—Single-and multi-scaled movement

Movement ecological simulations typically regard rule-based kinds of dynamics, which determine successive displacements. The current direction and speed are successively updated in accordance with how the model individual is defined to respond to local conditions inside the perceptual field. The response may additionally be influenced by the individual's rules for internal state (for example, hungry or not). This kind of motion is scale-specific, since a given spatial scale (the perceptual field, albeit fuzzy in real terms) and time (the defined response time to adjust movement to new conditions). Consequently, the movement is single-scaled, demarcated by the perceptual field in space and the response interval to new events in time. Both scales are obviously varying both locally, temporally, and with influence of the individual's present behavioural state. However, in generic model terms these variations may be considered scenario-specific constants, based on true or assumed average values. These constants then represent the statistical scale-specificity of the movement process in space and time.

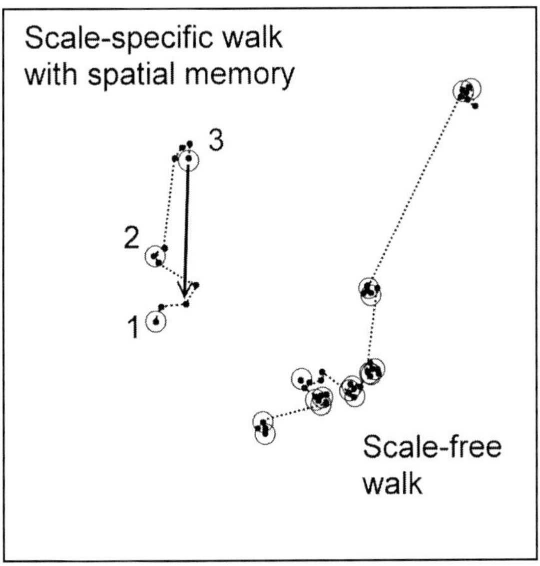

Figure Box 4

In simulations the movement rules may be simple—typically reflecting basic attraction and avoidance responses—or they may be more extensive and complicated. For example, a given number of time increments may be set to represent handling of a food item prior to further movement. If the movement rules include influence from spatial memory, occasional returns towards previous locations may take place, exemplified by an arrow from the neighbourhood of the fix marked as "3" in the illustration. Under such returns the influence from environmental input will typically have to be disabled to allow for strong directional persistence until the target is reached. Over time a path may be drawn by connecting high-frequency fix sampling (sufficiently to make the collected path appear "smooth," with little additional jaggedness expected at even smaller sampling intervals), marked as small dots in the path on the left-hand side of the illustration. Inter-fix line segments represent step lengths at the unit time resolution for the simulation, but the local habitat conditions determining these steps are not shown in this sketch.

A path may also be sampled at a coarser time scale (called lag; see Box 1), as shown by circles in the illustration. Since the movement is scale-specific and single-scaled in this example, a lag that is sufficiently large to embed several successive unit-scale movement events per circled dot will approach a Brownian motion path, owing to the transition from more or less deterministic movement events at unit time scale ("biological-mechanistic" scale, t) to stochastic step vectors at coarser lag $t_{obs}>>t$. This coarsened time resolution, leading to a Brownian motion-compliant sampled path under specific conditions, is in this book referred to as invoking the hidden layer.

In contrast to Brownian motion, the path marked as "Scale-free walk" in the illustration produces a qualitatively different kind of path in statistical terms. This kind of pattern may emerge from at least two kinds of multi-scaled processes (belonging to different universality classes, see Box 3). Multi-scaled movement is often referred to as "area-restricted search," "composite random walk," or "intermittent movement behaviour." Most steps at unit time scale are spatially fine-grained (short displacements; fine-scaled moves). However, occasionally some very large steps take place. If the mode switching happens in a manner which—based on a sample of distances between successive GPS fixes—leads to a distribution compliant with y times larger displacements happening $1/y^{\beta}$ as often ($y>1$, $1<\beta$), the distribution describes a power law. Power law statistics characterizes a scale-free pattern, but not necessarily a scale-free *process*—a fact that is a central theme of this book. A scale-free pattern is in practical terms always limited to a scale range (no animal is able to jump infinitely long between two observations). Thus, power law compliance is similarly bounded.

Box 5—Statistical mechanics of animal movement

The physics of animal movement will be described in greater detail in chapters that are dedicated to this theme. However, a nutshell overview is provided here. Basically, consider the micro-resolved path of an animal (black dots in the illustration in Box 4). This path represents high-frequency fix sampling to reveal all movement-influencing events ($t_{obs} \approx t$, where t is small). It represents the proper scale to directly observe or dynamically model animal behaviour with respect to its similarly micro-scale environment in its neighbourhood (inside the trailing field of perception). At this resolution behaviour should be expected to be relatively deterministic, with sensible responses to environmental input and internal state. In contrast consider the same path but coarse-grained to the temporal scale of circled dots in Box 4. A given step at this scale will then represent the resultant vector for respective behavioural/biological micro-steps at finer resolutions that have taken place over the latest sampling interval of size t_{obs}. At this coarser temporal resolution, we are not studying the movement behaviour directly, but a statistical-mechanical representation of this behaviour. Between this representation and the fine-resolved movement behaviour we have the hidden layer referred to in this book. For example, the distribution of inter-fix step lengths is an example of such a statistical-mechanically interpretable pattern. However, this transition from micro-scale biology to meso-scale biophysics depends on two premises. First, the temporal scale for path sampling, t_{obs} (circled dots in Box 4), needs to be sufficiently large to assume that many movement-influencing events have taken place during each of these time increments. In other words, a hidden layer of ignorance with respect to observed path details is invoked over the temporal scale range from micro-scale (t) to the scale of system analysis (t_{obs}). The magnitude of the hidden layer, t_{obs} - t, sets the magnitude of system entropy for the given time extent (sampling time, T). Sufficiently large entropy embedment is required for a statistical-mechanical level of analysis. Second, in a non-equilibrium system entropy is also enlarged by increasing the total time period, T, for a series that is sampled at time resolution t_{obs}. T should be sufficiently large to ensure large sample size of circled dots in the illustration. Under these two premises the sampled path is physical in a statistical-mechanical sense, with the biological interactions hidden at finer resolutions. At this coarser-resolved level, stochastic modelling and calculation of various system descriptors belonging to this level (e.g., emergent properties like step length distribution, fractal dimension, etc.) is a sensible approach when observing/studying movement and space use.

Later in this book I will describe how the complexity at the behavioural/biological level may be expected to fall neatly into one of eight main statistical-mechanical universality classes of movement, where Brownian motion represents one of these classes. Multi-scaled walk and memory-influence of various kinds represent other classes (see Box 3-4). The statistical-mechanical (physical) level may still start at quite small scales, allowing a statistical-mechanical analysis at intra-home range scales from somewhat larger than the expected field of perception to home range size and beyond. Over this scale range the various stochastic models that characterize the respective movement classes are expected to emerge from a wide range of respective micro-scale behavioural details. For example, ten similar-sized samples of fixes as in the illustration in Box 1 will show different dispersion of fixes, but the statistical-mechanical representations like the distribution of step lengths and accompanying parameters may be similar for all ten samples (a simulation example is shown in Figure 67 in this book). Deviances then become part of ecological inference at this meso-scale level of observation. In this manner phenomena observed at the statistical-mechanical level are important to understanding animal behaviour at finer scales—indirectly rather than by direct observation of micro-scale events. For example, the hypotheses H1 and H2 (see Box 1) regard analyses at the statistical-mechanical level.

The biological/behavioural and the physical approaches to study movement ecology are complementary. Statistical-mechanical approaches will never replace direct behavioural studies, but may supplement them for the sake of improved understanding of the complexity of animal ecology. Many space-use aspects are better resolvable on one side of the coin rather than the other. In the present book I am focusing on the dynamics and the accompanying statistics from the statistical-mechanical perspective.

0. INTRODUCTION: THE BROAD PICTURE

0.1 Complex space use from memory utilization

An animal that utilizes its environment under influence of a memory map creates a pattern of space use (for example, as observed from the spatial scatter of GPS relocations superimposed on a GIS map) that needs to be analysed by models that explicitly account for this cognitive power. Unfortunately, statistical methods in movement ecology have generally not satisfied this criterion. Simulation models with memory influence on movement have occasionally been proposed and explored over the last thirty years or so, but mostly on a conceptual and descriptive level. However, more realistic models for ecological inference are now emerging at a rapid pace (Dalziel *et al.* 2008, Fryxell *et al.* 2008, Moorcroft and Barnett 2008, van Moorter *et al.* 2009, Boyer and Walsh 2010, Nabe-Nielsen *et al.* 2013, Merkle *et al.* 2014, Vergara *et al.* 2015).

Despite these recent developments, theoretical progress has generally been slow with respect to solving some core challenges surrounding realistic modelling of memory and multi-scaled space use by animals. In this book I argue that part of the problem arises from a strong reluctance among theoreticians to explore bold steps away from basic process assumptions. These assumptions are inherited from the "memory-less" modelling paradigm (a subset of the eight universality classes that were introduced in the preface), basically due to lack of a well-developed framework with respect to movement under memory influence. How can memory-including models be designed from a framework that is basically memory-less? This will be elaborated on in later chapters, where memory effects are split into several main classes of space-use processing. In particular, I will be presenting an extended biophysical framework that implements memory in a less constrained manner than under the limitations given by "stretching the string" under the more traditional biophysical frame of reference.

Not all animals utilize memory either because of lack of this cognitive capacity or because they use it to a lesser extent as a function of current conditions. The theory for memory-less space use represents an extremely well-developed and coherent model system both at the individual and the population level (Skellam 1951, Okubo 1980, Turchin 1998, 2003, Hanski and Gaggiotti 2004, Moorcroft and Lewis 2006, Ranta *et al.* 2006). Thus, in the following chapters I will drill into this part as a background for what is to follow: the theoretical outline for a spatial memory-driven system under alternative model assumptions and embedding an extended set of universality classes of movement and population kinetics. This quest brings up the dual aspect of all ecological research: theoretical challenges with respect to model realism, but also practical opportunities if the challenges are resolved in a coherent manner. If spatial memory effects are modelled in a sufficiently realistic manner for a given biological/ecological context of space use, one should expect results from GPS-based analysis with less statistical noise in comparison to application of the more constrained theoretical framework. One should also expect reduced chance for confusing or even paradoxical results. Perhaps most thrilling, an extended theoretical platform allows for testing well-known hypotheses from new methods that are consistent with the best applicable framework (universality class) for the given system.

Over the following chapters a critical evaluation of biophysical model design of individual space use and population kinetics takes the centre stage. At the individual level kinetics regards the physical aspects of behavioural responses to the actual environment and how this process translates to movement (Box 5 in the preface). In a population context the simultaneous movement dynamics of many individuals within a defined area are considered, called an ensemble. At first sight such a statistical-mechanical approach may appear quite narrow and even peripheral for general animal ecology and wildlife management. However, consider these three aspects that exemplify some of the space-use properties to be scrutinized throughout the chapters:

(1) Animal dispersion over the landscape is typically considered compliant with a classical diffusion process. Here I examine the general applicability of this approach and focus on theoretical variants of an alternative kind of biophysics—a mixture of diffusion (exploration) and "anti-diffusion" (site fidelity)—for situations where complex dynamics apply (Box 4 in Preface);
(2) Density is an important proxy variable for ecological inference. At the individual level typical density measures are accumulated number of GPS relocations pr. unit area. At the population level density represents number of individuals pr. unit area. Here I criticize the general applicability of this proxy variable if space use

is memory-influenced, and present an alternative variable with potentially great applicability for ecological inference;

(3) Most animals do not move unconstrained over the landscape, but tend to express site fidelity by occasional returns to known localities. The spatial extent of an individual's home range—an emergent property of space use from directed returns—is typically quantified by an area demarcation. Here I show how this area demarcation—the home range—may be misleading object without further specification. The spatial dispersion of a sample of relocations may be more correctly described as a statistical fractal pattern with dimension $1 <\approx D <\approx 2$ than as an area with D=2 (a spatial object). A similar fractal property also seems to emerge in a population's spatial dispersion (see later chapters). In a statistical sense, a fractal dispersion of GPS relocations or a snapshot of a population's dispersion of individuals will then show "aggregations within aggregations within…" in a manner that will tend to satisfy a power law (Box 4 in Preface). Power law compliance of data over a given scale range indicates a scale-free or at least a multi-scaled kind of movement.

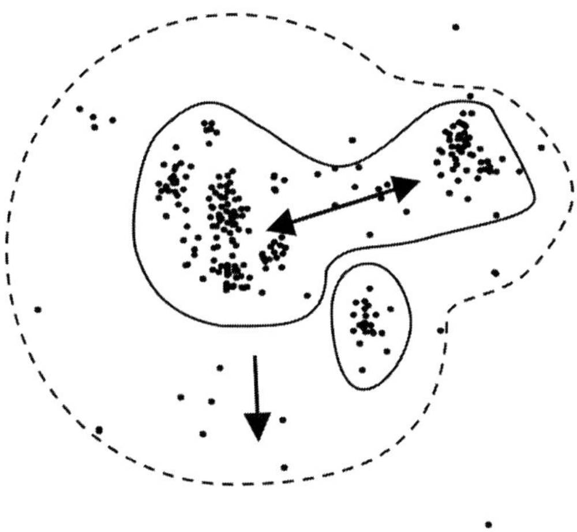

Figure 1. A hypothetical dispersion of GPS fixes, showing the emergence of a home range and intra-range sections of more intense use. The double-arrow line illustrates commuting between two core areas and the single-arrow line points towards less frequent use of the home range's peripheral parts.

Figure 1 illustrates these three aspects in the context of individual space use. In this hypothetical dispersion of GPS fixes (shown as dots), the animal's home range is roughly demarcated by the scattered line, while core areas of more intense use are shown as solid lines. First, the presence of a home range falsifies free dispersal and verifies site fidelity. This is not controversial, and is a core property of animal space use under a broad range of ecological conditions. Since site fidelity depends on spatial memory for targeted returns, Brownian motion and diffusion should *not* be expected to apply as a biophysical model for these kinds of data. This statement is definitely controversial, due to the actual dominance of diffusion theory as assumption for statistical methods for home-range analyses (much more on this theme later). Second, local density of fixes under the condition of site fidelity will depend not only on local habitat quality and other environmental attributes, but is also confounded by commuting effects (double-pointing arrow in Figure 1) and distance-from-centre effects (single-pointing arrow). Space use at the periphery of a home range will appear less intense due to distance *per se* and not necessarily an inferior habitat, and commuting between core areas will tend to inflate fix density between these locations. Thus, comparing local fix density with environmental qualities should be expected to produce very noisy results. This issue is generally acknowledged, but (here follows another controversial statement) in my view cannot be expected to be resolved coherently within the present theoretical paradigm. Third, in a statistical sense real data seems to show fix aggregations over a range of spatial resolutions partly as a consequence of self-reinforcing revisits. Also this aspect confounds habitat use/availability studies.

The challenge is thus to develop alternative system descriptors for constrained area use, fix density, and fractal-like space-use patterns. For all three aspects above—and many more—you will be confronted with the respective premises that need to be satisfied in order to apply the diffusion framework, the density variable, and the area demarcation protocol. *The diffusion template is not abandoned*; it is modified under a wider set of movement classes to account for specific properties of animal behaviour under influence of spatial memory where this condition applies. *The density variable is not abandoned*, but is implemented with extensions that are formulated within a modified system design, to be applied for ecological inference where the complex system condition is satisfied. *The home-range concept is not abandoned*, but its quantification is modified to account for non-trivial sample size effects on its demarcation, to be expected from memory-influenced space use of a particular kind. Four of the eight universality classes for space use are described for the particular context of memory map utilization.

When simulating home-range movement, you will be presented with an alternative proxy variable to the local fix density. This alternative quantification, in a homogeneous environment—in contrast to expectation from the classical fix density—will show a similar magnitude of space-use intensity at the centre and the periphery of the home range, and a similar magnitude for locations between core areas. Thus, when applying this alternative variable on data from a heterogeneous environment, you should expect the observed variation to reflect this environmental variability in a more clear-cut manner than the traditional density variable. In other words, the effects from distance-from-centre, core area commuting, and self-reinforcing use of some patches are "filtered away" when applying a proxy variable derived from theory of site-fidelity influenced space use of a particular kind. As a result, expect a clearer statistical picture (less noise) with respect to ecological inference in the context of space use, like habitat selection, for systems where the assumptions that accompany this framework are met. In addition to providing a potential for improved statistical power, the alternative framework may also resolve long-lasting issues—and even paradoxes—that have haunted specific aspects of ecological research for a long time. Crucially, you will also be presented with empirical support for this alternative framework.

If you have digested the preface prior to starting here in the introduction, you already have the background for invoking a critical evaluation of the mechanistic modelling paradigm: animals with the cognitive power of spatial memory utilization may under a broad range of ecological conditions defy mechanistic principles. An alternative *non*-mechanistic kind of process framework may be needed. This book explores this path from various angles.

This book presents two statistical-mechanical challenges in a quest for a better understanding and more realistic modelling of memory-influenced animal space use and population dynamics: (a) outline for model extensions and (b) outline for model applications. The first challenge implies an attack on the dominance of the holy grail of movement ecology—the Brownian motion paradigm and its many facets—by proposing an extended biophysical system description that contains a range of additional movement classes. In this extended framework Brownian motion is forced to a corner position in a multi-dimensional continuum of universality classes that covers varying degree of memory-less, scale-specific movement and memory-influenced, scale-free movement. The second challenge—model application—illustrates by simulations and tests on empirical GPS data how an extended model framework may provide powerful albeit mathematically simple (at least at this stage of development) statistical protocols when connecting space-use models to ecological inference.

Brownian motion may appear from a movement path sampled at a regular interval. The interval should be sufficiently large to produce a Poisson-compliant (negative exponential-like) distribution of displacement lengths and a random direction for successive steps along the observed path (Gautestad 2013b, Tilles and Petrovskii 2015). Observe "may appear," since the emergence of a negative exponential step length distribution requires specific conditions to be met by the actual movement process. A negative exponential implies that steps in the range L_i+dL (i=1,2,3,...) in the total sample becomes *k* percent less likely relative to the foregoing length interval (called a bin) L_i-dL, and *k* is a constant. The bin size dL is the defined increment on the x-axis of the distribution of step lengths. In the following a coarsened time resolution is in the present context called meso-scale sampling. This coarse temporal scale contrasts with high-frequency path observation at a resolution where successive behavioural responses to environmental conditions are revealed (micro-scale path sampling). The meso-scale, representing system ignorance of details, is in this book linked to the hidden layer concept. The hidden layer is an approach to make a principally deterministic movement of a particle at micro-scales (obeying Newton's laws of motion, as an example from physics) appear stochastic and potentially Brownian motion-compliant from the perspective of the coarsened observational scale.

Why introduce the hidden layer concept? In short, at a meso-scale of observation, the statistical-mechanical system abstraction may be invoked, with a huge potential for space-use analysis. At this level the actual space use process has to be explicitly classified, and consequently this approach forces us to consider some biophysical assumptions that traditionally have been ignored or taken as granted by ecologists. Consider first the classical class of space use, embedding Brownian motion and standard diffusion principles. The path that has brought the individual to its current position is under this hidden layer sampling condition—when combined with this mechanistic movement assumption—not influencing what the animal does next. "Next" then refers to where the animal is located after a given time interval. At finer temporal resolutions (implying a more condensed hidden layer, potentially violating a statistical-mechanical system analysis), the path is typically driven forward by deterministic behavioural responses and goals (Box 5 in the preface). The animal's internal state—like hungry or not—and new environmental input within the reach of its senses will determine the next micro-scale move. However, owing to sequences of intermediate and inter-independent influenced micro-scale movement events, the path from drawing line segments between successive relocations under the meso-scale sampling condition needs to be described by a stochastic rather than a deterministic model (Gautestad 2013b). The stochasticity is an emergent property from the respective sets of fine-grained and inter-independent moves taking place between successive relocations of the individual. Thus, the statistical-mechanical framework for Brownian motion depends on a Markov-compliant process at finer resolutions than the sampled path.

Brownian motion thus reflects a mechanistic process below the hidden layer, and the process has to be Markov compliant. As a consequence of the Markovian process premise, Brownian motion is *memory-less* and may be considered both temporally and spatially *scale-specific*; the very two aspects that have been undermined by recent analyses of animal space use. Animals from many taxonomic groups have a cognitive capacity both for spatio-temporal memory utilization—leading to non-random self-crossing of the path—and a scale-free (hierarchical) mixture of both tactical and strategic decisions on when and where to move. Such behaviour obviously violates the premises for Brownian motion and diffusion statistics at meso-scale sampling. This is very unfortunate, since it forces upon us two choices; conveniently ignore the empirical fact and continue using the traditional toolbox, or accept the shortcomings and start looking for an extended theory.

Such an extended framework may produce many unexpected ideas and results. For example, spatially explicit memory utilization may invoke an explanation of the scale-free property of movement. Memory and scale-free space use may show up to represent two sides of the same coin (to be described later). In this manner, scale-free and memory-influenced space use undermines the Brownian motion paradigm as a convenient representation of individual-level dynamics in a statistical-mechanical frame of reference; i.e., observing the system with a sufficiently deep hidden layer as in typical GPS-based studies. One thus needs to ask: if the Brownian motion abstraction is unrealistic, does it matter in practice with respect to ecological studies on animal space use? If yes, how critical is the error?

To conclude, time seems to have passed for assuming *a priori* movement to be realistically formulated by Brownian motion- and correlated random walk (CRW)-compliant diffusion over the landscape (Fagan and Calabrese 2014, Fleming *et al.* 2014):

> *While great strides have been made in building biological realism into CRW-based models, it is now clear that this framework is nearing its limits. For instance, many animals exhibit movement behaviours that repeat at regular intervals (e.g., daily, seasonally). Animals may also use memory to navigate, or may avoid recently exploited areas when foraging. All of these biological realities and many others violate the Markovian assumption (...) Memory, avoidance, repetition and other biological complexities introduce long-term autocorrelations into the movement paths of individuals, which the CRW and other Markovian movement models cannot accommodate or utilize. A new frontier of movement ecology is to relax the first-order Markovian assumption such that movement models can use the information contained in long-term autocorrelations to identify critical behaviours (...).*
> Fagan and Calabrese (2014)

For ecologists studying animal space use much is at stake if one accepts that the Brownian motion paradigm fails as a premise for the system under study. Some core ecological aspects will become impossible to implement realistically in a classical biophysical model construction. With reference to Figure 1, individuals often accumulate local experience, leading to more effective foraging and reduced risk from predators. One of the sites may thus accumulate more GPS relocations at the expense of the other, as an emergent statistical property from such self-reinforcing site fidelity. This behaviour may lead to a situation of increased experienced level of resource availability and reduced level of risk (Stamps 1995, Gautestad and Mysterud 2010b, Piper 2011). Thus, utilization of spatial memory and multi-scaled space use may under a broad range of ecological conditions provide a potential for fitness advantage relative to purely opportunistic movement, which may be considered compliant with memory-less space use. Two memory-utilizing individuals overlapping in space-use extent (and thus sharing a common resource base) may cope substantially better than a single opportunistic-moving individual in the same area. The former two may from initial chance effects gradually build different cognitive representations of the resource map and local risks, including remembered escape routes (Stamps 1995).

How to realistically implement this kind of interesting space use ecology in a dynamic model based on classical mechanistic principles? You cannot, since you have to accept that the behaviour is strictly mechanistic and thus Markov compliant. How to statistically infer these behavioural properties from spatial analysis of GPS data, based on Markovian-based statistical methods? You cannot, since you have to accept the behaviour is strictly mechanistic and thus Markov compliant! If you accept Markov, you must also be prepared to generate simulated space use that may be statistically incompatible with real behaviour under a broad range of conditions. You may also experience unnecessarily noisy statistical results when testing your hypotheses and you may even find support for paradoxical relationships. These provocative statements obviously need a book to explain, since they need to be formulated from many angles in a biophysically coherent manner.

The difference in space-use statistics between opportunistic (Markov compliant) and site fidelity-based dynamics is profound when mimicked by simulations, under condition of what we have termed habitat auto-facilitation (Gautestad and Mysterud 2010b). Positive feedback loops from intrinsic processes may be intermingled with negative feedback loops from dynamic responses to shifting environmental conditions over space and time, producing complex space use (Gautestad and Mysterud 2010b, Merkle *et al.* 2014). As formulated in a recent review by Piper (2011), *an inhabited space holds a unique "private value" to an animal based on its familiarity with the space.* As a consequence, the intuitive appeal of density-dependent regulation at the population level may not be easily transferred to the dynamics of individual movement. And vice versa: local aggregations of GPS fixes, reflecting "strong space-use intensity," may emerge from a mixture of self-reinforcing site fidelity and local habitat quality (Wolf *et al.* 2009, Gautestad and Mysterud 2010b). This complexity obviously creates challenges for modelling how population density is expected to vary over space, time, and respective scale ranges. Consequently, the predictive value of Markov-based models that are applied in contexts where and when the population is represented by individuals that utilize spatial memory in a multi-scaled manner becomes challenging to the degree that I tend to call it questionable.

0.2 Spatially explicit patch selection in behavioural ecology

Aspects of memory map utilization are now penetrating behavioural-ecological theory at an increased pace (Dalziel *et al.* 2008, Wolf *et al.* 2009, Nabe-Nielsen *et al.* 2013, Merkle *et al.* 2014), leading for example to extension of the Marginal value theorem (Charnov 1976), which describes how individuals, for the sake of optimizing energy gain from foraging, are expected to leave a patch when the rate of energy gain falls below a critical level (a function of deterioration of patch quality from foraging and energy expenditure when searching for alternative patches). In the model's classical formulation, the process of food consumption and search for patches with above-average resource levels is opportunistic in the sense that previously visited patches are re-encountered by chance. This is an unrealistic assumption in many instances, and the behavioural-ecological theory has now been extended and novel aspects explored for animals that are utilizing spatial memory to optimize foraging (Spencer 2012, Grove 2013). By reformulating and extending central ecological models and concepts like the Marginal value theorem at the individual level and diffusion processes at the population level, contemporary theory on the biophysics of movement will obviously also rattle some aspects of evolutionary ecology. In short; movement dynamics percolates both ecology and evolution.

0.3 Optimal foraging in stochastic movement models

Above I described progress in movement ecology to be basically divided into two camps, one camp focusing on the scaling aspect of movement complexity (scale-specific versus scale-free space use) and the other camp basically studying spatial memory aspects. In the following part of the introduction I summarize the scaling aspects, in particular with respect to foraging and statistical space use patterns that are expected to emerge from different strategies. From this starting point I gradually turn towards population aspects of movement – population kinetics – before drilling into a broader framework that involves both scaling and memory properties of complex space use.

Foraging theory aims to identify the complex mixture of behaviour and morphology most efficient to gather energy in different environments (Schoener 1971). Identifying characteristics of efficient search is one important component in such an effort. The Lévy flight foraging (LFF) hypothesis extends the theory of optimal foraging by bringing a closer attention to the distinction between scale-free and scale-specific movement (Viswanathan *et al.* 1999, Viswanathan 2010, Viswanathan *et al.* 2011, Tilles and Petrovskii 2015). It has arisen as a result of a constructive interplay between simulation models identifying optimal space-use tactics under given environmental conditions (Bartumeus *et al.* 2005,

Viswanathan 2010) and empirical testing of these predictions on animal movement data (Sims *et al.* 2006, Humphries *et al.* 2010, Humphries *et al.* 2012, Sims *et al.* 2012). In short, Lévy walk (LW, which may be considered synonymous with Lévy flight in the present context) exemplifies scale-free foraging. In this case, many short movement bouts are stochastically interspersed by some long moves and occasionally some very long ones, owing to varying duration of directional persistence of successive displacements. The Lévy distribution is more long-tailed than the exponential distribution that characterizes scale-specific movement like Brownian motion. The occasional very long steps of scale-free LW are identified as the most efficient in a resource-poor environment where food items or patches are randomly scattered and relatively unpredictable. Resource detection is then enhanced by a total movement path covering a large area for a given movement speed (path length per unit time).

In heterogeneous landscapes that are not strictly fractal (i.e., a condition that embeds almost all real habitats), LW has generally been described either as an emergent statistical property from the interaction process between the foraging animal and its landscape, or representing an adaptive behaviour at the fine-grained scale of the movement path (see Box 5 in Preface). With respect to the latter, Bartumeus and Levin (2008) postulate the presence of a "fractal re-orientation clock," a neuronal mechanism that may generate a Lévy distribution of step lengths (Cole 1995), enabling improved search success. Tilles and Petrovskii (2015) have presented an alternative, mechanistic model leading to Lévy distribution at a micro-scale of path sampling as an emergent property from a superposition of successive move-influencing events. This pattern may, under a specific model premise (lack of "dampening" of movement speed from respective events), be expected to survive also towards coarser sampling scales.

LW contrasts with the more scale-specific and "fine-grained" kind of movement, Brownian motion, which has already been described. Brownian motion may—from the classic assumptions under meso-scale path sampling—be expected to arise in environments with more frequent responses to events and conditions within the perceptual range, resulting in low degree of directional persistence of the path (Gautestad 2012, Tilles and Petrovskii 2015). These events successively take place below the hidden layer, at micro-scale temporal resolutions. One such event could, for example, be detection and handling of a food item, followed by a new displacement that is independent of the foregoing ones. The medium-range steps are in this case rarer than for LW, and the longest ones are completely absent. Consequently, the area covered during a given time period is small relative to LW under similar movement speed. A Brownian motion will result from sampling an actual movement path at sufficiently large time intervals to ensure that successive relocations (fixes) provide inter-fix distances that reflect a sum of independent and intermediate movement events (Bartumeus *et al.* 2005, Gautestad 2011, 2013b). Whether the behavioural response to specific local situations (like the emergence of prey within the perceptual field) is determined by simple or complex rules, or whether the respective rules are characterized by a small or large degree of stochasticity, the direction from one fix to the next will, under sufficiently large sampling intervals—i.e., by invoking the hidden layer—become Brownian motion compliant. Even LW will appear Brownian motion-like under sufficiently large path sampling intervals, owing to limitations on large displacements (Gautestad and Mysterud 2013).

On the other hand, the LFF hypothesis defines Brownian motion in a more direct and mechanistic manner, linked to actual search behaviour. The model explicitly describes both Brownian motion and LW as alternative modes of movement (movement rules at the behavioural/biological fine-grained scale of a path; see Box 5 in Preface), while Brownian motion is not considered a consequence of fix sampling at coarse temporal scales as described above. In other words, when an animal is foraging in Brownian motion mode, it is assumed to perform successive steps in more or less random direction with a Poisson-compliant (negative exponential) step length distribution. When in LW mode, the successive directions are still random, but the step length distribution is scale-free (power law compliant). The fractal reorientation clock postulate by Bartumeus and Levin (*op. cit*) may represent this mode conceptually, and the mathematics of continuous time random walk seems to be suitable for its theoretical formulation (CTRW; to be described later, in *4.3 Site fidelity under continuous time random walk*).

The realism of stretching a stochastic model like Brownian motion or LW from the meso-scale statistical-mechanical temporal resolution towards the micro-scale resolution of the actual path may be questioned (Gautestad and Mysterud 2013, Pyke 2014, Tilles and Petrovskii 2015). However, at present there seems to be a strong consensus among theoreticians to accept this premise at least in the context of the LFF theory. Unfortunately, this view will beyond doubt contribute to widen the gap between field ecologists and theoreticians. Animals are not drunken walkers.

The mechanistic modelling approach in movement ecology requires as much detail of the actual paths as possible (high-frequency GPS sampling) to get material for studying directional change, movement speed, staying times in local patches, and other aspects related to respective classes of behavioural mode of movement (Humphries and Sims 2014). The LFF hypothesis, which is part of this tradition, is thus based on a qualitatively different premise than what is outlined in the chapters to follow. Both Brownian motion, LW, and other stochastic movement models are in this book described and studied by observing more or less deterministic movement—whether foraging-related or not—at sufficiently large inter-fix intervals (invoking the hidden layer) to ensure transition towards a statistical-mechanical representation of space use. The theoretical and practical advantages from this approach will become clearer over the following pages. Behaviour and ecological inference is thus studied indirectly by proxy variables rather than by direct observations. Different classes of dynamic (behavioural) rules and environmental interactions at micro-scales will from this scheme lead to—for example—compliance with scale-free or scale-specific movement when analysed at meso- and macro-resolutions (Gautestad and Mysterud 2010a, Petrovskii *et al.* 2011, Gautestad 2013b, Gautestad and Mysterud 2013). The arguments for applying a statistical-mechanical frame of reference as opposed to the more standard micro-scale "behavioural mode" approach percolate all the chapters to follow. Hence, I criticise the LFF hypothesis for applying movement classes from the statistical-mechanical level (Brownian motion and LW) to describe rule-based movement modes at the behavioural/biological micro-scale level (Gautestad and Mysterud 2013; Gautestad 2013b). Recently the LFF hypothesis has received similar critique from the behavioural/biological scale perspective (Pyke 2014). As already mentioned a statistical-mechanical model able to mimic Lévy walk-like behaviour at micro-scales has also been proposed (Tilles and Petrovskii 2015), but this approach should not be confused with the present hidden-layer (meso-scale) representation of statistical-mechanical modelling and system analysis. In later chapters I will describe an alternative model leading to LW-compliant step length distributions, based on a specific statistical-mechanical postulate.

0.4 From individual space use to population kinetics

A critical evaluation of model premises for individual space use also has consequences for the population level, which traditionally is based on a similar set of model assumptions. Consequently, at the population level the applicability of the mean field approximation (Ovaskainen and Cornell 2006) is similarly undermined in scenarios where Brownian motion cannot be considered a realistic approximation of the statistical mechanics of individual movement. The mean field regards the condition where local immigration and emigration are assumed to cancel out at finer scales than a spatial and temporal resolution that reflects local equilibrium in biophysical terms (strictly, it may rather be called steady state, but in the following I keep the equilibrium term for intuitive simplicity). To provide this strength of "particle" mixing (multi-individual movement in the same area), the temporal scale for a model simulation (length of time interval under condition of discrete time steps) has to be assumed to be sufficiently large, given the choice of spatial resolution. Further along this cascade: if Markovian-compliant population mixing is not satisfied, the use of continuous time differential equations—whether ordinary or partial—to represent population dynamics also needs to be scrutinized, since the equations are based on the mean field premise being satisfied both at the scale of the demarcated system and locally at the system's grain size (for example, in coupled map lattice models).

In this manner a theoretical progression with respect to understanding the complexity of individual movement presents a need to develop a matching framework also for the population level, in order to ensure biophysical coherence between these two levels. Theoretical development towards such coherence is an important part of the following chapters. I pay particular attention to how memory-influenced space use may force upon us an extended biophysical framework for modelling population dynamics, both on a local scale within an intra-generation time extent and beyond these scale extents.

Within a local scale and a fraction of the generation time span, movement rate (immigration and emigration from virtual subsections of space) has a substantially stronger influence on population dynamics and dispersion than variations in birth and death rates. The spatial scale domain, where movement kinetics dominates the pattern of individual space use, may for many species and in many environments be quite large. For example, a red deer *Cervus elaphus* occasionally moves three to four kilometres within a time span as short as one hour, even outside the migration seasons when these animals are rather stationary and show strong site fidelity (Rivrud *et al.* 2010, Gautestad *et al.* 2013). At a scale of square kilometres, it is clear that spatially explicit population dynamics may be studied in this species; not only individual space use, since many individuals typically are overlapping their movement paths and home ranges at these scales. Intra-generation (for example, seasonal) migration may contribute to additional importance of movement aspects to understand population behaviour

(Bischof *et al.* 2012), in particular since long-term memory and varying strength of site fidelity often seem to be present as a supplementary driving force in addition to extrinsic influence from shifting environmental conditions. Spatial memory (cognitive map utilization) obviously allows for inter-season and long-distance commuting between respective familiar areas, as found in—for example—roe deer *Capreolus capreolus* (Mysterud 1999), moose *Alces alces* (Bunnefeld *et al.* 2011), and red deer *Cervus elaphus* (Bischof *et al.* 2012).

0.5 Reflection of individual space use at coarse scales

As will be illustrated by model descriptions and simulations throughout this book, complex movement (defined as non-mechanistic in Preface) may potentially influence dispersion and population range at large spatio-temporal resolutions. In the following I will first give some examples from behavioural ecology, followed by examples of how these individual-level aspects are violating the mean field premise for population dynamical modelling.

What may be called "intra-specific cohesion" (Gautestad and Mysterud 2006)—a tendency for individuals to seek living space (home range) in proximity of conspecifics—will have obvious fitness value. Individuals that express such conspecific attraction (Stamps 1988) improve their mating success and/or survival relative to space use without attraction-involving site fidelity. For example, many vertebrate juveniles like passerine birds return to their birth place after a natal dispersal phase (Pärt 1995). One possible benefit of returning is the advantage of choosing a site that has proved suitable for breeding. The animals that exhibit this behaviour are often the ones that require precise conditions for breeding. For example, sea turtles must find a beach with sufficient sand depth, close to feeding areas. Surrounding habitat may be muddy and unsuitable. Similarly, salmon return to streams where they were born, as these streams are guaranteed to have the required accessibility and conditions far upstream that the fish require.

Such philopatric behaviour—"homing" for short—is a good example of coarse-scale population kinetics from migration that is strengthening population gregariousness from non-random self-crossing of individuals' respective paths. Another behavioural mechanism for gregariousness is social interactions in more general terms. For example, contrary to what is described in the literature and expected for solitary animals, Eurasian otters *Lutra lutra* showed positive interactions with individuals, associating more often than expected by chance (Quaglietta *et al.* 2014). Moreover, otter movement patterns were correlated. Finally, otters shared diurnal resting sites more often than expected. Adult males and females with cubs overlapped spatially and temporally, even sharing resting sites when the males had no paternity. In another study, Fletcher et al. (2013) applied network analysis on individual movement data and verified strong population modularity (stronger patch sharing than expected from random choice) among species as disparate as cactus bug *Chelinidea vittiger*, snail kite *Rostrhamus sociabilis plumbeus,* and bullfrog *Rana catesbeiana.* Similar examples could be collected for many species, to the extent that intraspecific cohesion in one form or another is the rule rather than the exception.

This kind of intra-population "glue" means that the population as a whole gains a reduced risk of Allée effects (extinction risk owing to low population density). Intra-specific cohesion at the population level from conspecific attraction at the individual level counteracts the risk for extinction from diffusion: in an open environment a lack of these forces will ultimately drift the population into the Allée regime. As a consequence, the old assumption that individual space use can be confined to fine spatial scales while population dynamics regard coarser scales should be revised. Similarly, the mean field assumption of standard population-dynamical modelling—that individuals in a population are self-crossing their respective paths only by chance as a consequence of memory-less movement—should also be critically evaluated.

0.6 Challenging the mean field assumption

One important mechanism for conspecific attraction in particular and individual-specific site fidelity in general is the individual's cognitive capacity to memorize local experiences and utilize this information both tactically and strategically (Gautestad and Mysterud 2006, Börger *et al.* 2008, Dalziel *et al.* 2008, Boyer *et al.* 2012, Fronhofer *et al.* 2013). In populations and under ecological conditions where this behavioural premise is satisfied, space-use processes at the individual level and the population level may be intermingled over a very large range of scales (Senft *et al.* 1987, Johnson *et al.* 2002), and will thus require a synthetic approach to gain the full picture. However, the traditional thinking in theoretical population dynamics allows movement rates at finer spatial scale than the defined system demarcation to be ignored, under a combination of a mean field assumption and the Occam's razor principle:

population density and rates of change over time and space are assumed to relate primarily to birth and death rates, while movement only contributes to averaging out spatio-temporal density variations. This is the mean field assumption; local immigration and emigration are assumed to cancel out, leading to local equilibrium in statistical-mechanical terms (Ovaskainen and Cornell 2006). From classical premises, stronger mixing rate (strong diffusion from Markov-compliant movement) thus means better coherence between population dynamical equations and standard premises. Less effective mixing (movement rates smaller than birth- and death rates at the scales for the study, and/or net positive or negative migration at the spatial grain size of the population system) will require more extended models; for example, by applying partial differential equations, coupled map lattice models, or metapopulation-structured system descriptions. In short: spatial extent is resolved at finer spatial grain, for the sake of reaching a finer-grained resolution where the mean-field principle (and local equilibrium) is satisfied within each "cell" in this lattice.

However, both classical and recent theory and empirical results show that the individual movement aspect of the system dynamics may generate strong and complex density variation of individuals over both space and time in its own right. In other words, the density surface may be undulating over space, time and respective scale ranges as a consequence of both extrinsic (environmental) conditions and intrinsic, complex behaviour. With reference to the foregoing paragraph, pinpointing a transition scale for which the mean-field principle is satisfied may be difficult or even futile. The concept of philopatric behaviour—exemplified above—may underscore this dilemma. The more general term "conspecific attraction" and the even more general term "site fidelity" will tend to violate some central assumptions for common population models. In particular, individuals may tend to show density heterogeneity that does not cohere in a straightforward manner with the resource map (Figure 1). Conspecifics become part of the resources image in its own right, and lead to intrinsically driven positive feedback effects on population dispersion. As proposed in this book, owing to memory and scaling effects, this intrinsically driven heterogeneity may potentially even tend to become scale-free and thereby statistically fractal-compliant. The general dependence on the mean-field approximation in population dynamical modelling means that alternative model designs also need to be formulated and explored. In this respect I pay particular attention to the duality between individual movement and population kinetics (the integrated study of motion and reaction rates).

0.7 From mean field to multi-scaled dynamics

In summary, the main goal for the present focus on the individual/population level duality of animal dispersion in the landscape is to make some widely accepted and generally implicit assumptions more explicit. On one hand this will involve drilling deeply into the biophysical processes behind individual space use, and on the other hand it requires critical evaluation of the translation of these properties to a framework for population dispersion processes (Ovaskainen and Cornell 2006). I also propose a solution to specific issues and paradoxes when constraining the system design to these classical assumptions. In particular I argue for a synthetic system formulation including aspects related to scale, scaling (multi-scaled dynamics), and memory influence. This is the extended toolbox referred to in the preface.

There are specific conditions where Brownian motion may be expected to fit the actual system even where/when the actual animal has capacity for complex, spatial memory. Some examples of this are referred to in respective contexts in this book. However, the present consensus in the mathematical-theoretical field of population kinetics is deeply anchored in the Markovian framework (what else to use even if mean-field-based equations are inadequate?). Some are now anyway beginning to define limitations to this paradigm, as already cited in the first paragraph of this chapter (Fagan and Calabrese 2014). Here follows another recent demarcation statement:

> *It does not escape to us that the assumption of a foraging animal as a "brainless" random walker (i.e., with no internal states nor sensory or memory capabilities) should be viewed as a first-order approximation to the actual dynamics. Hence it does not represent the ultimate description of animal information use and movement complexity. Nevertheless, memory-less models can be realistic when the searcher looks for dynamic targets that move away from their original location on time scales shorter than typical revisiting times by the researcher.*
> Bartumeus et al. (2013)

I agree with this statement [see also Pyke (2014)], which *de facto* reduces the applicability of the Markovian framework from a generally accepted premise to a special case where conditions for opportunistic/mechanistic foraging may be considered optimal.

A specific population model with explicit consideration of a memory-driven and multi-scaled kind of space use at the individual level is outlined in this book. In short, this approach is proposed as a starting point to the condition where site fidelity in combination with multi-scaled habitat utilization may be expected to be a better strategy than purely opportunistic space use. Simulation output from a basic (generic) version of this design—the Zoomer model—verifies the crucial theoretical coherence between the statistics of spatio-temporal variability at the individual and the population level. Within the proposed design the system may be tuned from classical system assumptions towards the extended system reflecting intrinsic complexity. This tuning illustrates how the respective levels' compliance with classical or complex dynamics may be explored and empirically tested even at a generic level.

The core premises for the Zoomer model will appear unfamiliar relative to what you find even in contemporary theory on population dynamical modelling. Consequently you will definitely be (and of course should be) critical of the model's design and biophysical geometry. However, before shooting from the hip, consider that the survival of—or death sentence for—the Zoomer model should depend on:

- how it may resolve empirical and theoretical paradoxes by loosening the grip from the classical model premises; and
- how toggling between classical framework's memory-less processing and the Zoomer model's parallel processing premise (see below) may be achieved in a more gradual manner by "turning knobs" of the model conditions (bridge-building towards Markovian dynamics).
- Critically, you should also consider how these two alternative statistical-mechanical system designs may be tested on real space-use data, like large sets of GPS fixes at the individual level and spatio-temporal variability at the population level. Some pilot tests on empirical data are provided.
- You should pay particular attention to biophysical (statistical-mechanical) coherence between the two levels of system abstraction.

The four Zoomer model aspects above will be of particular importance in later chapters, following the introductory chapters, where modelling principles with respect to individual-level space use are particularly targeted.

0.8 The statistical-mechanical approach

In addition to the focus on scaling and memory challenges when studying animal space use, another aspect characterizes this book: the statistical-mechanical framework from invoking the hidden layer for system observation and analysis. With respect to analysing individual-level movement, the advantage of the hidden layer premise is obvious: GPS fixes are typically collected at a temporal resolution that ensures a displacement between two fixes is the result of many movement-influencing events at finer (and thus hidden) resolutions. As already described in the preface this brings the material into the domain of a biophysical description. Rather than analysing successive step lengths and turning angles as expression for a specific behaviour at their respective points in time (many intermediate moves are hidden from view), the dynamics appear in a statistical sense by studying the properties of collections of steps or the properties of the spatial dispersion of a large number of fix locations. For example, whether the distribution of step lengths comply with a Poisson (negative exponential) function or a power law function may be an indicator of scale-specific versus scale-free kind of movement.

What is the strength of the statistical-mechanical level? To explain it simply, consider the hypothetical and trivial example where N independently moving individuals are released simultaneously from the same location. First consider that each individual moves unidirectionally and with a fixed movement speed. After time T their respective positions will all be located at a virtual circle with the origin at the point of release and a radius R that reflects the given movement speed. R will trivially increase proportionally with square root of T on average. Then consider that the individuals stop and change direction now and then (not necessarily randomly!) before moving on until a new obstacle is met. In this scenario the circle after time T may reflect the average net displacement for the N individuals during T. If N is small and/or T is small, the magnitude of R is in this case quite uncertain. Repeating the experiment under similar conditions will lead to quite variable radiuses. However, consider that if N and/or T is large, one will experience that R from repeated experiments becomes less uncertain. In fact, R will tend towards satisfying the diffusion law, where R increases proportionally with the square root of T. In other words, given a sufficiently deep hidden layer of system details, the statistical-mechanical observable R becomes a statistically “sharp” representation of net displacement during T. Such “sharp” observables are exactly what we need to make ecological inference. A deep hidden layer is what a statistician and a physicist would characterize as containing "many degrees of freedom." The stronger the degree of ignorance of micro-details (represented by each individual's successive moves), the more exact in statistical terms the representation of the movement process becomes at the statistical-mechanical meso-scale.

Next, consider the situation where N = 1. In this case R represents a statistical-mechanical observable for individual-level space use rather than diffusion at the population level. Even for N=1 it is possible to invoke a statistical-mechanical representation. Instead of measuring R at time T, consider smaller intervals t_1, t_2, t_3, …, t_n; where n is large, each interval equals T/n, and N=1. In this manner we apparently build an individual-level system with a depth of the hidden layer dependent on n for a given T. As always in statistics it is nice to have a large sample size n, meaning that n*t=T. However, observe the conceptual difference between the depth of the hidden layer (inversely proportional with the size of the fixed interval for a given T) and sample size n.

In this individual-level scenario, the respective displacements during t_1, t_2, t_3, …, t_n do not have a common starting point but represent the distance ("step length") from one interval to the next. Consequently, the total set of steps covers different parts of space and will thus reflect both the temporal and local conditions during T in overall (aggregated) terms. Still, like R did for the diffusing population scenario, the average step length for the individual at a given temporal "grain" scale t will vary and reflect local conditions during the extent T. Observing the animal's location at t=0 and t=T does in this manner represent the individual-level meso-scale (biophysical) observation of diffusion at time lag T. The deeper the hidden layer, the sharper the representation of the observable R and the better it may represent a proxy variable for ecological inference. For example, R (under the assumption of a sufficiently large number of independent displacement events during T) will typically vary both locally and temporally. The null hypothesis is no variation, and represents movement in a homogeneous environment.

However, animals typically are expected to move in a more complex manner than obeying a simple diffusion law. Verification of constrained space use (non-random self-crossing of the path) indicates site fidelity at the individual level and thus utilization of long-term spatial memory. In this scenario the diffusion law where R is expected to increase proportionally with square root of T does not apply, whether we consider an individual or a population. Alternative statistical-mechanical observables are needed to regain the "sharpness" that is needed for ecological inference. Such inference—based on a Non-Markovian assumption—could for example reveal strength of habitat selection during T, sub-ranges of T, sub-ranges of space, and so on. Hence, the ecological hypotheses may be similar, but the methodology is changed due to a change of universality class of the movement. Further, by analysing how the spatial scatter of individual GPS fixes is distributed over a range of spatial resolutions, the estimate of fractal dimension (another statistical-mechanical concept) may provide similar information and potential for ecological studies. Like the radius R in the diffusion example above, the fractal dimension is also a variable that becomes sharper with increasing depth of the hidden layer. Ecological modelling already applies the hidden layer approach extensively in statistical studies, but normally not in a manner which makes the statistical-mechanical foundation explicitly acknowledged. Consequently, the applied framework is very restricted, typically limited to Markov-compliant and scale-specific variants of statistical-mechanical movement classes. Hence, memory effects on space use may in this case violate hidden assumptions for applying the given model.

In short, the statistical-mechanical framework potentially offers a much broader base for ecological studies than the classical limitation to a Markov-compliant model system. Later in this book I describe the "Scaling cube" to capture this extended perspective: a three-dimensional system characterization comprising (a) degree of scaling, (b) degree of spatial memory implementation, and (c) degree of hierarchical space utilization (the parallel-processing aspect of the dynamics). A statistical-mechanical representation of the system is explicitly assumed. The x- and y-axis for scaling and memory may be characterized as the Markov "floor" of the cube. This floor should embed all statistical-mechanical (hidden layer-invoking) representations of mechanistic space use, whether deterministic-mechanistic, stochastic-mechanistic, or intermediate variants (to be explained in 2. *Basics of movement physics: the standard models*).

In other words, choose *any* mechanistic model in movement ecology, often collectively referred to as individual-based models (Grimm and Railsback 2005). Then produce a large series representing the movement path under a given set of model conditions. Next, sample the path at frequency every nth step of the original path, for the sake of transforming the micro-scale representation of space use to a meso- or macro-scale representation. As already exemplified above, the latter invokes the hidden layer and brings the model behaviour into the statistical-mechanical regime. Depending on specific statistical attributes, which characterize the x- and y-axes of the scaling cube and can be quantified from analysis of the sampled data, the actual model can then be pinpointed at a specific location or region of the x-y floor of the scaling cube. The actual behaviour's statistical-mechanical properties at sufficiently coarsened temporal sampling resolutions will determine the values of the coordinates x and y. If the individual-based model is mechanistic, z=0. Brownian motion is the movement class where z=0, and where x=0 and y=0 are also satisfied. If z>0, it would imply additional influence from non-mechanistic dynamics, represented by what is explained as parallel processing in later chapters.

While ecologists traditionally have chased an ever-increasing temporal resolution of their GPS tracking devices, there is in fact a critical limit to be expected where the inter-fix interval may become insufficient to allow for a statistical-mechanical analysis. Simply stated, too little dynamics (too few events) take place on average during respective intervals to allow for sufficient process coarse-graining. The hidden layer is too narrow to make statistical-mechanical principles applicable. In this case there are not a sufficient number of degrees of freedom to satisfy a proper probabilistic modelling of micro-states and the accompanying macro-state. As already indicated, making sense of complex space use (as opposed to dynamics obeying classical, mechanistic space use) may *require* a sufficiently deep hidden layer when observing or modelling the system. To conclude, it should be quite apparent both from the preface and from this introduction that a proper theory for individual and population ecology not only requires extensive application of tools from mathematics, computing, and statistics; the models should also be coherently resolved from the perspective of biophysics.

0.9 The next step

The general theory of animal space use is old and voluminous. However, in this book some key aspects of it will come under scrutiny. In this chapter I have sketched the first elements of what will gradually and hopefully appear an alternative path through the jungle of understanding the basics of individual movement and habitat selection, and how these complex processes translate in a coherent manner into the level of population kinetics and dynamics. The path is termed "alternative", because I challenge a well-established paradigm, transcending both individual and population processes. Thus, in order to lead the reader through my line of arguments, and the logics, models and empirical tests to back them up, a key prerequisite for success requires one hand finding a balance between simplicity of basic illustrations and on the other hand drilling into hidden model assumptions. In the next chapter you will learn how a little detail in a data analysis at first appeared as a statistical nuisance but soon grew into a theoretical monster. My world view of animal space use theory had been rattled.

1. A LARGE MYSTERY IN A LITTLE DETAIL

1.1 The first glimpse

Who could know on that particular summer day in 1991 that what we saw on the PC screen would determine the major path of my research for the next twenty-four years, and still counting? What emerged on the screen was a particular diagram from a re-analysis of some telemetry data on movement of ten free-ranging sheep, *Ovis aries*, provided by associate professor Ivar Mysterud, head of the research group I had joined a few months earlier. Previous results in Ivar's group had confirmed the common knowledge that these potentially free-roaming individuals formed home ranges; they basically constrained their movement to an area of approximately 3x5 km^2 over the summer season. But now the re-analysis seemed to indicate that the area used by the sheep was more complex than just confirming site fidelity. The home range apparently expanded with an increasing number of relocations included for the demarcation, collected successively every six hours through the season, without showing any tendency to approach space-use saturation with respect to area extent.

On the surface it looked like a little technicality issue; the home ranges were apparently temporally unstable. However, under deeper scrutiny a more fundamental problem with classical model assumptions should appear, growing in strength and complexity. New layers of the challenge were successively peeled off, but the issue persisted. Finally, the inner layers should reveal a feasible solution to the sheep data paradox, but only after invoking an extended kind of biophysics. As the fragments for this system architecture had been merged together in a piecewise manner, a coherent integration of physics and movement ecology where memory map utilization was involved started to emerge. Below I describe the evolution of this exploration of the Russian doll. The walk-through simultaneously gives a condensed overview of the classical system descriptions of animal movement where some central but rarely acknowledged model assumptions are scrutinized. In later chapters I will also show implications for population systems, based on the revised premises. Some empirical support for this revision is also provided.

By necessity, as will be revealed through the sections, this critical evaluation of both the individual-level and population-level systems will involve a stronger focus on physical aspects of an animal's path than you may be accustomed to as an ecologist. Along an animal's zigzag path through its environment, myriads of interesting biological events are taking place; the animal's internal state at successive points in time determines its responses to shifting environmental conditions from one moment to the next. However, what are in focus in this book are the system properties at a coarser temporal resolution than this instance-to-instance analysis. The data available from an individual's path is assumed to consist of a sequence of dots on a map—typically a set of GPS (Global Positioning System) plots—and line segments connecting these dots to form a sampled path. The dots are thus successive relocations of the animal—"fixes"—that are collected, for example, every hour to reveal movement and space use. At this meso- or macro-level (above the hidden layer), the space-use process is described in statistical terms, like the average step length between fixes, average fix location in space, and so on, towards more sophisticated system descriptors. To infer about ecology, these coarser-scale descriptors are compared to environmental attributes.

In a sampled path the micro-level biology is apparently hidden at finer resolutions of time. However, these coarse dot-and-path representations of movement and habitat use may in fact indirectly reveal a lot of biological and ecological information about the micro-scale process of movement. Hypothesis testing may be performed at these higher levels of system representation, owing to the many degrees of freedom that characterize large sets of fixes. However, while all ecologists are well aware of the myriad statistical protocols that make up the toolbox for such analyses, there has traditionally been less focus on these protocols' dependence on respective sets of assumptions. Some of these assumptions regard the *physical* aspects and properties of the process. For example, if the spatial scatter of fixes has revealed a denser accumulation in one habitat class relative to another, there are many methods available to quantify selection strength for this type of environment. The comparison of local densities of fixes then depends on assumptions about biophysical aspects of the space-use process, to be explained in detail in following chapters. Some of these premises are well known and explored both theoretically and empirically, but others—those focused on in particular in this book—are less broadly acknowledged. Some even seem to have remained undetected and neglected by theoretical ecologists, despite their importance to understand a movement process from a physical perspective.

These "hidden" (implicit) premises are described here, perhaps for the first time in detail in the context of animal movement. But this presentation goes even further by conceptualizing system properties and a proposal for an accompanying model architecture that go beyond even what physicists have been occupied with and developed theory for. For example, if specific conditions are met by the space-use process of a given individual, local density of fixes from a period of sampling may not only be a poor descriptor of selection strength ("intensity of use") in this locality, but it may even become a misleading one. To understand the line of deductions leading to this hypothesis and how to test it, it is necessary to start at a very elementary level and then successively move into deeper physics of movement.

> *In every field find the strangest thing and then explore it.*
> John A. Wheeler, Physicist (1911–2008)

Now, returning to the 1991 event. The actual method to demarcate a sheep's home range predicted that, after passing a statistical "small sample size artifact," the observed home range size from a given number of fixes should tend to reach an asymptote with increasing fix sample size. This area asymptote would then represent the approximate size of the accumulated space use during the sampling period (White and Garrot 1990, Börger *et al.* 2006). The apparently missing home range demarcation in the sheep data was by itself somewhat odd given the previous verification of home ranges for these specific individuals, but might still be explained as simple statistical trivialities.

Now the first of these plausible explanations had been tested. The hypothesis was that the non-asymptotic expansion could simply be due to what was known as a "home-range drift," which at that point in time had been described for red fox *Vulpes vulpes* (Doncaster and Macdonald 1991). Statistically, spatial drift would introduce so-called autocorrelation in the data. In this case, the aggregated area demarcated by successively increasing the number of fixes as time goes by is trivially expected to expand. Thus, an asymptotic approach towards a "true" home-range area is not to be expected under this scenario. For example, in the hypothetical spatial scatter of fixes in Figure 2a, the subsets of relocations on a map (dots) from successive sub-periods 1-4 through a total period T show a clear drift tendency. Two randomly drawn fixes from, say, one of the sub-periods (embedded by circles to illustrate the point) will then tend to be closer in space than two fixes that are drawn from the total area of space use during T. Thus, by adding the area from period 1 and 2, and then 1-3 and finally 1-4, the observed space use from an increased sample of fixes will not level off.

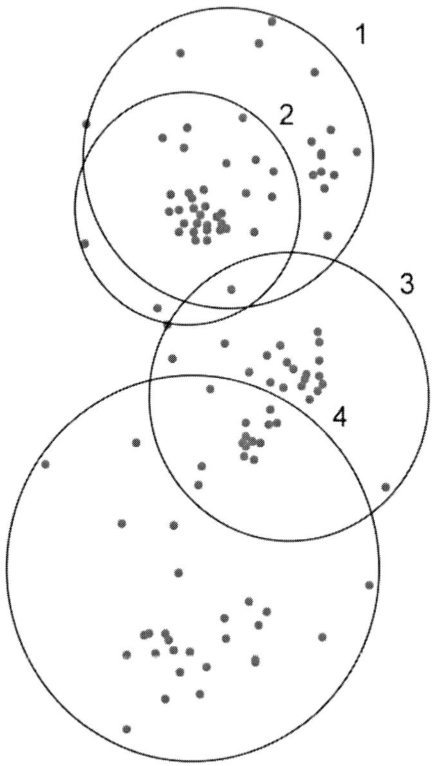

Figure 2a. A hypothetical collection of GPS fixes, where each circle embeds subsets of fixes from successive time periods of data collection.

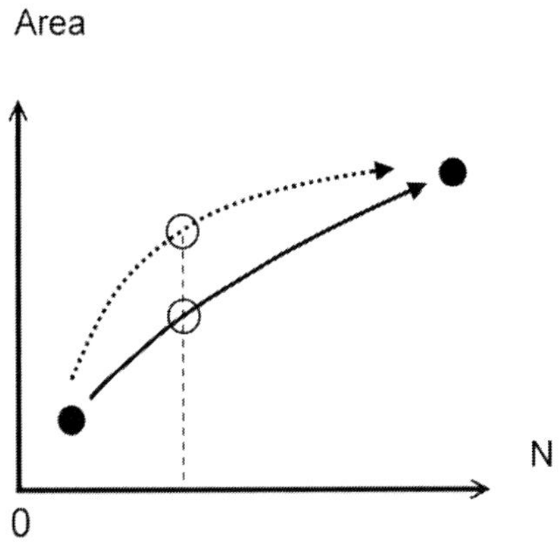

Figure 2b. As the number of relocations (N) increases, the area that demarcates N fixes typically increases. The continuous line illustrates time-proportional sampling, and the dotted line shows frequency-proportional sampling. The latter regards increase of N uniformly spread within a given total sampling period.

A simple protocol to test for such an autocorrelation effect in the sheep data was applied by plotting the demarcated area from N fixes as a function of N and changing N from two sampling methods. The first method was to increase sampling intensity within the given total time period T. A successively smaller interval between the fixes in the sample, spread uniformly over T, should then under influence of autocorrelation be expected to show a larger area demarcation in comparison to the second method: increasing N successively from the start of the sampling period (N increasing proportionally with time). This approach to verify an autocorrelation effect is indicated by the vertical difference between the two circles in Figure 2b, where the dotted regression line indicates the expectation from uniform sample size expansion within T. Absence of autocorrelation—indicating a stationary home range—would then be seen as approximately overlapping regression lines. Under this condition an area asymptote should be expected to appear: adding additional fixes should reveal little or no extra gross area of space use.

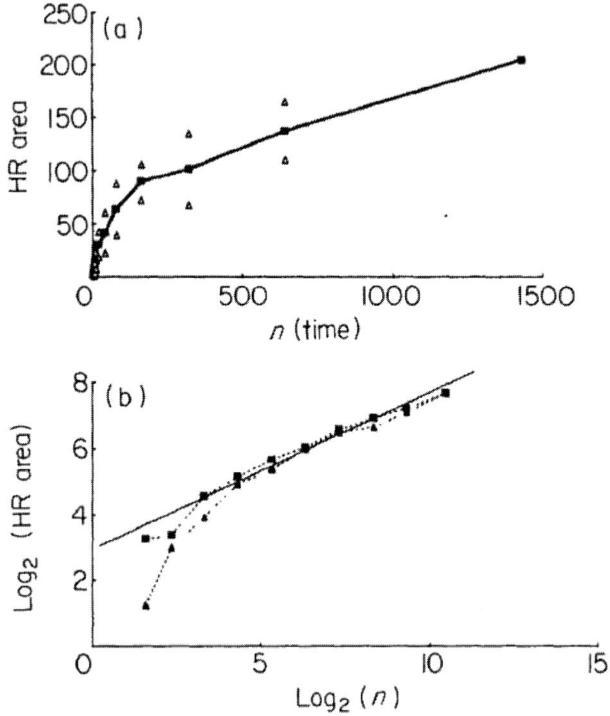

Figure 3. Home-range area demarcation (minimum convex polygon, average of 10 individuals) as a function of sample size of relocations of free-ranging sheep, with arithmetic and log-transformed axes. The latter includes both time-proportional and frequency-proportional sampling.

But the result from the sheep data that particular day in 1991 was not supporting the autocorrelation hypothesis. For example, increasing the sample from fifty to five hundred fixes revealed—on average—a similar mid-range area of demarcated space use regardless of which of the two sampling methods was applied.

If the underlying cause for the expansion was drifting space use (the autocorrelation hypothesis), the demarcated area embedding, for example, 20 percent of the total sample of fixes would logically have been much larger from the first protocol—increased sampling intensity over a fixed period—than the latter. Instead, the data behaved as if the sub-period demarcations in Figure 2a were approximately overlapping and thus confirming absence of spatially drifting space use. As indicated by the lower part (b) in Figure 3, we thus seemed to have a potential paradox in front of us. Area expanded non-asymptotically with increasing N, despite approximate absence of autocorrelation. Thus, this preliminary test, published in 1993 in *Journal of Applied Ecology* (Gautestad and Mysterud 1993) after having been rejected in two other journals, apparently verified that the area expansion seemed time-independent but observer-dependent. The harder we looked—by increasing fix sampling intensity—the wider space use emerged, irrespective of sampling time as such.

"So what?" you may ask. If you are an expert in the field of animal movement, you may swiftly come up with several explanations for the observed pattern. For example, in the actual study we had applied the minimum convex polygon containing all fixes at the given sample size to demarcate the area. This method has repeatedly been criticized owing to sensitivity to the location of the outermost fixes. However, as would be clear from the follow-up tests, this statistical challenge could not explain the odd result in Figure 3. Over the years a range of demarcation methods and ideas to explain the pattern were successively tested. Some are referred to below, and some remain "*in prep.*," but the nuance from the time-independent and non-asymptotic expansion of observed space use still seems to hold.

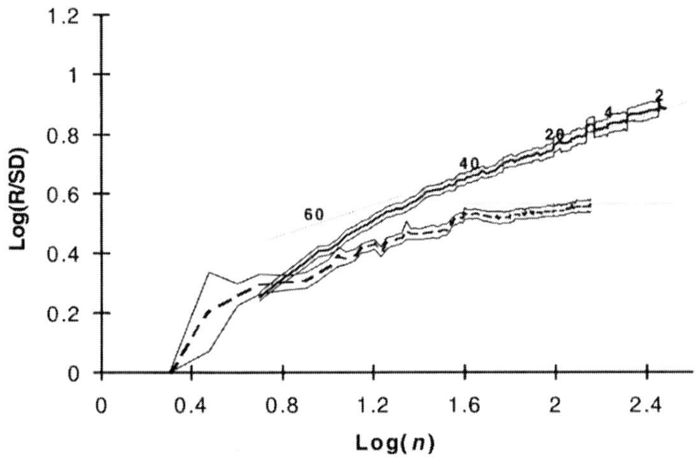

Figure 4. Sample size-dependent home-range expansion by black bear, calculated from rescaled range analysis. For comparison the scattered line describes simulated space use from classical model premises, implying an area asymptote at large sample sizes (grey line). See main text for details.

In cooperation with Professor Michael R. Pelton, University of Tennessee, we analysed his research group's large database on movement of eighty-four black bears, *Ursus americanus*, collected during 1976–83. The bears lived in Tennessee, in the Great Smoky Mountain National Park and Cherokee National Forest. The analyses were performed in 1994, but journal acceptance for publication was not achieved until 1998 (Gautestad *et al.* 1998). Compared to the sheep data, this new analysis consisted of a larger data set (6,487 fixes), and regarded a species at a higher trophic level. We also replaced the minimum convex polygon method applied on the sheep data with a more sophisticated statistical protocol, rescaled range analysis (R/SD; range divided by standard deviation) (Hurst 1951, Mandelbrot 1983), which had never been applied previously in the context of animal space-use data. Again, the main results from the 1993 paper were confirmed: in a subset of thirty-three female bear home ranges with relatively stationary centres of activity, the ranges seemed to expand non-trivially with a similar rate as we found in the sheep data.

In Figure 4 the bear result is shown by continuous line (with standard deviations), and the null model result is shown by dotted line. The latter regards the results from applying the R/SD method on simulated data under classical premises, where an area asymptote is expected. Stationary spatial centres of activity for the bears in Figure 4 were tested for and confirmed, which indicated absence of home-range drift and therefore absence of autocorrelation effect. Again the expansion seemed both N-dependent and time-independent, and the paradox remained. The remaining bear series not included in Figure 4, owing to drift of the centre of activity (and a significant level of autocorrelation), showed—trivially—a steeper slope for the area expansion, in closer compliance with what was conceptualised in Figure 1-2. Thus, the paradox regarded the consistent area expansion—absence of area asymptote—in non-autocorrelated space-use data. Other results also kept the paradox alive. For example, a similar non-asymptotic expansion of home-range area when averaging over all series was found in a meta-analysis of forty-five literature data on demarcated space use as a function of N. The data covered a wide range of birds and mammals, and a very heterogeneous ensemble of area demarcation methods (Gautestad and Mysterud 1995). Consequently, only the average result from area demarcations was reflected upon in this study.

Neither I nor Ivar believed in a metaphysical explanation, so open-minded discussions and a continued hunt for trivial reasons for the strange result determined the path forward. Fortunately, Ivar not only acknowledged the oddity in the initial results, but also supported a continued hunt for an explanation. His open attitude and broad field experience with terrestrial mammals in general and ungulates in particular were critical to maintaining our endless attempt to bridge empirical observations to a realistic theory on animal movement. One conclusion stood clear: We had to think out of the box to find solutions. In short, the current paradigm had to be scrutinized. In this process Ivar also helped secure me a four-year doctoral fellowship, including office space over many years following, to explore and hopefully explain the issue from the perspective of an extended or alternative theory for complex space use.

The need to invoke some kind of complexity theory (which will be defined in a later chapter) was supported also from another angle of analysis. Strangely, the spatial scatter of fixes seemed to satisfy the property of a statistical fractal. In short, the sheep seemed to utilize space in a manner that resembled fix aggregations within fix aggregations within…over a range of spatial resolutions. The spatial dispersion of food sources in the sheep habitat was heterogeneous but not fractal-distributed, meaning that environmental forcing could not explain the fractal and thus scale-free dispersion of fixes.

The concept of fractal patterns had recently percolated to the field of ecology in 1991, so an initial test of fractal dimension of the fix pattern was also included in the 1993 paper. This part of the result was also in conflict with the existing home-range paradigm and thus contributed to increase the statistical nuance: the fractal dimension D was too small to satisfy the theoretical expectation. In the original analysis we concluded with $D \approx 1.5$, but in a more extended analysis we found D closer to 1 on average, over a range of spatial resolutions (Gautestad and Mysterud 2012). This magnitude of D implies a strongly heterogeneous dispersion of fixes at various levels of spatial resolution. The expected strength of spatial patterning from classical assumptions would give D between 1.5 and 2 (more on this later). Entry-point introductions to fractal analysis in an ecological context are found in Sugihara and May (1990) and Gautestad and Mysterud (1994).

The sheep data confronted us with two paradoxes: non-asymptotic expansion of home-range area even in absence of autocorrelation, and a fix aggregation pattern that could be described as a scale-free pattern. The fixes seemed to organize themselves, statistically, into an aggregative pattern that looked similar at different spatial resolutions. The bear data additionally indicated that the oddities of non-trivial area expansion of space use might be more general among mammals at least, and not linked to a specific area demarcation method.

After having tried in 1993 to explain the strange results from the sheep data to another group working on animal space use at our department, we were kindly allowed access to one of their own data sets. They suggested that free-ranging sheep for some reason probably were non-typical representatives of animal space use. The data provided was based on movement by the root vole *Microtus oeconomus* in relatively homogeneous field enclosures. I immediately analysed the samples using a third method, grid cell counting of visited cells rather than minimum convex polygon or R/SD. However, again the strange pattern appeared with similar statistical characteristics as for the sheep and bears.

Unfortunately the response from the group was silence and ignorance. Thus, the root vole's potential support for the generality of the sheep result is still unpublished. In retrospect I can't blame them for dismissing the issue in Figure 3 and the similar result from root vole home ranges. It *was* really odd at the time, to the extent that sound scientific practice would imply dismissal until strongly verified in broader terms. At that time we had not performed the bear analyses in Figure 4 and the meta-analysis from literature data, so the root vole results were just test number 2 in the queue. On the other hand, progress is of course best achieved in a constructive dialogue where arguments from various directions are tested and new proposals might emerge. We were disappointed by the lack of interest.

Similar responses were often experienced also in other contexts, like workshops and conferences. In general, it seemed, people were comfortable within their present model paradigm and saw no reason to spend time on what may have appeared as metaphysics. A home range was by default considered *a specific area* (fractal dimension close to 2), albeit with fuzzy borders, which unfortunately would have to invoke some level of subjective judgement of borderline definition (Börger *et al.* 2006). Our analyses showed fractal dimension closer to 1 rather than 2.

By not considering space use's fractal properties, a non-asymptotic expansion (if it appeared in specific analyses) could easily be dismissed by defining peripheral relocations with strong influence on the space-use oddity as "statistical outliers" and thus ignorable. In fact, for the bear data in Figure 4, we had already removed a few extreme outliers as a conservative precaution, so the result should be less influenced by this aspect. Despite this, R/SD showed non-trivial expansion. An alternative workaround when confronted with such odd results is also common: implying one of many area demarcation methods that—by fine-tuning specific "smoothing" parameters—could effectively depress the effect from such sample size dependence.

1.2 Pinpointing a feasible reason for ignorance: KDE

The last statement in the foregoing section may provide a key to understanding the lack of interest in the paradoxes that we identified from analysis of sheep, bear, root vole, and other data sets. This may simultaneously explain why the oddity had not already been widely reported and acknowledged in the community of ecologists—despite tens of thousands of range demarcations and thousands of published papers in this field. It may all coalesce onto a specific acronym: KDE.

The widely applied kernel density estimate (KDE) method calculates spatial contour lines (isopleths) for the locations with the highest fix densities. KDE has repeatedly been proposed to be the best practice for home range demarcations (Börger *et al.* 2006, Fieberg 2007) and thus superior to, for example, the minimum convex polygon method. However, it has been shown repeatedly in applications of KDE on simulated home-range data that this method tends to overestimate the demarcated area for small sample sizes relative to the contour area for larger samples [see for example (Fieberg 2007)].

This type of bias has been documented also in analyses of real home-range data, for example in studies on white-tailed deer *Odocoileus virginianus* (Fieberg and Börger 2012), cerulean warbler *Dendroica cerulea* (Barg *et al.* 2005), black bear *Ursus americanus* and brown bear *Ursus arctos* (Belant and Follmann 2002). When the KDE method was tested on an extensive sample of fixes from lion *Panthera leo,* it was found that repeated returns to familiar locations created substantial bias in the area estimates, and this "artifact" increased in strength with increasing sample size of fixes: larger samples led to stronger "peaks" (high-density fix locations at specific locations) and thus narrower contour lines from the KDE method (Hemson *et al.* 2005). Strong local aggregation of fixes violates a basic statistical premise of the KDE approach. According to KDE's deeper statistical assumptions, two fixes may happen to overlap by chance, but many such instances will conflict with the conditions the method is made for—leading to too-strong narrowing of the area demarcation in the neighbourhood of such aggregations. However, such aggregative space use—intra-home-range site fidelity in short—is typical of space use by animals, as was reflected in, for example, a low fractal dimension of the fix dispersion of sheep: lower magnitude of the dimension means stronger clumping of fixes, in a scale-free manner, where fix aggregation is dispersed statistically over a range of spatial resolutions. KDE statistics is not compliant with fractal fix patterns, which violates the premise of "standard" statistics of kernel functions.

Standard statistics means compliance with a spatial utilization distribution (UD) function that may be decomposed into a superposition of Gaussian-distributed space-use kernels (summing up the spatial utilization) with respective spatial centres, widths, and magnitudes. The UD is a smoothed surface which describes the core and fringe areas of space-use intensity. In other words, the respective kernel components comprising UD are in sum assumed to represent respective local intensities of space use (fix density) where these spatial variations are the result of inter-independent revisits and staying periods in respective patches. Observe the key word *inter-independent*, which implies random re-encounters.

If this assumption of independent revisits to a given site is not satisfied, the KDE output may be influenced by a statistical artifact. Rather than showing a similar-sized home-range area for small and large sample sizes as ideally expected from the KDE method (whose purpose is to compensate for under-sampling the true home-range area), a larger N can be expected to successively underestimate the area, owing to a specific kernel truncation effect. Conversely, a small N will lead to an overinflated area within the contour lines.

The issue is illustrated (somewhat exaggerated) in Figure 5. When sample size of fixes is small, the kernel contour line will reflect a larger uncertainty (larger intrinsic variance level, reflected by a contour line called an isopleth) with respect to calculating the most probable area where the animal has spent most of its time. As the sample size increases, the uncertainty aspect decreases and the contour line consequently becomes more convoluted. Thus, the area within the contour may very well either be found to stay the same, or even decrease with increasing N. Ideally—from the premises of the KDE theory—the contour-delimited area should in fact stay the same with increasing N (thus home-range area could be estimated even from small N). However, a fractal fix aggregation will tend to create too strong of a "pull" on the contour line as sample size of fixes increases.

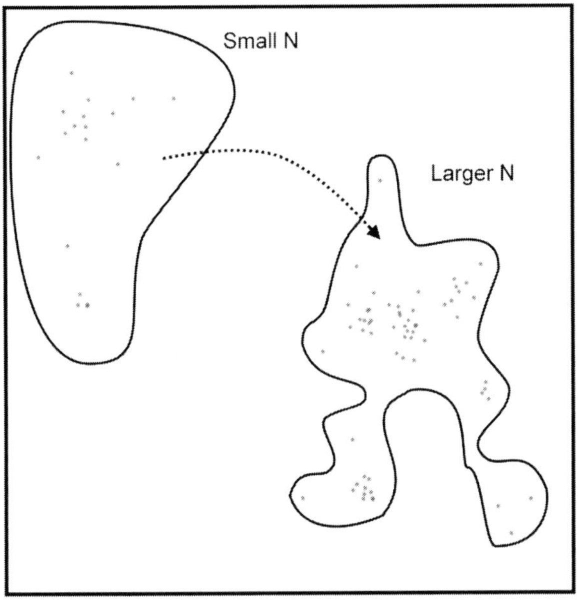

Figure 5. Hypothetical set of GPS fixes, showing area demarcation from a KDE estimator where the area becomes more convoluted as sample size, N, is expanded.

This issue, related to the widely accepted KDE method, draws attention to the complexity aspect of animal space use: scaling and memory. The statistical artifact from applying the KDE method can thus be understood from the animal's space-use behaviour. Non-random self-crossing of an individual's path owing to memory map utilization and site fidelity leads to self-reinforcing use of some patches at the expense of other patches. In other words, a home range is emerging from this positive feedback process.

As documented by our empirical analyses and coherent model simulations, site fidelity leads to a spatial pattern of the dispersion of fixes that satisfies a statistical fractal (soon to be explained in more detail), where there is heterogeneous and so-called scale-free distribution of fixes over a wide range of spatial resolutions. In comparison to a multi-modal "standard-statistical" distribution, assumed for KDE, the peaks in space-use intensity are steeper and the widths surrounding the peaks are much wider, satisfying a superposition of Cauchy distributions rather than Gaussians, leading to a UD that is rugged over a large range of resolutions. In this case the distribution cannot be approximated by a smooth, functionally defined surface as assumed for application of the KDE method.

Owing to the expected kernel truncation artifact, a home range that appears from a spatially continuous isopleth at low N may become a set of space-use "islands" in larger samples owing to a better representation of the fractal aspect of the fix dispersion when N is large. If this statistical "nuance" is compensated for by tuning a specific "smoothing" parameter, the fragments are better merged as in small-N samples. However, as a consequence the area demarcation by the modified isopleth increases. In this manner, parameter tuning to maintain a continuous area demarcation with increasing N is expected to expose the home-range ghost paradox from non-trivial N-dependence on home-range estimates also from applying the KDE method.

1.3 Leaving the comfort zone: confronting memory effects

Ivar's never-ending agreement and optimism that something theoretically novel and interesting was lurking in the data managed to keep my spirit afloat. From his wide empirical experience on animal movement, he immediately resonated well with our 1993 paper's working hypothesis to explain the pattern in Figure 3: animals with the ability to memorize events, like local habitat properties and resources, might utilize this information in a more complex manner than anticipated by current theoretical conceptualization of animal movement. Thus, without adjusting the model to account for these system conditions, paradoxical results—when interpreted under the established paradigm for space use at that time—could emerge in analyses.

In particular, animals might relate to their environment over a scale continuum, as indicated by the observed fractal property of the fix pattern:

> *In reality, the animal probably relates to the fine-to-coarse scale continuum in parallel, so that the actual set of movement steps becomes a randomized mixture of many small and a few large steps. ... A hierarchically structured movement process...*
> Gautestad and Mysterud 1993, p. 524

The general concept of hierarchical space use was not new. For example, a 1987 paper by Senft et al. (Senft *et al.* 1987) is often referred to in this respect, but many others have also struggled with similar insight. A review of the scaling challenges is nicely reviewed by Walters in a book chapter (Walters 1992):

> *The basic problem is that animals interact with the habitat at several spatial and temporal scales, using behavioural cues and experience (memory maps) in complex and poorly understood ways.*
> Walters (1992), p. 119

Our 1993 paper's Eq. 1 seems though to have been the first attempt towards formulating an explicit "translation" of memory-influenced scaling complexity in animal movement into a statistical-mechanical model. This model raised the hypothesis that the emergence of a fractal-like pattern of fixes was a statistical consequence of memory-utilization involving a range of spatio-temporal resolutions—covering a complex mixture of movement tactics and strategy. The complexity aspect in our initial working hypothesis implied that "something"—related to the animal's cognitive capacity and information processing, spanning a range of spatio-temporal scales of a memory map—was influencing movement and thus the pattern of spatial fixes in a non-trivial manner:

> *In a complex environment, individuals are continuously affected by factors which influence their movement. Thus, the distance moved and direction between two points in space and time are a function of the integrated (summed) influence during this period, in addition to a complex influence from coarser scales (memory).*
> Gautestad and Mysterud (1993), p. 525

This was our first conceptual attempt to connect multi-scale space use to spatial memory—utilization of a memory map. As will be shown in later parts of this book, an explicit consideration of such memory and scaling effects within a specific biophysical framework may explain the paradox of non-trivial area expansion with sample size of fixes. In fact, I will argue that this property is *expected* to appear in general terms and under a wide range of environmental conditions, as an emergent property of complex space use under the specific biophysical premises. The influence of long-term memory on animal movement has been well documented for a broad range of animal taxa (Morris 1981, Collett *et al.* 1986, Garber 1989, Rodriguez *et al.* 1994, Bailey 1995, Ostfeld and Manson 1996, Sherry 1996, Mendl *et al.* 1997, Healy 1998, Beekman and Ratnieks 2000, Wohlgemuth *et al.* 2001, Burt de Perera 2004, Boyer *et al.* 2006, Merkle and Wehner 2009, Wolf *et al.* 2009). The mammalian brain's capacity for self-location in space for a coarser scale than its current field of perception, a prerequisite for path-planning and homing, has also gained increasing support from recent experiments on rats (Solstad *et al.* 2008) and humans (Doeller *et al.* 2010).

Despite this non-controversial insight that spatial memory may be an important driver on animal space use, researchers in wildlife ecology demarcate home-range sizes based on, primarily, the KDE method, which assumes a *memory-less* movement process, owing to dependence on assumptions from standard statistics. For example, an individual showing site fidelity, which may be defined as an increased chance for self-crossing of its path relative to a purely opportunistic movement, will violate central premises of many statistical methods in ecology, including the KDE method. A consensus is always difficult to break out from, not at least because leaving the comfort zone in this case would require a stronger focus on physical aspects of animal movement. Traditionally, ecologists have been more inclined to borrow elements from the theoretical mind-set of statisticians than physicists. Thus, in my academic field as a zoologist and ecologist, a focus on advanced protocols from statistics has been the traditional direction to seek solution to difficult pattern analyses. In this book I advocate an extended approach: analyses involving animal space use may have even more to gain theoretically by reaching out towards statistical-mechanical aspects of movement, which more explicitly seeks to understand the *process* behind the statistical patterns. Statistical mechanics, where the term "mechanics" refers to the physical nature of this theory, is in fact the major premise provider for statistical methods in movement ecology. However, these premises are normally not recognized explicitly. Hopefully the following pages will contribute to bringing some of these premises closer to the surface.

To summarize the initial working hypothesis in the 1993 paper, with regard to collecting an individual's position every Δt minutes or hours, the statistical pattern that appears from plotting successive locations on a map could—according to our working hypothesis—reflect a statistical mixture of coarse- and fine-grained habitat utilization in an exploratory sense, intermingled by occasional returns to previous locations as a simultaneous reflection of site fidelity. Both aspects—multi-scale habitat use and memory-dependent returns—obviously depend on memory.

Owing to the typical interval between successive fixes in a series of magnitude minutes or even hours, which is large relative to the time to respond to a specific movement-influencing event, it was clear from the onset that a kind of stochastic model would be needed for system description in process terms. In other words, since a specific displacement during (for example) Δt =30 minutes, one hour or more represents a sum of many more or less deterministic responses during this interval, the net distance and direction—according to the paradigm of animal space-use modelling—should be expected to be better represented by a random walk type of movement than a deterministic algorithm (Turchin 1998). As will be explained in detail in the following sections, random walk does not imply that animals are supposed to move like drunken walkers. It is simply a statistical property that emerges as a consequence of studying a path by connecting successive fixes at sufficiently large temporal scale Δt (Box 5 in the preface). In other words, movement is observed and analysed in the *physical* realm of space use. However, despite the long history of applying classical and Brownian motion-like random walks in the field of movement ecology (Skellam 1951, Okubo 1980), we needed something that was capturing the novel aspects we were chasing; presumably linked to scaling complexity from memory influence where short-term tactical behaviour was mixed with—or constrained by—coarser scale strategic goals. One such goal could, for example, be the return to a resting place after some period of excursion involving foraging and other activities. It could also involve a return to a patch that previously had been found to be rewarding in foraging terms. In short: strategic goals could regard a wide range of movement optimization in behavioural ecology terms, given that the animal has a cognitive capacity for this kind of reasoning, which may manifest itself as strategic moves.

In retrospect the 1993 paper now appears a bit tumbling and fumbling in both style and content, perhaps as one might expect due to the bewilderment of the strange results and the struggle to interpret them. For example, the proposed simulation algorithm, "random walk on a random walk," was in 1993 proposed to be synonymous with the concept we called multi-scaled random walk (MRW or "random walk with memory") that was introduced in the same context. However, in successive theoretical developments, the random walk on a random walk algorithm (Kehr and Kutner 1982) was replaced by an alternative method to simulate MRW.

In retrospect the proposed working hypothesis to explain non-asymptotic and time-independent home-range expansion as a mixture of many short-term and a fewer longer-term moves seemed both simple and intuitive. Any ecologist with practical experience with space-use data would agree with this statement. However, the model was at this point rather hand-waving and intuition-based, owing to lack of a proper spatially explicit process theory that could explain it in detail at that time. Anyway, application of fractal geometry to describe hierarchical space use as scale-free and even revealing a specific rate of area expansion as a function of sample size represented novel dimensions to the modelling challenge. Since the voluminous random walk theory (Turchin 1998) is connected to the physical rather than biological aspect of space use, we stressed this distinction explicitly by naming the 1993 paper "Physical and biological mechanisms in animal movement processes."

Over the years I was constantly toggling in and out of a nagging feeling: were we chasing a ghost? As a reflection of this mood, the 1995 paper—containing the meta-analysis of forty-five data sets—even got the heading "The home-range ghost." Would the next testing method or data set make the initial oddity vaporize, like all crackpot ideas tend to do under closer scrutiny? Further developments raised additional paradoxes when movement was described under traditional model premises, and made it increasingly clear that a theory of memory-influenced movement needed a line of development that started at a very fundamental level.

1.4 The basics of studying individual space use

Following up the issue of home-range demarcation in the previous section but generalizing the perspective somewhat, consider that you have fitted a GPS transmitter to an animal. You have let the individual calm down for a period, and then started collecting records of its location every hour. How is it moving around? After five hours the spatial scatter of locations (called fixes) is contained within the small circle in Figure 6. After ten hours the area is larger as indicated by the larger circle, where the five initial fixes are marked as grey dots. After some time of fix sampling, you draw a diagram of area (A) as a function of number of fixes (N). There is a large menagerie of area demarcation methods available, each with its own strong and weak properties with respect to demarcated space use. Four of these have already been defined above. Let us keep it simple, and not specify the actual method to quantify A at this point. Further, to reduce random effects and statistical error, let us assume that you repeat the sampling procedure above for many individuals (under similar conditions), and calculate the average A for each of the respective magnitudes of N.

Typically, three results may appear. First, the most basic and trivial result, marked as **1** in Figure 6. In this case the animal has moved around, responded to local habitat conditions in a sequential manner. Some events lead to attraction, others to repellence; occasionally the animal is taking a break, and sometimes it is moving slowly and sometimes moving faster.

Since we in this example are assumed to be collecting fixes at intervals of $\Delta t = 1$ hour, any given displacement from one fix to the next will reflect the accumulation of a large set of unobserved displacements at finer time resolutions like a minute or less. Fitting the function A(N) after averaging over many independent series of (N,A) plots will lead to further removal of the effect from specific events on the fitted A(N) line. In the case of **1**, the best-fit function is linear—the accumulated area is (on average) increasing proportionally with N—and thus confirming a paradigm model for animal movement, Brownian motion. Each direction-influencing micro-scale event is in this case similar to a collision event between two gas molecules, or a small particle being shuffled around by water molecules. While each collision or reshuffling of movement direction may be totally deterministic (for example, in accordance to Newtonian mechanics in case of molecules and particles), sampling the fine-grained path at a coarser time scale than the average direction-influencing interval—for example, one hour—will make the observed path stochastic and thus Brownian motion-compliant. For improved realism you may expand the generic model with terms for some persistence of path direction and more or less spatial drift, but it is still basically a Brownian motion. In an open environment, Brownian motion will show an extreme form of spatial autocorrelation, which was explained conceptually in Figures 1-2. Two successive locations in a set will tend to appear close together, relative to the average two-location distance in the entire set.

However, many animals will show a constrained kind of space use, a home range, where "home" indicates site fidelity and "range" traditionally indicates a specific area—but with fuzzy borders—where the animal is utilizing its environment for a smaller or larger period of time. Much research has over the years been devoted to studying home-range area under various conditions, to reveal biological and environmental relationships with predictive power for other sites, times, and individuals. For example, home-range size may be expected to vary as an inverse function of the magnitude of available resources (more food, smaller home range), and larger animals typically have a larger home range than smaller species. However, implicit in this home-range concept—as it has been modelled over the years—is the assumption of a given area size under stationary habitat conditions (albeit with a diffuse and statistically difficult border delineation). Hence, as indicated by the line **2** in Figure 6 and also shown by simulation results already shown in Figure 4 for the bear data, area is expected to increase with a slower and slower pace with increasing sample size.

Figure 6. Variants of area expansion as a function of sample size of relocations.

Since peripheral parts of the range normally have lower visitation rate (observed as smaller local density of fixes), the gradual approach towards an area asymptote should be expected to be slow. Typically, theoretical simulations of area-constrained movement have shown that fifty to three hundred fixes, collected at sufficiently large interval Δt, may be necessary to reach closeness to the asymptote from common area demarcation methods, and thus approximating the "true" home range in a statistical sense.

1.5 The home-range ghost

Setting various well-known complications from environmental conditions, sampling methods and area demarcation methods aside, the mystery is represented by the function **3** in Figure 6. The pace of area expansion with increasing N is slowing down, but there is no indication of an asymptote even for N much larger than 1,000 fixes or more. In fact, for the data sets that are studied so far and including data sets beyond those already described (to be presented later in this book), the expansion—for non-autocorrelated fixes—typically seems to comply with a so-called power law where area expands proportionally with square root of N:

$$A(N) \approx cN^{0.5} \qquad (1a)$$

The bear data in Figure 4 R/SD expanded proportionally with $N^{0.25}$ (slope 0.25 with log-log transformed axes), implying that area, which changes proportionally with the square of R (the one-dimensional "range") increased proportionally with $N^{0.5}$. The power exponent for the bears was thus similar to the sheep data in Figure 3. The method to demarcate space use, A, has in our own research been varied from minimum convex polygon (sheep), R/SD (black bear), and grid cell counting (root voles and all subsequent data analyses). The kerned demarcation estimate (KDE) method has not been applied, owing to the (unfortunately rarely acknowledged) statistical artifact described in 1.2 *Pinpointing a feasible reason for ignorance: KDE*. A power law as in Eq. 1a contains no asymptote. The slowing area expansion due to an exponent smaller than one in Eq. 1a is easy to calculate: for example, while increasing N from fifty to two hundred is expected to double the area observed, an increase from N=200 to N=800 is needed for the next area doubling, and so on.

Of course, in practice this limitless expansion has to have an end. Any animal has traversed a fixed area within a given period, for example, a summer season. However, Eq. 1a is expressing a power law expansion within a large range of N, much larger than can be explained by "small-sample" statistical artifacts.

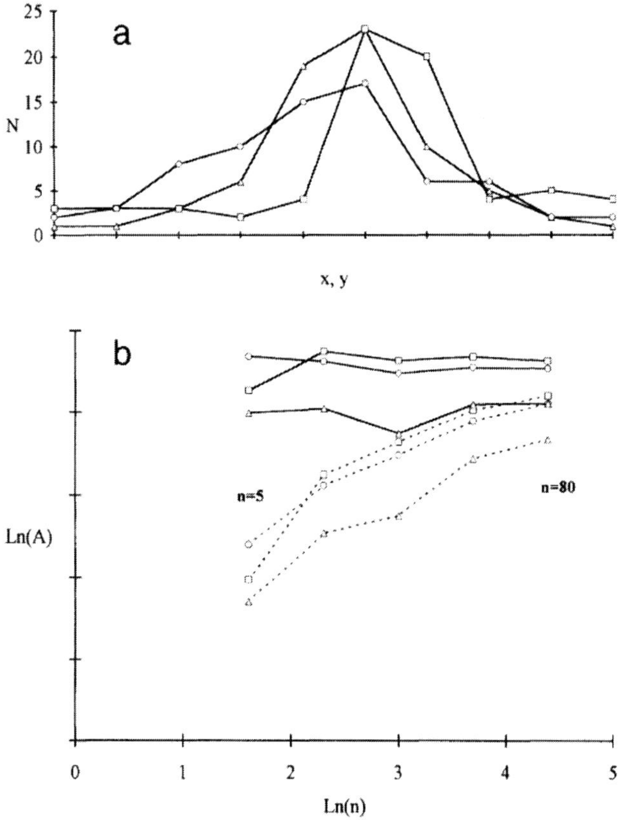

Figure 7. Variants of static area utilization distributions (a), sampled by the Monte Carlo method to simulate sample-size area expansion (b). Dotted lines represent the raw result, while the continuous lines show area that is adjusted for small-sample effect. Thus, large and small samples show similar home-range size after this adjustment.

Depending on area demarcation method, there is therefore a compensation factor to account for to neutralize the small-sample artifact. This factor is not included in Eq. 1a for simplicity and to stress its generality (Gautestad and Mysterud 1993, 1995). For example, if the minimum convex polygon is embedding all fixes at sample size N (a method that was applied for the sheep data only), the right-hand side in Eq. 1a should be multiplied by a function Q(N), which takes a value larger than one for small N but then approaches unity for larger N and thus can be ignored in large samples. Therefore, Eq. 1a will in this case read $A(N) = Q(N)cN^{0.5}$. Non-autocorrelated fixes are assumed. Q(N) is a "rule of thumb" to neutralize the expected small-N artifact from this particular demarcation protocol, under the null hypothesis that an asymptote exists (Gautestad and Mysterud 1995). Q(N) represents an adjustment for sample size effect on area estimate from a non-parametric method (minimum convex polygon). The neutral model (null hypothesis) is derived from the following simple expression:

$$A(N) = Q'(N)K \qquad (2)$$

K represents the area asymptote, and Q'(N) varies between zero and one. This latter function reflects how demarcated area A is expected to start out with a small, positive value for small N and then successively grows with N with a smaller and smaller pace. This implies a positive, but successively smaller first derivative of A(N) while Q'(N) grows asymptotically towards one. Thus, Q'(N) is the inverse function of Q(N), which starts with a large number for small N and then approaches the value one asymptotically with increasing N.

The principle to compensate for the small-sample effect is illustrated in Figure 7, reproduced from ref. (Gautestad and Mysterud 1995). It shows in pane (a) three examples of simulated utilization distributions (expressing the varying spatial density pattern of fixes inside the home ranges) under the classical premises, and in (b) how a minimum convex polygon area asymptote after N=80 is gradually approached but not quite reached under these somewhat varying conditions. However, by multiplying respective (N,A) plots with the calculation for Q(N) [the function is specified in (Gautestad and Mysterud 1993, 1995)], the small-N artifact is cancelled out. Q(N) has the largest adjustment effect for samples where N<50 or so. For larger samples Q(N) continues to approach the value one asymptotically and loses its influence. Hence, if Eq. 1a applies even after adjusting with Q(N), "something else" is needed to account for compliance with Eq. 1a.

A simple method to reveal the difference between an asymptotic function and a power law function as described in Eq. 1a is to log-transform both the x- and the y-axis, like the presentation of sheep data in Figure 3b. In case of a power law and pre-adjusting for small-sample artifacts by the function Q(N) (Gautestad and Mysterud 1995), the function becomes linear—in the present case with regression slope ca 0.5 over the range of log(N) where this relationship holds:

$$\text{Log}(A) \approx \log(c) + 0.5*\log(N) \qquad (1b)$$

The classical model is given by replacing the exponent 0.5 with zero and ignoring (or compensating for) small-sample artifacts. Log(A) then becomes N-independent and equal to log(c), which represents the area asymptote. However, the power law in Eq. 1 has exponent larger than zero. Thus, unit home-range area is found by interpolating to N=1 to find *c* (area pr square root of N). When the exponent is larger than zero as in Eq. 1a,b, this unit area is not represented by an asymptote but a substantially smaller spatial unit than the area found from N>>1.

Without a theoretical foundation, it is just expressing a statistical relationship. In the start, it was an empirically derived regression function, accompanied by relatively hand-waving and concept-based preliminary hypotheses. To fully understand the relationship, it was obviously necessary to turn to the process behind the pattern in a more rigid manner. The need to focus on physics was apparent. Why did a power law emerge, and not the conventional asymptotic law? A power law indicates a scale-free space-use process, complementing the spatially fractal dispersion described previously. Why an exponent close to 0.5 (typically in the range from 0.4 to 0.6), rather than close to zero even after adjusting for small-sample artifact? Why does Eq. 1 hold both for sample sizes that change proportionally with time and—crucially—when sampling is performed uniformly over a fixed period T (defying autocorrelation as a simple explanation)? What are the theoretical transitions between Eq. 1 and non-constrained Brownian motion? And clearly most interesting from an ecologist's view: what may a realistic physical system description with these properties provide with respect to understanding animal movement and space use in a specific environment? What is the potential to extrapolate this local insight to other places, times, and individuals, in particular when comparing these protocols to the strength of traditional model premises?

As became increasingly clear while the doctoral fellowship period progressed, it was very difficult to find answers to these questions. In 1994, two directions of research stood out. First, I could simply have left the issue on the shelf, accepted it as a nuisance, and moved forward with research on animal space use under the current paradigm. This was obviously the safer lane with respect to maximizing my chances for a career in the field. Second, I could increase the stakes dramatically and explore the issue towards the bottom. Metaphorically, my career might then take the same downward direction if a solution was not backed up by a sufficient number of referee processes and papers in demanding journals. I took the risky lane.

In the following I provide a guided tour through the basics of studying the biophysics of complex movement and space use, where "complex" in the present context implies cross-scale influence from a combination of temporal and spatial memory. Where other books and papers on animal movement tend to take some basic system properties as granted, I follow another approach. This book is primarily focusing on these unspoken premises. From my experience with analysis of movement data (and not forgetting actual field experience in younger days), I have found it necessary to dig deeper. Since no acceptable solution to the already mentioned paradoxes (and more to come in later sections) was found within classical models, I had to start building novel elements for an unorthodox foundation.

Thus, most of this book regards a review of this process, with some new and unpublished elements added. The presentation is mathematically very basic relative to current model spin-offs from the classic space-use paradigm that has matured over hundreds of years. However, *conceptually* the new approach and model elements may appear cognitively demanding. I have struggled continuously to avoid "classical" perception of how movement-influenced systems are supposed to tick along over space and time, to keep focus on how the process appears under the alternative or extended premises for the dynamics. Over and over again I've had to arrest myself in thinking along classical paths while trying to conceptualize and formulate the new approach. Most of all the physical aspect of movement—mechanistic and statistical-mechanical properties of particle kinetics—needed to be revealed and evaluated. For example, what were the basic prerequisites for applying Brownian motion—a classical random walk model—as a representation of movement?

1.6 The next step

You have now been introduced to The Home Range Ghost, the "statistical nuisance" that changed my line of research in a fundamental manner. In many respects this individual-level paradoxical pattern complemented the empirical paradox I had already discovered in the population-level study on aphids at the university campus a couple of years earlier. In my quest for coherence I decided to make a bold turn towards biophysics. Next chapter explains why.

2. BASICS OF MOVEMENT PHYSICS: THE STANDARD MODELS

2.1 Transitions from deterministic to stochastic movement

Many ecologists would say that to understand random walk, one should start with its statistical properties. I oppose that view. To understand random walk, we need to start with physics—the process or driving force that generates the statistics. In particular, random walk and Brownian motion regard the pillar of physics embedding statistical mechanics. However, before entering that particular scene, I start at a lower level in the hierarchy of system conceptualization: at temporal micro-resolutions where we find the playground for deterministic—rather than statistical—mechanics. Here the analogies between physics of motion and forces on one hand and movement-generating behaviour may be compared. Following this elementary entry-point, I coarse-grain mechanistic movement paths to the level of statistical mechanics, which gains the major focus throughout the following chapters. From simple illustrations it should be reasonably clear why I, in the preface, claimed that (a) modelling of animal movement and population dynamics is basically based on mechanistic principles; (b) these principles provide the statistical-mechanical foundation of individual-based and population models; and (c) if spatial memory and multi-scaled habitat utilization are involved in the process, the mechanistic principles are violated to the degree that a qualitatively different kind of statistical mechanics may be needed.

As a starting point consider the plain and simple dynamics of the bouncing billiard ball in Figure 8, moving with constant speed from **A** to **B** during a time period T. It is trivial to predict the future location **B** from the position **A** in the given simplistic environment, knowing the speed and direction. If you actually find the ball at **B** after time T, additional intermediate relocations (circles) at intervals Δt would be a waste of effort to pinpoint the ball's real path. Instead, you could simply calculate these positions by interpolation. In fact, in this simple system you would construct a deterministic-mechanistic model to describe it. In a first simplistic definition (to be expanded and rattled in a stepwise manner below), a mechanistic system implies that it may be completely explained through the motion of matter under physical laws, as the prediction of movement from **A** to **B** under a premise of constant speed in the given environment. If the prediction does not hold, it is a matter of expanding the equation of motion with more terms to capture a more extensive range of factors determining the path from **A** to **B**. Thus, under this premise the system is not only mechanistic, but *deterministic-mechanistic*.

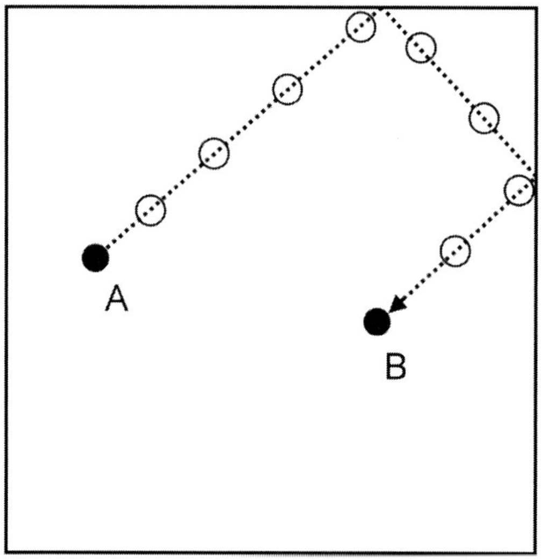

Figure 8. Movement by a billiard ball from A to B on a flat surface, with intermediate locations marked by open circles.

Next, let us send the ball into a more complicated environment. Consider that the floor of the container is rugged and also includes some smaller, randomly located obstacles that lead to reflections of the ball's direction, similar to hitting the wall. Unless you increase the observation frequency dramatically (i.e., Δt is strongly decreased to $\Delta t' << \Delta t$), and you are able to expand the movement conditions by adding realistic terms for local variations in the environment (adjusting for the fine-grained ruggedness and obstacles), the deviation between your prediction of position **B** and the true location **B'** after period T will make the prediction valueless. **B** could end up anywhere within a large spatial zone of uncertainty. Even a largely expanded mechanistic equation for the movement, accounting for at least the major path-influencing obstacles, may not be sufficient. This means that you in practical terms must reduce the observation period T dramatically to regain any predictive power at all from your deterministic-mechanistic model. Theoretically, in this manner you could maintain a deterministic-mechanistic system description, but realistically you will have to replace it by a *stochastic-mechanistic* model, in this case a system resembling Brownian motion.

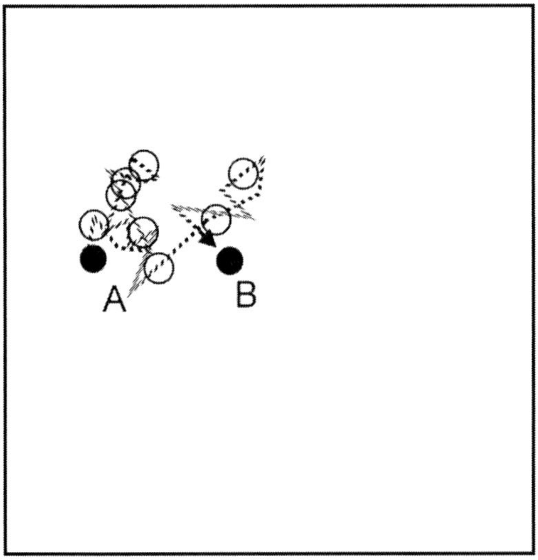

Figure 9. Billiard ball movement on a rugged surface.

Brownian motion implies that the deterministic micro-resolution mechanistic details are ignored and replaced by a more coarse-grained representation of the series of movement-influencing events. In this more complicated scenario, movement will lead to a Gaussian-formed ("bell-shaped") probability function for the individual's whereabouts, F(x,y,t), with the spatial axes as x and y. Time progression of the area covered by the Gaussian function (its variance) should also be considered. The probability of observing **B** within the small "box" of local range ($x+\Delta x;y+\Delta y$) at time t is given by F(x,y,t) with values at the z-axis. Thus, the bell-shaped probability distribution function with **A** at the origin describes the expected distance moved—and the variance of this distance (mean square deviation)—from a starting point as t is increasing. The variance increases over time as the Gaussian function becomes extended and "flatter," since its volume has to remain constant (the summed probability for B's whereabouts sums to one). Simultaneously, the local probability for finding **B** within any given location (x+Δx;y+Δy) is diminishing as time goes by.

If the ruggedness is strong and collisions with obstacles are frequent, the path of a given length becomes convoluted, and may progress in a new direction many times during even a small interval from a given point in time t to t+Δt. Owing to the stochastic nature of the path in this scenario, **B** will probably end up very close to **A**. This happens despite a constant movement speed (or constant average speed) and thus constant total along-the-path distance travelled during a small increment Δt or within the total period of observation, 0<=t<=T (Figure 9). The path will follow a more or less random inter-step direction for successive displacements during Δt, given that this interval is larger than the average direction-interrupting event interval due to the rugged and obstacle-containing surface.

Directional persistence (maintenance of direction from one step to the next) will increase if Δt—consider it as your observation scale for fix sampling (temporal grain)—is decreased, leading to what is called a correlated random walk representation of the process (Turchin 1996). In this case, with high observation frequency on the path, two successive steps will tend to have a similar direction. The hypothetical deterministic-mechanistic model with potentially "a zillion" terms to account for micro-scale obstacles will gradually reappear at even finer resolutions. In Appendix A, I show a correlated random walk that is sampled at different intervals, showing transition towards Brownian motion-compliant path at larger intervals. Conversely, starting with the large sampling interval, the Brownian motion will change to a correlated random walk at finer sampling intervals.

A billiard ball is a perfect example of Markov-compliance. If the ball is at rest or moving with constant speed (and disregarding friction), it will stay in its current state until a force—a change of conditions—changes its speed by invoking acceleration or deceleration. Animal movement is of course more complicated even under the simplifying Markov premise: successive displacements are the results (resultant vectors) of a superposition of internal state (intrinsic dynamics) and environmental conditions (external input). On the other hand, a billiard ball has a constant internal state (weight, composition, form), and its motion is in the hands of the environment only. Thus, the main quest to understand complexity of animal movement involves a realistic modelling of the transition of an individual's internal state to actually moving, where to move and when to execute the respective goals. In particular, spatial memory processing and scale-free space use may be regarded as core aspects in this scrutiny. Consequently, this is a completely different ball game in comparison to the memory-less billiard ball's behaviour.

In ecological theory of movement, it is a customary approach to consider an intermediate kind of system abstraction between deterministic- and stochastic-mechanistic model designs. For example, one rule in the space-use model may be represented by correlated random walk, while another rule is added to account for various forms of spatially directed movement. This may be invoked as more or less deterministic responses to specific local events and conditions (like discovery of a food patch and attraction towards it) (van Moorter *et al.* 2009, Nabe-Nielsen *et al.* 2013). Such events will distort the otherwise mainly forward-moving correlated random walk path. Increased level of mechanistic details in the movement algorithm should then bring the model prediction into stronger compliance with the local system conditions. For example, the deterministic set of rules could be expanded, or there could be introduced some habitat-specific variations in degree of directional persistence in the stochastic component of correlated random walk (say, stronger directional persistence in a suboptimal habitat, leading to shorter staying time in such patches).

There are also fully stochastic-mechanistic models that mimic a mixture of free dispersal by Brownian motion and a simultaneous attraction towards a given centre (directional bias), leading to emergence of a home range (Moorcroft and Lewis 2006, Fieberg 2007). For a relatively non-technical review, see Smouse et al. (2010). These kinds of *mixed mechanistic models* represent the dominating approach to illustrate concepts of movement in an ecological context, like optimal foraging and space-use strategies (Turchin 1998, Grimm and Railsback 2005, Börger *et al.* 2008, Revilla and Wiegand 2008, Viswanathan *et al.* 2011, Nabe-Nielsen *et al.* 2013).

An important property of correlated random walk at a given scale Δt is the transition towards classical Brownian motion if the resulting path is sufficiently "coarse-grained" by sub-sampling the original path at larger intervals. The stronger the directional persistence at temporal scale Δt, the stronger the path needs to be sub-sampled by increasing Δt to reach a level where directional persistence is lost and the coarser path consequently describes a Brownian motion (Turchin 1996). It is in fact a nice property—statistically—that Brownian motion is maintained for all coarser sampling resolutions than this transition scale (given otherwise identical conditions for the walk). And finally, the reason for applying Brownian motion theory for animal movement is that micro-scale correlated random walk may also be represented by a more or less extensive set of deterministic movement rules. Regardless, a transition towards Brownian motion will appear (according to classical premises that are scrutinized below) if the resulting path is sub-sampled at a sufficiently large Δt relative to the micro-scale where the deterministic rules are executed. Thus, application of Brownian motion physics in movement models does not imply that the actual individual is a drunken walker. The deterministic details are simply hidden at finer temporal resolutions than the sampling rate of the path.

Despite the stochastic nature of a Brownian motion and correlated random walk in mechanistic models, there is still a potential for strong system predictions in a statistical sense, thanks to the properties emerging from coarse-graining a path. For example, the Gaussian probability function described above for the spatially explicit location of **B** after time Δt is an emergent property owing to the central limit theorem; "the law of large numbers," where each "number" is a movement-influencing event in this case. The term "emergence" implies that this distribution is not expressed explicitly by the micro-scale movement events but appears as a side effect of the process. In a non-homogeneous environment, the Gaussian form of the space-use probability distribution will of course be distorted, owing to locally varying strength of directedness at the micro-scales of correlated random walk. An important field of research will then seek to reveal the reasons for such distortions from the generic model. Theoretically, this process was clarified with great detail a long time ago.

The Gaussian probability distribution for pinpointing **B** after time t is spatially explicit. The probability for finding **B** regards a specific location (or more precisely, within a small neighbourhood of the location). Thus, it is called an Eulerian system description. Alternatively, one may study the process in a spatially implicit manner, like collecting displacement lengths between successive relocations along the path, regardless of the actual spatial location of each displacement. This approach regards the so-called Lagrangian system description. In this case the expected step length function describing the distribution of distances moved during a large set of intervals Δt is expected to fit a negative exponential: a Poisson distribution (Benhamou 2007, Gautestad 2013b). According to this distribution, larger steps than the average displacement during Δt are rare. In particular, the probability for a step in the range (x + Δx) is smaller than the probability in the range (x - Δx). If the relative ratio P(x + Δx)/P(x -

Δx) for successive step lengths is maintained with similar magnitude for respective increments of x ("bin size" Δx), the distribution is negative exponential-compliant and describes a scale-specific process. The probability for longer steps rapidly falls towards zero, since $P(x + \Delta x) < P(x - \Delta x) < 1$. In summary, the spatial explicit probability function F(x,y,t) will in a homogeneous environment describe a Gaussian probability distribution, while a spatially implicit analysis of step lengths is in a complementary manner expected to describe a Poisson (negative exponential distribution).

2.2 From mechanistic to statistical-mechanical modelling

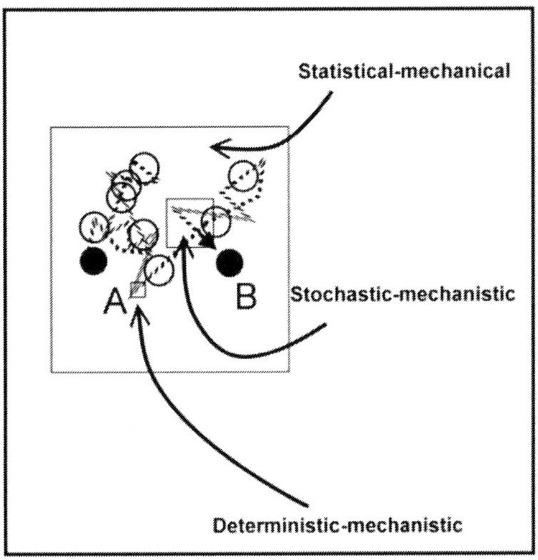

Figure 10. Invoking the hidden layer by sub-sampling a deterministic path, leading to observational shift from mechanistic to the statistical-mechanical level.

The observant reader will now correctly have a feeling from the foregoing section that we have moved even further away from the deterministic-mechanistic system we started with in Figure 8 and complicated to a stochastic-mechanistic system in Figure 9. Changing focus to the emergence of Gaussian and Poisson distributions in large samples has in fact moved our system abstraction closer to the realm of statistical mechanics! At this "holistic" level (Figure 10), referred to in the preface as a consequence of invoking a "hidden layer," we can also look for other, novel system properties that are not explicable at the mechanistic level. For example, the expected distance from **A** to **B** (proportional with the magnitude of the standard deviation of the spatial explicit Gaussian distribution) is expected to increase proportionally with the square root of T—the standard diffusion condition. Again we are ignoring a range of complicating factors to account for in real system parameterization, but these are skipped for simplicity in this walk-through. In some environments and under some conditions, the diffusion will be faster or slower, and the difference will be clearer the larger the sample of fixes and replicates from start position A that are included to estimate the diffusion rate.

Thus, Brownian motion (which will emerge from sampling a correlated random walk or a mixed mechanistic movement at larger Δt, where directional persistence is lost) represents a dual kind of coarse-grained "single-particle" system representation. First, it is a feasible approach both in stochastic-mechanistic and mixed mechanistic models to capture the random aspect of complicated movement. Second, Brownian motion is also an implicit premise for a statistical-mechanical system description—in this case classical, non-equilibrium statistical mechanics. Theoreticians in the mechanistic camp—including modellers in movement ecology—may therefore claim that Brownian motion (including correlated random walk and other variants) are just deterministic motion with more or less statistical noise added. At the other side, physicists in the statistical-mechanical camp may claim that Brownian motion is the single-particle manifestation of diffusion, and thus clearly inside their domain of system abstraction. In short: Brownian motion presents itself as a coin with two faces. Its head describes micro-scale mechanistic details (which may be deterministic, stochastic, or a mixture), and its tail describes the more "holistic" statistical-mechanical properties that emerge in—for example—the functional form of a collection of step lengths. This duality between micro-scale mechanics and meso/macro-scale statistical mechanics makes Brownian motion an ideal meeting ground where ecologists may reach out for new modelling tools on the physical side of the coin.

Putting on the statistical-mechanical glasses when observing Brownian motion, new model parameters specific for this level of abstraction will appear; for example, the diffusion rate (basically, area covered pr. unit time); the functional form of the distribution of relocation lengths at temporal scale Δt; and so on. At this level the system properties *indirectly* reflect aspects of the underlying biology and ecology. However, at the statistical-mechanical level, single observations like a specific displacement from **A** to **B** during T make no sense. Only a large set of such observations gives meaning for quantification, in contrast to the mechanistic level at micro-scales where each displacement may be modelled and interpreted individually, by the actual movement algorithm. In contrast to the mechanistic level, the more movement-influencing and successively independent events that take place on average during Δt (the micro-scale depth) and the larger the sample of moves (macro-scale degrees of freedom), the better the chances for proper model parameterization in statistical-mechanical terms.

Parameters and functions at this coarser level of abstraction may in fact tell more about the moving particle than a fine-grained deterministic-mechanistic model in a complicated environment (Gautestad 2011, Petrovskii *et al.* 2011). Thus, some aspects of the system may be better analysed—and provide a better prediction power when extended to other systems—at the statistical-mechanical level than at the temporally fine-grained mechanistic level. As will be described later, in complex systems involving spatial memory, the statistical-mechanical approach may even turn out to be a prerequisite for a model's predictive power.

Figure 10's conceptualization offers a nice starting point to illustrate the important concept of mean free path in statistical mechanics. The mean free path is the average distance travelled by a moving particle (such as an atom, a molecule, a photon) between successive impacts (collisions) which modify its direction or energy or other particle properties. Translated to our initial billiard ball arena in Figure 8, consider the mean free path to be determined by the average distance to the walls of the arena from a random position inside the arena. In the more complicated area with ruggedness and obstacles, the mean free path will be smaller. Thus, the diffusion will consequently be slower, which explains the statement above that the distance **A-B** during T will typically be much smaller than in the initial billiard ball arena, despite a constant movement speed at the mechanistic resolution. The mean free path obviously resonates well with important aspects of movement ecology. For example, an animal will typically perform a more convoluted path in a resource-rich habitat (area-restricted search, ARS). Consequently, the residence time in this habitat will be larger. In other words, while a mechanistic system description may focus on local residence time as a function of local ecological conditions, a statistical-mechanical description will study the same aspect from the perspective of the inter-related mean free path, mean square deviation as a function of time and diffusion parameterization. In this case correlated random walk or Brownian motion represents the model for the physical "particle" experiencing movement-influencing events.

An important property of a statistical-mechanical system description and simulation is the *ignorance* of whether mechanistic micro-scale details at the lower levels in Figure 10 are deterministic-mechanistic, stochastic-mechanistic, or a mixture of these properties. For example, in Gautestad (2013b) I described how an object called Markov the robot moves around in accordance to deterministic-mechanistic rules. Similar to the billiard ball in Figure 9, but in a more explicitly heterogeneous terrain, the robot was set to move in a unidirectional manner until an obstacle or other direction-influencing event appeared. The robot's algorithm then decided the actual response and thus the direction for the next move. The robot and the given environment together described a stochastic-mechanistic system when an observer is sampling the path at sufficiently large scale Δt within a given period T to ensure many successively independent direction-influencing events during Δt. Thus, the deterministic-mechanistic path turned into a single representation of a statistical-mechanical path at the coarser time resolution. Many such path realizations (typically under similar conditions) would be needed to provide sufficient material for a statistical-mechanical description.

Compliance with the statistical-mechanical law of diffusion at the abstraction level of statistical mechanics may then be verified by testing for this level's observable quantities. The mean free path that was experienced by Markov the robot during T is one example where the magnitude of this quantity tells a lot about how the creature experienced and responded to its environment during T. For example, what strength of spatio-temporal resolution power was the robot set to respond to? An estimate of the mean free path owing to the Brownian motion theory would be able to reveal this. Knowing the environment's various attributes, the mean free path estimation may indirectly tell something about the robot's transition zone of ignorance and responsiveness. On the other hand, knowing the robot's inner working (the movement algorithm) and its settings that determine responsiveness and resolution power, the mean free path estimate may indirectly reveal properties of the environment. A strong responsiveness and a great ability for perception of micro-details in the environment will translate to a smaller mean free path estimate from the movement data, owing to larger frequency of interrupts on the otherwise unidirectional movement. Thus, the diffusion rate will then also be smaller.

Testing the robot's compliance with classical diffusion means to test whether the Gaussian probability distribution function fits the local density distribution from a large set of **B**-locations after time T from starting repeatedly in the close neighbourhood of **A** (with random start direction). Wider distribution (larger mean square deviation) means stronger diffusion rate (larger area/T) in the given system. This would then express and quantify weaker influence from floor ruggedness and local obstacles on linear, deterministic movement at mechanistic micro-scales. If such a set of replicate moves are repeated from another start location, one could then—based on smaller or larger differences in the respective diffusion rates—conclude something about the micro-scale system details in general terms. Thus, the mean free path parameter is coherently related to the spatially explicit diffusion parameter. It could be calculated—for example—in the Lagrangian context of a Poisson distribution of step lengths, owing to study displacements regardless of specific location within the defined area and sampling period.

2.3 Behavioural ecology at the statistical-mechanical level

To summarize, other types of observable quantities and system details are analysed at the statistical-mechanical level than in mechanistic representations of movement. In a movement ecological context, while compliance with a specific set of foraging rules is explored and simulated at the mechanistic level, the general outcome from such rules in more macro-level terms is studied at the statistical-mechanical level. For example, does a specific set of movement rules (successively executing a "what to do if…" algorithm as response to local conditions and internal state) lead to a similar diffusion rate, and how does this rate correlate with forging success in the given environment? Do these rule variants converge to a similar probability distribution function—for example, the Gaussian (spatially explicit) or Poisson (spatially implicit)—at the statistical-mechanical level? If so, what is the distribution parameters' sensitivity to details of the respective rules or the complicatedness of these rules (the difference between complicatedness and complexity was defined in the preface)? In particular, may a very complicated and extensive algorithm be simplified and still lead to similar parameter values at the statistical-mechanical level? What rules tend to influence these parameters and which do not (sensitivity analysis)? In the following I will first elaborate on the duality between the traditional mechanistic system approach in movement ecology and the complementary system description in a statistical-mechanical frame of reference. Following this, I will show how the latter approach may have much to offer, in particular where data consists of GPS series of individual relocations.

In fact, many kinds of statistical analyses are in fact statistical-mechanical, despite the fact that this is rarely explicitly stated. This often leads to what—in my view—is an unrealistic conception of animal movement and space use in mechanistic terms. I will return to this later. First, in a concrete example, consider a simulation where the individual's direction of movement and stopping probability are sensitive to the appearance of local food patches within a small detection distance (called the perceptual field). When hungry, the animal tends to approach the food, capture it, and stop for ingestion. Perhaps it performs an area-restricted search for additional food after having consumed the first item or patch, but with smaller probability the less hungry it is. The degree of hunger and thus environmental responsiveness is easy to specify by the deterministic-mechanistic rules for simulating the movement.

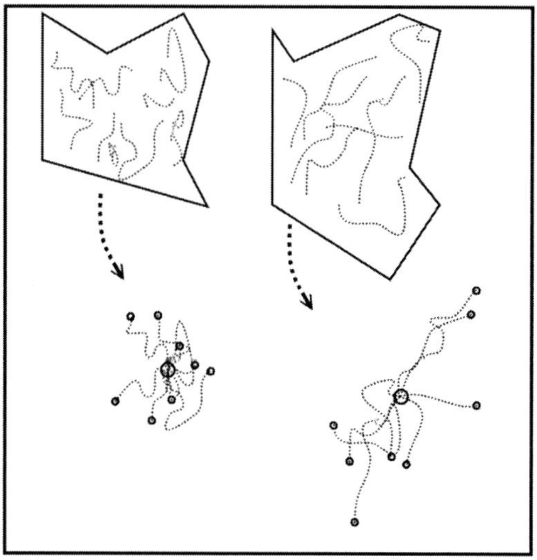

Figure 11. A random collection of path segments from a fixed time interval (upper left) are spatially shifted to be superimposed on a common spatial point that represents their respective start positions (lower left). The similar example to the right illustrates a condition that leads to larger path lengths on average (larger time segment or faster movement on average).

After having collected a large set of fixes at temporal scale Δt—which is much larger than the time increments for sequential execution of the movement rules (to ensure correlated random walk or Brownian motion compliance)—and repeated the simulation many times under similar conditions, one may then analyse the system as outlined above for the statistical-mechanical level. The overall conditions that were experienced by the animal during T=N*Δt would then be reflected in a given standard deviation of the Gaussian distribution. At this point in the thought experiment, the fix-collected path at scale Δt is assumed to show strong spatial autocorrelation. In other words, two consecutive fixes will on average be closer in space than two randomly selected fixes during T.

With only one data set for the given environment, there is no room for hypothesis testing. However, the number of replicate paths could be increased by estimating standard deviation of path segments where the animal traversed the given area, after normalizing the set of series. Normalization implies that dispersion from each start position among these path segments is shifted to a common start position—their weighted mean (x,y) position for the first fix in each series—before calculating the standard deviation for the set. According to diffusion theory the magnitude of the standard deviation under a given set of system conditions would then tend to stabilize as a function of increased number of path segments.

The principle is illustrated in Figure 11, showing two habitat types that are demarcated by polygons. An individual is tracked with small Δt during a given number of revisits to respective sites, leading to a start and an end position for each tracking series while still inside a given habitat type. Each of these series is assumed to represent an inter-independent revisit to the respective polygons; otherwise, the movement would not comply with basic premises of classical diffusion. The tracking time T is also constant for each tracking event. The dotted-line arrows point to the normalized sets of respective path segments, where the start positions are overlapping (larger circle) and the end positions are located at the net distance traversed from the start position from respective path segments (smaller circles). Clearly, the animal has—on average—displaced itself a longer distance during T in the right-hand habitat. Hence, the standard deviation of the end locations of the centre-normalized paths is larger, and the diffusion rate is larger. The difference could then be subject to biological-ecological interpretation, like correlating it to relative differences in resource density between the two sites.

From the diffusion law it should be easy to adjust for variations in T, if the simulations were run for different lengths T. The standard deviation will increase proportionally with square root of T—an intrinsic property of a Gaussian probability distribution. Alternatively, one could study the diffusion strength from another approach, by quantifying the mean free path-related parameter in the Poisson distribution for displacement lengths at temporal scale Δt. This could be found by studying the distribution of distances from start to end positions in the respective sets of paths. Anyway, in both cases a small diffusion rate and a small average displacement length (implying a small mean free path) will be found in a relatively food-rich environment. In this manner, fine-grained mechanistic simulation and macro-scale statistical-mechanical analysis complement each other in an elementary study of a given movement algorithm or a set of variants.

However, there is a transition zone to account for, since we specified for the conditions in Figure 11 that Δt was large, but still small enough to produce spatial autocorrelation at this sampling scale. Under such high-frequency tracking, the observed path will appear relatively "smooth," with jaggedness appearing upon zooming into some segment of the path. If Δt leads to correlated random walk, the direction of the line segment from fix A to fix B will tend to be similar to the direction from B to C. The series of fixes is then both spatially autocorrelated and positively direction-correlated. Consequently, when paths are observed at such fine resolutions (like in Figure 11), the standard deviation will tend to increase with a higher rate than square root of T—closer to increasing proportionally with T (Turchin 1996). On the other hand, Brownian motion—and thus compliance with classical diffusion laws already at Δt and towards larger intervals—requires statistical independence between successive directions along the path. This can be achieved by increasing Δt sufficiently (Turchin 1996)—given that the process is mechanistic and therefore Markov compliant. To achieve Brownian motion a larger Δt is obviously needed in the habitat to the right in Figure 11. This important concept of transition between the mechanistic and statistical-mechanical level is central to the theory of movement ecology, and will be explained next.

2.4 Markov compliance and the characteristic scale

Consider a path comprises a chronological series of location points, *P*, at discrete time intervals associated with diverse context-varied attributes, which can be represented using a finite set of triples:

$$P = \{(P_{\tau 1}, \tau_1, A_{\tau 1}),(P_{\tau 2}, \tau_2, A_{\tau 2}),\ldots(P_{\tau n}, \tau_n, A_{\tau n})\} \quad (3)$$

where $P_{\tau n}$ denotes the spatial position at time τ_n and $A_{\tau n}$ denotes the associated attributes, such as movement speed, direction, cognitive state (e.g., hungry, scared, or tired) and other factors that at time τ_n contribute to a given displacement. Among the triples, “position” and “time” are compulsory elements and are the syntactic properties of a path, whereas “attribute” is optional and varies among different scenarios along the series.

The difference $\Delta\tau=\tau_n-\tau_{n-1}$ implicitly defines a *characteristic time scale* for the process, and this scale is assumed to be of similar magnitude for the respective terms in Eq. 3. With respect to movement, this characteristic scale is the finest time resolution where the individual is “close” to dynamical equilibrium with its extrinsic and intrinsic environment (external conditions and internal state, respectively). Being close to dynamical equilibrium means that the movement at a given moment has absorbed previous movement-influencing events. Thus, in absence of new environmental changes, no further change in the individual’s current direction and speed is expected (speed zero at this point implies that the animal is still staying put until a new influence emerges). In practice, the condition for the characteristic scale for a system (like a moving individual in a given environment) being close to dynamical equilibrium is defined as *the correlation time*. For exponential decays of perturbation influence, it is related to the correlation length, which has both a temporal and a spatial extent: the spatial distance and time interval where only 1/*e* (ca 37%) of the response to a given perturbation remains (*e* is the base of the natural logarithm *e*=2.71828...).

Eq. 3 describes a memory-less process. In fact, all the mechanistic examples above—from the billiard ball in Figure 8 and Figure 9 to the references to modelling opportunistic search for food—regarded memory-less mechanistic processes. In short, when the billiard ball or an animal starts from position **A** in Figure 9, the next displacement during a small time increment is determined by the current environmental conditions and successive responses to these. Except for an influence from the foregoing (or a few foregoing) time increment(s) to allow for determination of directional persistence of the path in absence of disturbance, all previous events are already reflected in the present location **A**. Thus, regardless of the sequence of events that brought the ball animal to **A**, this may in the case of memory-less movement simply be disregarded when determining (with a larger or smaller degree of stochasticity attached) the next position. This kind of process will then reflect Markovian system compliance or a low-order Markovian if a few historic steps are influencing the current behaviour. In the latter case, sampling the path at a coarser temporal scale n*Δt, where a positive integer n sets the magnitude of the sliding time window of historic influence, will bring in two modifications: (1) the coarser path becomes fully Markovian, but (2) the sampling of the path at this temporal scale means that a deterministically generated path becomes mixed-mechanistic (semi-stochastic-mechanistic) or fully stochastic-mechanistic. The latter is typically modelled by Brownian motion.

In this case of Markovian-compliant mechanics at micro-scales, the theoretical coherence between deterministic mechanisms, meso-scale mixed model representations, and macro-scale statistical mechanics is mathematically and statistically well-behaved. Markovian dynamics represent a key premise for classical diffusion processes. Hundreds of books for each level of system description have been published. In ecology, though, the focus has generally been on the mechanistic levels and statistical inference at these levels, under the umbrella of individual-based modelling (Grimm and Railsback 2005, Moorcroft and Lewis 2006). In this tradition there is generally little or no focus on an explicit consideration of statistical-mechanical system aspects and premises beyond generally assuming compliance with statistical properties of classical diffusion and Brownian motion (if the path from the mechanism is sufficiently sub-sampled).

2.5 Markov from another angle: the principle of superposition

Before we dive deeper into statistical mechanics of memory-enhanced systems, it is important to describe the Markovian system—a memory-less process—in more mathematical and conceptual detail for the sake of further illumination of the distinction between scaling simplicity and scaling complexity. In particular, I will highlight the transition from a mechanistic system description to a statistical-mechanical description from additional angles. Why such fine focus when the foregoing presentation may appear sufficiently detailed? Many may even say unnecessarily detailed! My answer is clear: movement ecological theory generally does not describe (for example) Brownian motion as a meso/macro-sampled statistical-mechanical representation of micro-level mechanistic movement. Rather the recent advances of the theory increasingly seem to describe Brownian motion as one of several movement *modes* at mechanistic scales (Humphries *et al.* 2010, Sims *et al.* 2012, Bartumeus *et al.* 2013, Humphries and Sims 2014); i.e., in the context of behavioural ecology rather than statistical mechanics (Gautestad and Mysterud 2013). Clearly, opposing this mechanistic-behavioural interpretation of Brownian motion (and other variants soon to be explained) will require a very detailed and many-faceted line of biophysical arguments. Some of the concepts laid out in the following sections will be unfamiliar to most ecologists, simply because they—to my knowledge—have not previously been explicitly described in the literature on movement ecology. In the following I will also provide some points that connect common system assumptions at the individual and population level.

The classical paradigm for individual movement mechanisms is based on the physical concept of serial processing; in other words—a mechanistic system as conceptually expressed by Eq. 3. In other words, it represents a classical Markovian process. A local event in the environment is assumed to trigger a local response, whether the response involves an individual moving from one site to another (Stephens and Krebs 1986) or a population adjusting to a new density (Jørgensen 1988, Yodzis 1989). A new event triggers a new response, and so on. Some events may take longer to respond to (like a migrating bird moving from its winter to its summer territory when the season changes) than other events (e.g., fleeing from an attacking predator). However, the central principle is that a given movement speed dX determining the distance moved from one location to the next during a small interval in time, dt, equals the sum—*a superposition*—of independent movement-influencing factors ($F_i(x,t)$) affecting the individual from a small neighbourhood (x) around the individual at a given time (t):

$$dX/dt = a_1F_1(x,t) + a_2F_2(x,t) + ... + a_nF_n(x,t) + \varepsilon \qquad (4)$$

All the respective influences $F_i(x,t)$ contribute in sum to "pushing" the individual around in its living space. For simplicity I express X as a scalar, and not a vector in space. Non-specified influences are summarized by a component ε, expressing statistical noise. A realistic model will require ε to be small, which may require the number of components, n, to be large. When we focus on movement mechanisms in a deterministic context during a small time interval, some of these influences may be relatively negligible—like the migratory drift, with a_i close to zero in comparison with the rest of the set (Hirata *et al.*), j=1,2,...,n; j≠i—meaning that this equation for instant motion can be simplified or *condensed* (smaller n) (Wiegert 1993). Other influences may be so spatially patterned (locally variable, but spatially autocorrelated) that a partial differential equation $\partial X/\delta x\delta y\delta z\delta t$ (where x,y,z,t represent spatial coordinates and time) would be appropriate (Brown and Rothery 1993). A trivial example would be an animal reacting to a locally variable habitat with varying movement speed dX/dt (Skellam 1951, Turchin 1991).

The overall distribution of step lengths from solutions of a vectorized (direction-pointing) form for Eq. 4 will in the long run describe a spatially two-dimensional Gaussian distribution with inter-independent displacement events, and the distribution of the direction of steps will in overall terms be uniform over 0-2π radians. Thus, from a holistic, statistical-mechanical perspective where condensation of Eq. 4 in a sense is maximized, Eq. 4 is expressing a process that is compliant with a Brownian motion (Brown and Rothery 1993), even if all the defined terms in Eq. 4 are totally deterministic. Details on this transformation from a deterministic process like Eq. 4 to a totally stochastic process of Brownian motion were described in earlier sections.

In addition to the magnitude of condensation—as expressed by the size of n in Eq. 4—there is another important simplification in this kind of mechanistic system description; the *linear* form of the movement equation. If all the $F_i(x,t)$ components in Eq. 4 are linear (i.e., they do not invoke a non-proportional change in local movement speed dX/dt for a given total change in $F_i(x,t)$), dX/dt complies with the principle of *linear superposition* [see Campbell (1989), p. 6]. The principle of linear superposition specifies that the behaviour of the system can be described by linear and thus generally solvable differential equations. In simple terms, one can then say that one breaks the problem—a solution to dX/dt—into many small pieces with respective subscripts *i*, and then adds the separate solutions to get the integrated solution to the whole system (X). By expressing dX/dt as a sum of n terms, it is consequently assumed that each of the movement-influencing factors $F_i(x,t)$ is independent of the others at fine-grained scale dx,dy,dz,dt.

This linearization and inter-term independency is quite convenient in a modelling context, because it means that the limit value for dX/dt as dt approaches zero does exist in a well-defined way. In other words, the function is differentiable and analytically tractable [see pp. 3-4 in (West and Shlesinger 1990)]. When dt approaches zero, so does the spatial volume of dx,dy and dz. At these fine resolutions the system is assumed to be in local equilibrium, as already explained.

2.6 Linearization of non-linear mechanistic interactions

2.6.1 Spatial coarse-graining

Let us for a moment ignore volumes and return to one-dimensional distances, as in Eq. 4. For real individuals' responses to movement-influencing events are always of a *non-linear*, or non-proportional nature at small magnitudes of x and t. However, these kinds of non-linearity can under specific conditions—in short, Markov compliance—be assumed to become "linearized" when we consider an ensemble of independent movement-influencing events within a subsection of space and time (larger x and/or t). I will now explain these "specific conditions" in some detail, since this linearization assumption often seems to be violated in real systems where scaling complexity emerges. Scaling complexity will be described after the following example of scaling *simplicity*.

Consider a hypothetical situation, a scatter of poisonous flowers in an area where a grazing animal moves around. The system is assumed Markov-compliant, as already defined in a previous section. We define that the animal has a non-linear response when it encounters such a flower. According to our defined set of movement rules, if it sees or smells zero, one, or two flowers within what we can call the animal's field of perception, the animal's path during the present time interval is not influenced. When it sees three or more flowers at the same time, the animal runs away (with probability p=1 for simplicity), trying to escape this potentially dangerous environment (Figure 12). In other words, the response is strongly non-linear, or non-proportional, from a local and short-term (i.e., tactical) perspective. This implies a lack of response (p = 0) under the condition of low abundance of flowers in the neighbourhood and deterministic movement (p=1) at flower density above a given threshold. Simulating this movement and then regressing dX/dt versus local density of flowers for small values of the neighbourhood x, we can expect to find a good fit to the non-linear functional response for the animal and this plant. For simplicity, we ignore the complications from other movement-influencing factors, except for a random walk step of length x after (for example) two time increments in the absence of flowers within range x.

Next, let us superimpose a virtual grid on the area, using grid cell size about the size of the field of perception, and include local densities of flowers in each grid cell. The field of perception pertains to the idealized peripheral border for the individual`s senses. This is a trick we can use to be more specific about the spatial scale, the "immediate neighbourhood," x. Within this scale the individual can react directly to events and environmental conditions. In this spatial configuration of virtual grid cells with size x^2 we get a local variation of local densities of poisonous flowers in the animal's immediate surroundings (flowers pr. x^2), and the corresponding local movement responses to these densities (p= 0 or p=1) at unit time increment, plus occasional random displacement at a somewhat coarser time scale to avoid termination of movement in absence of flowers within x.

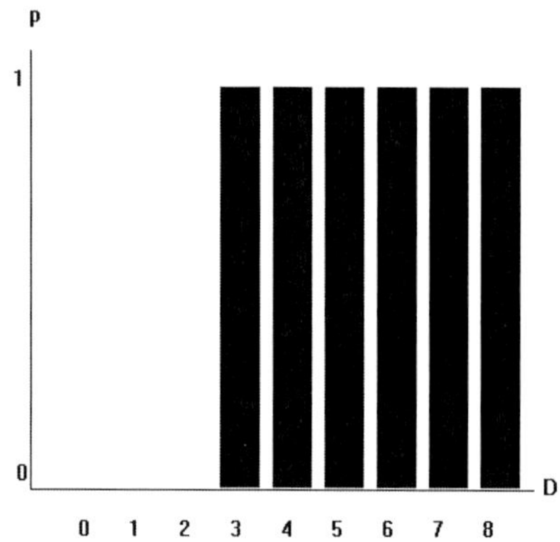

Figure 12. Non-linear fleeing response by an animal as a function of local density of poisonous flowers (0 = ignore; 1 = move on).

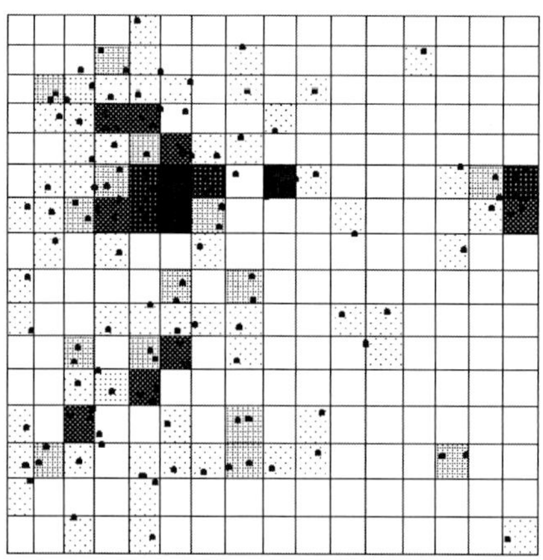

Figure 13. Local flower density at the given grid resolution.

Depending on the scatter of flowers, we get a distributional pattern of local plant densities, where density equals number of plants in a given grid cell. Thus, density includes an explicit reference to scale. In Figure 13, the local flower density at grid scale x^2 is expressed by a grey scale from white (no flowers) to black (three or more flowers).

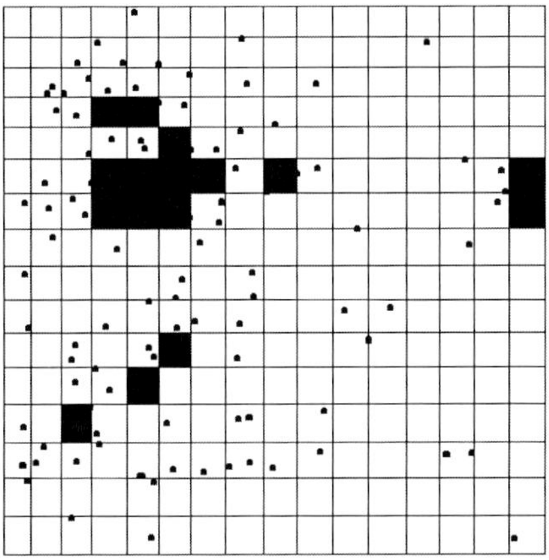

Figure 14. Local fleeing response at the given resolution.

Next, we make a similar spatial mapping for each of the grid cells of the average movement speed dX/dt by defining dt as the unit time increment approximately equal to the animal's response time to local conditions. The response time will of course vary from moment to moment and from place to place, but we simplify by averaging. In contrast to the mapping of local flower density, the latter mapping (Figure 14) should show a larger contrast due to the non-linear functional response: more black-and-white and less grey tones. In our ideal model, grey tones may be considered to be absent at this scale (x and t) with a resolution similar to the field of perception.

What happens with the pattern when we zoom out to a coarser spatial scale, by averaging the plant density and the corresponding local response of movement over groups of four adjacent grid cells? We will see that the black/white contrast from the non-linear movement response becomes less apparent, as shades of grey emerge also in the dX/dt mapping. If we continue to coarse-grain the pattern into even larger groups of cells (Figure

15), we will eventually end up with an approximately *linear* regression line between density of plants and the probability for fleeing (p) during dt with increasing spatial resolution x, close to the inflection point in Figure 16. It is easy to imagine that p is positively correlated with movement speed dX/dt in overall terms at these coarser spatial scales. Larger p means larger displacement on average. However, at these coarser resolutions we have lost additional details with respect to the actual location of the animal within a given cell. It could be anywhere inside the actual cell of size x. But from a statistical-mechanical perspective, the average movement speed in the respective sections of space at scale x during dt will increase with increased micro-scale probability for fleeing from local danger within the trailing field of perception. In other words, while the response is strongly non-linear at the (mechanistic) micro-scale that approximates the scale of non-linear, deterministic-mechanistic behaviour, the process transforms itself to a more linear and proportional response—more in accordance with the premises of Eq. 4—as we spatially coarse-grain the pattern. From that transition scale and upwards, we make no big mistake by assuming that *the principle of superposition* applies, owing to linearization of the functional response at these scales. At these coarser resolutions we are studying statistical-mechanical aspects of the system, while the deterministic-mechanistic non-linear process is hidden away at finer spatial scales.

In a statistical sense, it is easy to see why this tendency towards linearization (Toffoli 1990) appears. After sufficient coarse-graining, both the flower density and the response represent spatial *averages* with underlying probability distributions around these averages. Thus, a large plant density in a given coarse-grained cell may indicate that it embeds a relatively large number of high-density cells at finer resolutions with respect to flowers, which all initiate local and inter-independent "fleeing responses" from the animal. However, some of these smaller cells (of magnitude the field of perception) have density below the fleeing threshold. These cells contribute to lowering the mean fleeing frequency at coarser scales, below what we would expect from the strongly non-linear rule that we defined initially. It is reason to underscore once again: when an individual enters such a coarsened virtual cell, its local position within this cell should be considered homogeneously distributed (all micro-locations at finer resolutions are equally likely during t). This is a crucial assumption in statistical mechanics: all accessible micro-states under a given set of system constraints are equally likely (however, full compliance with this postulate also requires temporal coarse-graining; this will be described shortly).

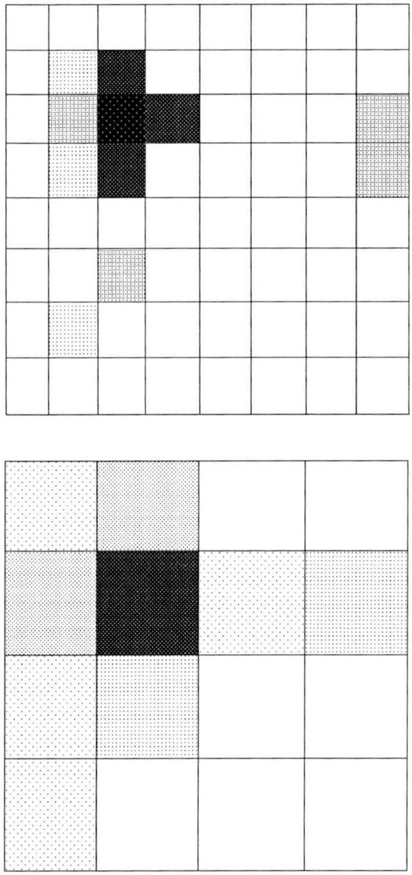

Figure 15. Invoking the hidden layer of detail ignorance by spatial coarse-graining of the virtual grid.

The mechanistic reason for this linearization effect at the statistical-mechanical level lies in the fact that the functional response in the present example was defined to be *scale-specific* to the scale of the original grid cell size, which was approximating the animal's field of perception. This idealized spatial resolution represents *the characteristic scale* for this element of the movement process (Norton and Lord 1990, Wiens 1990, 1995). The animal responded tactically to the local flower density at this scale. A tactical response also implies a sequential process, since it responded dynamically to here-and-now plant density at spatial unit scale x and temporal resolution dt. Under this premise the flower density the animal experienced (and responded to) a long time ago and a long distance apart is not influencing the local response during this unit increment dt, which equals the average staying time in a grid cell of unit size. In other words, the process is Markov compliant.

The modelling advantage from compliance with the Markovian principle and a mechanistic system design should now be obvious. The flower density and the functional response (say, a specific term $a_iF_i(x,t)$ in Eq. 4 where other terms are set to zero in our example) at coarser scales than x represent averages from a set of independent spatial units of this term at finer scales. Events within these micro-scale units are independent, since the local "push" on the animal in a given cell is independent of the magnitude of local push in any of the other cells that contribute to the average estimates [see With and King (1997) for a review of movement processes in a landscape modelling context]. Thus, we can assume that the *linear* form of Eq. 4 is based on an assumption about Markov-compliant movement mechanics at micro-scales. In other words, each term $a_iF_i(x,t)$ represents a coarse-graining of finer-scaled process events. In terms of the transition between mechanistic and statistical-mechanical frame of reference,

$$a_iF_i(x,t) = \{a_iF(x_1,t) + a_iF(x_2,t) + ... + a_iF(x_m,t)\}/m \qquad (5)$$

where x_1, x_2, ..., x_m are m embedded unit cells at the process' characteristic scale, and the scale in focus by the observer (modeller) is m times as large as these subunits. In other words, $a_iF_i(x,t)$ represents a macro-scale linearized term for local movement, in a system approximation based on classical statistical mechanical postulates and premises. Dividing by m "normalizes" the process back to the unit scale for the scale-specific process.

The central limit theorem applies at the macro-scale m, since $a_iF_i(x,t)$ is the spatial average response to a given component of movement-influencing factors over a large set of (assumed) locally independent functional responses to this component. Thus, the distribution of the expected *average* response (average local speed) to a given component (a term in Eq. 4) at coarser scales than the characteristic system scale approximates a Poisson (negative exponential) distribution, even if each local component at the micro-scale of behaviour processing is strongly non-linear. Further, the distribution of speed responses at a given set of coarsened grid cells (like shown in Figure 15) will become approximately linear in the neighbourhood of the average response speed (Figure 16). One of the statistical-mechanical system instances that allows for calculating the average speed at temporal scale dt is given by the average p at the y-axis in Figure 16, reflecting the average plant density at the x-axis in Figure 13. The average point (x,p) ≈ (6.5,0.8) represents the given movement speed term $F_i(x,t)$ of Eq. 4 in this particular realization of average plant density, when the dynamics are studied at the scale of the entire arena.

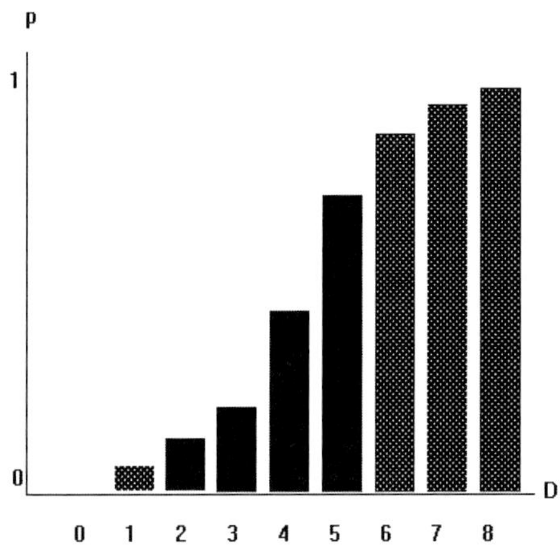

Figure 16. At the coarser resolution, the non-linear response function in Figure 12 becomes more linear (closer to proportional fleeing response as a function of flower density).

This property of serial, scale-specific processes leads to the linearization effect at coarser spatial scales than the characteristic scale. Linearization and thus compliance with the principle of superposition is for example a prerequisite for applying (standard or partial) differential equations to describe local system dynamics. Thus, this cornerstone mathematical tool for contemporary modelling of animal movement depends on Markov-compliance at micro-scales. When we in the following chapters turn towards accounting for memory effects on space use, this dependence on the linearization principle when applying differential equations (and other classical tools) will force us to consider other model designs.

2.6.2 Temporal coarse-graining

As a last but important twist of this example where the duality between a mechanistic and statistical-mechanical system description is defined, consider what happens as we coarse-grain in both time and space. I mentioned above that linearization "trick" with respect to Eq. 4-5 required that all accessible micro-states under a given set of system constraints were equally likely. Consider a coarser grid cell, where x is substantially larger than unit spatial scale, which was defined as the field of perception (representing the system's characteristic scale). When an individual enters this cell, the probability that the animal is close to this border location during the unit time increment dt is larger than a location closer to the opposite part of the cell. Thus, if dt is maintained at unit size, intra-cell homogenization of location is not satisfied at coarser scales than the unit cell size.

Within a cell of system-characteristic size, the small increment t for the process was assumed to be sufficiently large to homogenize the animal's probable location, but system coarsening towards larger cells will require larger dt as well. In a statistical-mechanical system, unit spatial scale is represented by the mean free path of the particle, which in our context is the individual (see *2.2 From mechanistic to statistical-mechanical modelling*). At coarser scale than the mean free path, the animal will move Brownian motion-like, owing to Markov compliance. Thus, the expected time to traverse an m^2 larger virtual grid cell will be m(dt) larger, according to the diffusion law. Consequently, system linearization of *de facto* non-linear processes at micro-scales—in order to satisfy the principle of superposition behind a mechanistic description like Eq. 4-5—will require unit time scale, dt, to be expanded proportionally with spatial coarse-graining. Therefore, coarse-graining to mx with m>1 will require coarse-graining to m(dt) to satisfy the premise that the animal is locally ("within-cell") randomized at spatial resolution mx.

What about coarse-graining in time only? Local plant densities have insufficient temporal variability in our example, so we now replace the poisonous plant with a predator population. Let us say that predators attack the model animal (now a potential prey) if three or more predator individuals by chance are present within the animal's field of perception. For simplicity, we assume that the predators move independently of one another, and that their field of perception is smaller than for the prey individual. So, even if a predator enters the prey's field of perception, there is a chance that it remains undetected by the predator. Our model individual is assumed to know this (or at least it responds as if it knows), so its functional response is similar to the plant example above: it simply ignores up to two predators in its present neighbourhood, but flees when more predators appear (the chance of remaining undetected seriously vanishes). Again, we have a situation where dX/dt represents a local "push" on the (prey) animal, with the new situation that the local intensity of this push varies in time as well as in space.

Let us now again superimpose a grid with cell size equal to the field of perception. This time, we will find that the functional response curve may become more linear (expressing proportionality of the response) even without coarse-graining in space, given that we consider a larger dt. The reason is that each time the prey individual revisits a given grid cell (during a sufficiently larger dt, this may well happen), it encounters a number of predator individuals which is statistically independent of the local number at previous visitations to this cell. In other words, successive "pushes" become independent but with variable intensity (so that a distribution of push intensity underlies the mean push for a given grid cell). Thus, if the mean number of predators in two cells were two and three during an enlarged time interval, the respective average fleeing response when visiting a given unit-sized grid cell (when we express this average response as probability p for fleeing) would probably not be p=0 and p=1, respectively, but something in between (say, p=0.3 and p=0.6) due to the central limit theorem. That is, the averaging procedure over a larger interval leads to coarse-graining over independent events in *time* rather than space in this example, so that the same linearization effect as I described for spatial coarse-graining will appear. Such temporal coarse-graining was in the preface referred to as invoking the hidden layer.

In fact, if predators come and go with a high pace relative to t, this unit time scale for the prey animal may even be sufficiently large to "homogenize" the probability of fleeing early during t relative to later during t. In this case, system linearization of a non-linear process may be satisfied even at the unit time scale for the prey. The linear form of Eq. 4 may thus be achieved even without spatial coarse-graining and then normalizing back to unit scale, as in Eq. 5.

2.7 The cross-scale homogenization assumption

The main takeaway messages from the examples in the foregoing section are that (a) for serial, scale-specific movement processes (and thus mechanistic by nature, and Markov-compliant!), any non-linear functional responses become locally linear around the average response value if we average the response in time and/or space beyond the characteristic temporal and spatial scale for the response, and (b) mainstream modelling of movement and dispersal at the individual and the population level use the cross-scale linearization approximation as a basic assumption so that the principle of superposition, implicit in Eq. 3-4, applies. If you accept this conclusion you may find it superfluous to read the following part of this section, since it only elaborates on these aspects in even more technical detail. On the other hand, if you find this conclusion too far-fetched in its ruthless attack on mainstream ecological modelling I advise you to study also the following paragraphs before moving on to 2.8 *From non-equilibrium to equilibrium*.

At this point the alert reader will probably recall a diverse menagerie of non-linear ecological models which apparently conflict with the introductory statement in the previous section that all space-use processes are inherently non-linear. For example, a simple (spatially implicit) logistic growth equation for a population of size P expresses how a population's rate of change, dP/dt, changes non-linearly. While P grows sigmoid-like, the first derivative is increasing and then decreasing with increasing P. dP/dt = 0 both when P=0 and when P reaches the population's carrying capacity, and it is positive in between (Yodzis 1989). P is here a macro-scale observable for number of individuals in a population. In other words, it represents density: P individuals pr. area unit. In a scaling context and with reference to Eq. 5, this implies that logistic growth also rests on the principles of response linearization and superposition: the specific growth rate, (dP/dt)/P, is assumed to be constant under spatial coarse-graining. This premise—called the mean field approximation or the law of mass action (Ovaskainen and Cornell 2006)—is satisfied if one assumes that (dP/dt)/P does not change qualitatively or quantitatively within a given defined population of P individuals, whether we calculate the average density from many small spatial subunits with smaller P per unit or from fewer larger spatial subunits with proportionally larger P per unit.

In mathematical terms, such assumed scale-independency of non-linear functional relationships of P in ecological models is based on *the principle of spatial homogeneity*. Homogeneity in statistical-mechanical terms means that over time there is a step-by-step immediate response over space from the ensemble of P individuals to a small change of conditions defined by the integration of the given differential equation over such a small time step (for example, see Nee et al. (1997), p. 137). In the context of statistical mechanics, the system is in a stepwise manner close to an equilibrium state.

A good example of a spatially homogeneous system is a sufficiently stirred bioreactor ("growth chamber"). A homogeneous system implies that the population is in a state of statistical-mechanical equilibrium at finer scales than the system's extent; the defined area (or volume) for P. In other words, immigration and emigration from a virtually defined subpopulation—consider again superimposing a virtual grid (or a cube in a three-dimensional system)—will tend to cancel out during a small interval t. A non-homogeneous system, as illustrated by the coarser cells in Figure 15, may still be considered *locally* homogeneous within cells at this coarsened scale. In other words, a partial-differential equation could be applied to predict transient dynamics of P at this scale. The local difference in P at this scale could be derived from different carrying capacity between cells at the given resolution. In a stirred bioreactor K would of course be cross-scale identical owing to homogenization. This *statistical-mechanical* postulate needs to be satisfied with respect to local-temporal homogenization of micro-state probabilities. The principle is the same as previously described for the individual-level system. Thus, there is coherence between individual-level and population-level system description. A relevant description of linear statistics and statistical-mechanical (thermodynamic) equilibrium for the present context can be found in (Kaye 1993), pp. 164-167 and in (Sklar 1993), pp. 210-212. A simplified presentation will be given in the following sections.

If the principle of superposition applies and the population system is (at least locally) homogeneous, this also means that the *kinetic principle* applies for the population. The kinetic principle concerns systems where the rate of change (dP/dt in the present example) is proportional with the concentration of the reactants (P in this example). Thus, dP/dt divided by P (the specific growth rate) is constant under coarse-graining when we assume "other things equal," since increasing the spatial grain size by a factor k means that we can assume k times larger P per grain. In other words, P increases linearly with grain size and in compliance with Eq. 5. Therefore, the principle of superposition, local homogenization of micro-state probabilities and the kinetic principle are expressions for the same process property, that the system shows scaling *simplicity* (i.e., state variables and their derivatives change linearly over scale). In short, classical theory thus applies, both with respect to the mechanistic and statistical-mechanical level of system abstraction. In such a system, spatial memory effects from complex individual movement are absent and can be ignored at the actual scale of analysis.

The kinetic principle relates to the chemistry of scale-specific and serial (Markov-compliant) processes at micro-scales at statistical-mechanical (thermodynamic) equilibrium. However, in a statistical mechanical sense, it is the same principle that underlies the use of differential equations in ecological modelling, whether a given model includes nonlinear interactions or not. With nonlinear interactions one has to assume—the moment one leaves pure mathematics and puts the model into a spatio-temporal physical context—that the population with P individuals exists in a closed system, or that the population is spatially homogeneous at the given scale of system representation. If not, the non-linear pattern may be averaged out at coarser scales, leading to a minimum scale (grain size) for homogeneity (de Roos *et al.* 1991), as was illustrated in Figures12-16. At coarser scales than the system's characteristic scale—given by the "particle responses," which means the individual level behaviour—nonlinearities can in statistical-mechanical terms be expected to be averaged out owing to the linearization principle already described—with a pace that depends on the degree of spatial synchrony of the population dynamics (a stirred bioreactor provides maximized synchrony).

Interestingly, I find no indication in the literature on modelling animal movement or animal population dynamics that the specific growth-rate is scale-dependent in a manner that violates the classical-mechanical postulates in the context of grain size within a given spatial extent. Thus, the principles of superposition and Markov compliance are generally accepted as premises for these models, *as they need to be*—given the actual mathematical tools applied. For example, the mean field approximation (Ovaskainen and Cornell 2006), which is a key premise for differential equations, is related both to the superposition, kinetic linearization, and homogenization principles. They are different formulations of the same elephant, with the name Markov. On the other hand, if the Markov compliance is not acceptable as a realistic premise for the given system, some alternative model architecture—with its own set of specific premises—should be invoked.

Before ecological theory adopts such a fundamental assumption like the principle of superposition (from compliance with the Markov principle and thus compliance with a serial and scale-specific process that allows for linearization) for the system in question, this assumption should be sufficiently validated. The superposition principle and Markov assumption may be feasible for population modelling of large classes of situations, or it may not. It may be applicable to some scale ranges in space and time, while not over other resolutions. If compliance is not satisfied, some alternative theory needs to be developed and explored for these situations. Interesting ecological laws could then emerge to classify these conditions for Markov versus non-Markov.

As will be the focus of the upcoming chapters, one such candidate for exception is a population consisting of individuals that utilize a memory map for their navigation and space use. If Markov is not applicable at the individual level, the mathematical structure of Eq. 3-5 is not applicable for a spatio-temporal system representation. The system—whether we consider the individual or the population level—may then defy the linearization property and the principles of homogenization and a kinetic process. If these aspects are not satisfied, the superposition principle and the classical kinetic principle are consequently also violated. Thus, the strange, paradoxical results in animal movement as already described for sheep, bears, and other species may lead to a need for scrutinizing models not only at the individual level but also at the population level. This will be treated in detail later in this book, but first we need further clarifications with respect to the statistical mechanics of movement; in particular its diffusion aspect, as it is solved in the context of non-equilibrium, Markovian-based statistical mechanics.

2.8 From non-equilibrium to equilibrium

The property of autocorrelation was described in *1.1 The first glimpse*. Clearly, the process of diffusion in *2.2 Transitions from mechanistic to statistical-mechanical models* regards autocorrelated sequences of fixes at observational scale Δt during T. Referring back to Figure 6, diffusion is connected to situation **1,** where space-use area is increasing proportionally with N and where N is proportional with T.

Consider the convoluted movement in Figure 9 progressing over a substantially larger period and within the same area constraint. Thus, T will then allow for a path that is repeatedly reflected from one of the walls. From a statistical-mechanical perspective, the system is far from equilibrium since the observed space use over the time scale range from Δt to T evolves as a function of time. If the original magnitude for Δt is maintained, it means that the observed area will first expand diffusion-like (if considering a collection of repeated simulations, and superimposing like in Figure 11), but eventually reaching a saturation with respect to the demarcated area for the actual dynamics. Whether we compare two successive fixes at temporal resolution Δt prior to or after this maximum area is reached, they will be much closer in space than two randomly selected fixes. Thus, the autocorrelation level is still high.

To achieve a sample of non-autocorrelated fixes, Δt should be increased sufficiently to ensure—statistically—that more than one border reflection has taken place during Δt. In this manner, two successive fixes at this temporal resolution will tend to have the same spatial distance as two randomly drawn fixes. Thus, to achieve a large N at this coarsened temporal resolution, T should be very large. At this scale range from enlarged Δt to similarly enlarged T, the system as a whole is—in statistical-mechanical terms—in equilibrium. Spatial autocorrelation is lost at these magnitudes of Δt within the closed system, but is still lurking at finer temporal resolutions.

Figure 17 illustrates this transition. In the top pane a particle (a model individual) has moved in a deterministic or stochastic (correlated random walk) manner, or a mixture, and the path has been fix-collected at a very high frequency. The area covered by the path is still expanding over available space, and the system is thus in a non-equilibrium state: a single-path representation of diffusion in progress. In the middle pane a smaller number of fixes is collected over a very large Δt and T in comparison to the pane above. In this case, owing to the substantially enlarged Δt, two consecutive fixes will appear randomly distributed within the arena, like two fixes that are drawn randomly from the series. Thus, spatial autocorrelation is absent and the system is in equilibrium within the arena. Since the arena in this example is homogeneous with respect to local conditions for movement (for example, obstacles and ruggedness are assumed randomly dispersed), the distribution of fixes appears uniformly distributed.

In the lower pane the diffusion process is slower close to the centre. For example, obstacles could be more abundant and ruggedness could be more enhanced in this zone (we are still considering a system with constant movement speed). Consequently, the local density of fixes will tend to be higher here, reflecting the average condition of path expansion-slowing events at micro-scales much smaller than the actual Δt. If this pattern is stationary over T and thus similar-looking for consecutive replicate periods of length T, the system is still reflecting equilibrium despite heterogeneous fix dispersion.

Figure 17. The effect from sampling a path at different time intervals, within a bounded area.

The concept of statistical-mechanical equilibrium is key to understanding the process of area-constrained movement from the perspective of classical premises. Under this condition we may take advantage of specific system descriptors for quantification and testing of biological and ecological relationships. The pattern in the lower pane could, for example, represent fix collection of an animal that on average showed a more convoluted correlated random walk (and higher local density of fixes at coarser Δt) where encounters with food items were higher—in this case this could be the centre parts of the arena. For non-autocorrelated fixes (sampling at sufficiently large Δt), the density pattern of fixes, called the utilization distribution, will then reflect local habitat selection strength. A classical use versus availability analysis (Manly *et al.* 2002) could then be performed, testing the strength habitat preference by comparing local fix densities in relative terms (N/(A*e(x,y)), where A represents the area of the given habitat type (Manly *et al.* 2002) and e(x,y) is a function to be described in more detail below. With reference to the middle pane in Figure 6, consider d(.)=1 everywhere for simplicity. In the lower pane e(.) is smaller farther from the centre.

In an alternative classical system that also describes statistical-mechanical equilibrium, the pattern in the lower pane could also reflect movement in an environment where the animal performs central place foraging, and the combination of correlated random walk and a centre-pointing "advection term" (bias)—a stronger tendency to drift back towards the centre the farther from this centre—are united in a single stochastic-mechanistic model (Moorcroft and Lewis 2006, Fieberg 2007). The combination of these two model properties, locally varying parameter values for directedness and step length (the correlated random walk component), and a simultaneous influence from a spatial "vector field" that represents biased movement towards the centre describes the archetypal model of home-range compliant space use. In stochastic-mechanistic simulations, a so-called bivariate Ornstein-Uhlenbeck process is often applied in this case (Fieberg 2007).

In the description above I've been faithful to the classical premises for movement in the context of modelling space use with respect to home ranges. Thus, to model constrained space use and site fidelity under the classical premises, statistical-mechanical equilibrium is important. It is equally important when this framework is applied when seeking to solve the apparent paradox inherent in Eq. 1—the home-range ghost. Recall that data from sheep, black bear, and other species showed non-trivial area expansion even based on non-autocorrelated fix series (the equilibrium condition); a result that invoked the ghost label. A classical explanation for the home-range ghost pattern will have to be able to demonstrate by simulation a series of spatially non-autocorrelated and area-constrained series that expand non-asymptotically as a function of N. If this is achieved, the simulations additionally have to demonstrate—under this equilibrium—that the expansion is both power law compliant and generates a spatially statistical fractal with $D \approx 1$ with respect to fix dispersion.

Assuming non-autocorrelated fixes, we are in a position to refer back to situation **2** in Figure 6; the classical area asymptote model for a home range. According to this model, statistical-mechanical equilibrium and asymptotic space use are mutually dependent: whether we increase the sample of fixes by collecting uniformly over a fixed period (collecting N fixes from the total set N_{max}), or we simply take a continuous section of fixes (meaning that N increases proportionally with time over which the sample was collected), the demarcated area size should be similar. This simply follows from the classical premises of statistical-mechanical equilibrium, which in traditional home-range modelling is categorized as absence of spatial autocorrelation.

Again according to the classical asymptote model, if the area utilization is uniform—as in the middle pane of Figure 12—a relatively small sample is needed to demarcate most of the true home range—albeit with subjectively defined rule for delineation of area owing to fuzzy borders. In contrast, if the animal has tended to move faster and/or respond by more unidirectional successive steps close to the borders, the distribution of fixes will tend to show higher density in the centre section relative to this periphery. This would emerge either from a larger mean free path, in statistical-mechanical terms, and/or a stronger influence of the drift term in the periphery (centre-pointing bias). This situation is typical for real home-range data. Fixes are typically wider dispersed the farther from the weighted centre of activity, as illustrated in the hypothetical utility distributions in Figure 7. In this case, a larger N is needed to be able to demarcate a substantial part of the true area utilized during T. It takes a larger sample to find fixes in subsections where utilization has been low (larger movement speed or larger mean free path). The compensation function Q(N), needed to estimate true home-range size from small N when applying the minimum convex polygon method, therefore should reflect this aspect. Q(N) was described in the paragraph following Eq. 1a. While Q(N) describes an adjustment for small sample size in a study of N-dependency of area demarcation, the spatially explicit function e(.) to account for distance effect from centre of activity regards adjustment of local density in the context of estimating habitat selection strength.

Figure 7 in a previous section contained both variants of heterogeneous space use, owing to the assumed condition of classical equilibrium. In that particular example, the utilization distribution was simply simulated by the so-called Monte Carlo procedure: fixes were randomly drawn from cells on a virtual grid that was superimposed on a defined area representing the home range. The probability to draw fixes from centre cells was set larger than for more peripheral cells. From a premise of a statistical-mechanical equilibrium condition (assuming sufficiently large Δt to ensure non-autocorrelated fixes), this simulation is coherent with the more sophisticated stochastic-mechanistic modelling of the utilization distribution just described, involving a vector field for bias. Crucially, both variants of mechanistic movement—with or without the bias term—lead to situations where an area asymptote is expected for observed space use, as N is increased at this level of spatially non-autocorrelated fixes: sufficiently large Δt and a premise of area-delimited movement to achieve equilibrium at this Δt.

However, non-uniform area utilization in the context of central place foraging creates practical problems for an analysis of habitat preference based on local density of fixes. In the lower pane of Figure 17, e(.) is smaller towards the periphery at equilibrium; thus, intensity of space use becomes larger (owing to smaller denominator and e(.)<1) when distance from centre-effect on intensity of space use is adjusted for in $N/(A*e(x,y))$ and e(.)<=1. A smaller e(.) is expected in the peripheral parts of a home range (since we are normally considering a two-dimensional space in terrestrial systems, one dimension z may be ignored). Thus, e(.) is a compensation function that has to be derived on a case-to-case basis. Habitat preference is then—under the premise of Markovian-compliant movement and density calculations at equilibrium scales—expressed by local density of fixes (ideally adjusted by the spatial function e(.)) as the proxy variable for habitat selection. The difficult task to estimate e(.) at sufficiently detailed spatial resolution in empirical data is another matter though. For example, how to disentangle the effect on local fix density from a smaller e(.) from the effect of a larger diffusion rate in non-preferred intra-home-range habitat sections, which could also be more dominant towards the periphery? This important challenge will be returned to later.

It is important to summarize that a *closed* system is a prerequisite for statistical-mechanical equilibrium under classical premises, and provides a potential for non-autocorrelated fixes in a Markov-compliant system. In the middle pane of Figure 17, this was achieved by the arena borders. In the lower pane it was this arena, a central-directed movement bias, or a combination of these two influences that maintained the area containment for the process and thus compliance with the premise of a closed system to achieve equilibrium. If the directional bias is sufficiently strong in peripheral parts of the space use, it may effectively function as a closed borderline at the distance where this bias on average cancels out the tendency for correlated random walk towards more distant locations from the centre of activity.

It is also important to acknowledge that a Markov-compliant process, a correlated random walk with or without a bias term (advection term) added, is inherently memory-less. Thus, a home range has to be maintained by environmental constraint: movement takes place within a space-limited habitat (bouncing back when the border is reached), or a local gradient of (say) resources makes it locally preferable to turn around when the resources become too scarce or difficult to utilize.

Under the classical equilibrium-based condition, *a movement path within a home range will only self-cross by chance* (the probability for self-crossing may or may not be higher towards the centre); not by goal-driven returns. Deterministic self-crossing would require long-term, spatially explicit memory utilization. From an ecological point of view, you would call it site fidelity. If an animal is expressing site fidelity, it does so owing to a cognitive capacity to utilize a memory map (some insects may navigate from another process, vector addition, but here I focus on the spatial map utilization by vertebrates). Interestingly, some recent Markovian-compliant mechanistic space-use models are also implementing long-term, spatial memory (Börger *et al.* 2008, Dalziel *et al.* 2008, van Moorter *et al.* 2009, Nabe-Nielsen *et al.* 2013). How this is achieved will be described and discussed shortly.

2.9 The next step

To understand the profound effect of memory on animal space use, I have now led the reader through the fundamental basics of memory-less particle dynamics and statistics. One of the reviewers of my book manuscript (a physicist) kindly hinted that this walk-through was perhaps a bit too basic. My response was that I needed it to lay the foundation of the more deep-drilling and off-piste framework that emerges in the chapters to follow. Unfortunately, the general audience of ecologists is not trained to consider animal space use from the perspective of biophysics. In the next chapter I show how memory effects will shake and potentially topple the existing paradigm, given that memory is realistically embedded in models.

3. APPROACHING MEMORY

3.1 Preparing to exterminate the home-range ghost

Classical diffusion, Brownian motion, and Markov compliance—the cornerstones of movement ecological theory both in 1991 and today—are apparently not sufficient to explain the home-range ghost pattern in Eq. 1. With reference to the lower panes in Figure 17 (non-autocorrelated fixes), increasing N by decreasing Δt within a given T should give a similar area demarcation as increasing N at a given Δt. That aspect was satisfied by (for example) the sheep results in Figure 3. However, despite this verified absence of autocorrelation effect on demarcated home-range area, Eq. 1 states that area grows lower law-like and not asymptotically with increasing N, even after having adjusted for small-sample artifacts. In the previous section I elaborated on this property from the perspective of classical statistical-mechanical equilibrium.

It stood clear from the onset of my fellowship period that I would have to extend my training beyond the traditional biological fields of expertise and move towards statistical mechanics. Though many ecologists would argue for some kind of path analysis when analysing fix data in detail, it is important to realize that these sets of spatial relocations—for example, GPS fixes—are typically collected with Δt of ten minutes, one hour, or much longer; i.e., they are not exactly tracking every mechanistic step along the movement path of an animal. Hence, a statistical-mechanical model framework should provide a feasible approach towards studying space use at these levels of system coarse-graining. However, closing in on this kind of biophysics was somewhat scary in itself, given a range of challenging concepts like entropy, microstates, ergodicity, energy levels, and so on. It definitely did not help that previous generations of ecologists who had tried this path to model population or community level systems apparently had got stuck. None seemed to have tried to model a single-particle (individual level) process from this angle, beyond applying stochastic-mechanistic or mixed model algorithms. Step distributions, diffusion compliance, etc. are frequently analysed from a statistical perspective using *premises* from classical statistical mechanics (inherent in the statistical methods that are applied)—rather than letting these premises become the prime target for their research.

Mechanistic modelling—and thus the Markovian framework—has always ruled the field of spatially explicit simulation of movement. However, as was indicated by the foregoing sections, a mechanistic and a statistical-mechanical system analysis are qualitatively different despite dependence on similar micro-scale processes. To make things even worse, back in 1991 both classical and contemporary statistical mechanics seemed to offer no help for a system consisting of a particle with spatial memory. In short: gas or water molecules do not suddenly decide to take a directed return towards a previous location, potentially far away from its present location. In classical statistical mechanics it is a basic and important premise that Brownian motion leads to self-crossing of a path by chance alone, not as a result of planning! In short, classical statistical mechanics describe spatially memory-less processes. In the previous sections this was specified as Markov-compliance.

3.2 Spatial memory from a mechanistic premise

The likelihood for a quick solution to the space-use issue of sheep, bears, root voles, and other vertebrates was not good, given the lack of off-the-shelf physical theory for particles under influence of spatial memory. On the other hand, exploring a white spot on the map stood out as irresistible. The answer to the initially strange results seemed increasingly clear to depend not only on the statistical-mechanical system abstraction, but a theoretical expansion of this field.

In this expansion, memory effects on movement should thus take the centre of the stage. Site fidelity, the tendency to return to familiar locations more frequently than expected from random self-crossing of a path, would then be hypothesized to lead to a home range as an emergent property. In contrast, a home range from a combination of Brownian motion (or correlated random walk) and advection (centre-pointing bias) is not spatially memory-based in strict mathematical terms. In physical terms the bias component in mechanistic modelling of movement is represented by an environmental *field* (Figure 18). Wherever the animal is, its next move is—according to this model architecture—influenced by two forces: the tendency to move forward in absence of appearance of new movement-influencing events (the correlated random walk component of the mechanistic equation) and the centre-pointing bias (the advection component of the mechanistic equation). At any point in space, the latter may be seen as a vector with a centre-pointing direction and a length that expresses the strength of the bias at that point in space. Thus, as expected by a mechanistic model like Eq. 4, the bias is determined locally, at the actual point in time and space along the particle's (the animal's) path, and recalculated over successive time increments in an independent manner relative to previous increments.

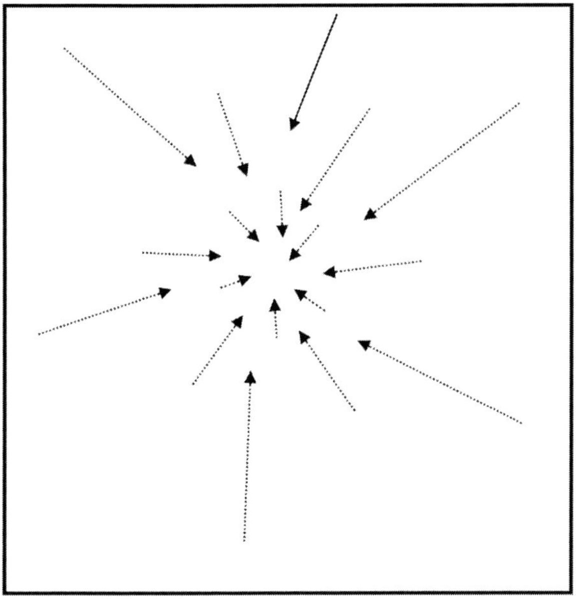

Figure 18. Illustration of spatial memory effect implemented under the classical biophysical framework, in the form of an advection field. Arrows illustrate vectors of field strength and direction at a given point in time.

In the next instance, the field strength is typically approximately similar (the field surface is differentiable and "smooth") and contributes—in superposition with the local strength of the correlated random walk tendency—to determine the next move. Spatial memory is not operationally formulated in this kind of process, where the system may be abstracted to a partial differential equation (Moorcroft 2012). Anyway, biased, correlated random walk is often applied mechanistically in ecology to represent site fidelity (Börger *et al.* 2008, van Moorter *et al.* 2009, Nabe-Nielsen *et al.* 2013, Benhamou 2014). Here we are again confronted with an apparent paradox: how can a Markov-compliant process—where a local response (like a move) is always determined by local causes—be realistically influenced by spatial memory of events that took place in distant space and distant time?

Various routes out of this challenge, but with varying realism, have been proposed over the years (Smouse *et al.* 2010). In the following I will start by describing the basics of a specific group of approaches to account for spatial memory within the framework of mechanistic and thus Markov-compliant movement (Börger *et al.* 2008, Dalziel *et al.* 2008, van Moorter *et al.* 2009, Nabe-Nielsen *et al.* 2013). The solution to the site fidelity challenge has been to conjecture that an animal is capable of extending its "field of perception"—the surrounding area where it relates to current properties and events directly—to include the accumulated information on its memory map. This information then provides a potential to target previously visited locations in the future. For example, the attributes of a distant location that was visited some time ago is assumed to potentially influence the individual's next-move calculations on equal footing with what it relates to directly through its senses. The more distant in space and time, the less influence ("weight") on the current decision where to move next (according to the typical algorithm in this approach). The dynamics are, in a statistical-mechanical frame of reference, truly Markovian despite spatial memory influence. The memory map is modelled as influence from a virtual field, which determines the local strength of directional bias in a moment-to-moment manner. The Markovian architecture in these approaches thus satisfies the principles illustrated by Eq. 4-5. Markovian compliance implies a mechanistic kind of dynamics.

In this mechanistic approach to implement memory effects, typically represented by biased correlated random walk or a more complex algorithm, a stochastic element is combined with deterministic rules for memory-based directed returns towards previously visited locations (van Moorter *et al.* 2009). This may sound like the model approach described in connection with Figure 13, but in the present scenario the walk is modelled explicitly as a mixed mechanistic model for a movement path. This implies a combination of intrinsic directional persistence of two kinds: a tendency to keep the direction of the foregoing step (correlated random walk) and a potential bias towards remembered locations with remembered properties (for example, expected food content, calculated from a given algorithm). The relative weight of the resultant vector for bias relative to the vector for correlated random walk at the current instance in time will determine whether the individual at this point in time is drifting in the direction of the current bias field vector or not. The resultant vector is summing influences from a potentially large set of remembered locations, meaning a similarly large magnitude of terms in Eq. 4 prior to some kind of system condensation for the sake of simplification.

In other words, the strength of the bias towards respective remembered locations depends on a moment-to-moment perception of the potential foraging gain from revisiting a specific patch versus just "moving along" in a correlated random walk manner. Physically, a correlated random walk will—when directional persistence is strong (close to what is called ballistic motion)—comply with *the principle of least action*: the path taken during a small time interval will be the one for which the action is stationary (no change) with respect to first order derivative. When slicing time into small intervals, the movement vector will tend to keep its direction. Referring back to the environment-determined vector field in Figure 18, the memory-originating bias field is undulating in strength in different directions from the animal's present position, as a function of the current set of bias vectors pointing towards known locations. As the animal moves along, the undulating field will follow along. Other factors than distance from present location might of course also contribute to weakening or strengthening the current attractiveness of various spots on the memory map.

To summarize this mechanistic model approach, memory is implemented in a Markovian-style manner: both directional persistence and memory-based bias are re-calculated moment by moment by the actual movement algorithm. Thus, the paradox of combining what in principle is a non-Markovian system (long-term memory influence) with mechanistic model architecture is cleverly resolved by applying modelling principles that are developed in physics for the influence from an external field. In other words, the bias will behave as a vector describing a local force, as described in Figure 18. The process is thus scale-specific with respect to the time dimension of the process, despite infinite memory of spatial attributes (like assumption about current food patch qualities outside the individual's present field of perception). The movement is in a moment-to-moment *equilibrium* with the actual conditions, whether these are perceived directly or cognitively calculated based on memory map information. The calculation of movement is therefore based (a) on local conditions at the model animal's present location at the given point in time (confirming that the model is mechanistic), and (b) on perceived gain from moving in the direction of the assumed most profitable patch at the present point in time if present conditions locally are less attractive (for example, after food consumption in local patch). The latter involves comparison that involves all historically visited locations and their respective attributes, deflated value-wise to the present point in time and present distance to respective historic locations.

Hence, this model structure implements memory of past patch visits by using spatially explicit memory to expand the individual's field of perception relative to its current location. The bias field is a composition of both intrinsic components (cognitive state, spatial memory, and neural processing based on this information) and environmental attributes of known locations and environmental conditions within the individual's current field of perception. With reference to Eq. 4, I propose that implementation of memory effects in a mechanistically designed model will look like this:

$$dX/dt = \sum_{i=1}^{j-1} a_i F_i(x,t) + \sum_{k=j}^{n} a_k F_k(x,t) + \varepsilon \quad (6)$$

The difference relative to Eq. 4 regards the specified range of terms with subscripts $1 <= i <= j-1$, describing influences on movement from conditions within the field of perception; and the second range of terms with subscript $j <= k <= n$ including influence from memorized locations outside the immediate field of perception. The latter range regards the mechanistic implementation of a bias field from spatial memory. Thus, the two ranges of terms in Eq. 6 operationally regard (a) opportunistic search and (b) strength of site fidelity. The latter represents an emergent property—leading to a multi-clustered home range—that will perform a stronger or weaker influence on moment-to-moment movement as a function of the strength of this set of terms (the directional bias) relative to the other set. Therefore, a small magnitude of the resultant vector from directional bias relative to the resultant vector from the set of search terms will lead to an expectation of a larger home range. If site fidelity becomes too weak relative to the summed search terms, the home-range property vanishes owing to reduced site fidelity (Börger *et al.* 2008, van Moorter *et al.* 2009).

Like Eq. 4, the process in Eq. 6 is also mechanistic by design, owing to the immediate execution (given by the dX/dt architecture) of the terms' relative strength from moment to moment. A mechanistic system representation implies that the system is in a continuous state of statistical-mechanical equilibrium (steady state) with actual conditions. Hence, from a statistical-mechanical perspective Markov-compliance is achieved by mimicking execution of influence from spatial memory as a physical field force. The strength of drifting along the resultant vector for bias relative to the correlated random walk vector is obeyed instantly, and these terms' relative influence is continuously updated. These principles were summarized in *2.5 Markov from another angle*. Mechanistic model design, Markov compliance, and statistical-mechanical equilibrium at the neighbourhood of the animal's current position are three attributes of the same elephant.

3.3 Coarse-graining the mechanistic memory model

While the mechanistic model (with variants) in the original papers was executed at a behavioural micro-scale, one could simply get a statistical-mechanical impression of the process by sub-sampling—"fix collecting" —every q'th relocation, where q is sufficiently large to allow for passing the mechanistic micro-scales and entering the meso- and macro-scale statistical-mechanical representation. Properties of this coarser representation of space use could then be studied to infer indirectly about the animal's behaviour as seen from a coarser and more holistic level. Alternatively, one could deduct some general statistical-mechanical properties of the model and apply these properties as the basis for a direct statistical-mechanical simulation rather than sub-sampling every q'th location from an actual mechanistic simulation. In Gautestad et al. (2013) we performed the latter approach, somewhat hidden as part of an extensive (51-page) online appendix of this paper. We called it Memory-enhanced random walk (MemRW) to distinguish it from other statistical-mechanical classes of movement, like Brownian motion. Here follows an excerpt of the simulation protocol and results.

Two primary aspects of a model following the design in Eq. 6 should traverse clearly to the statistical-mechanical level, where micro-scale mechanistic details are hidden from "view." A statistical-mechanical representation is achieved by sampling a given path every q'th time increment, where the original path is generated from a series of moment-to-moment events at unit time increments (the mechanistic representation of the movement). As already described, q needs to be sufficiently large to give a proper statistical-mechanical image of the process. Thus, both a large q and a large sample size are needed. The former provides successive net displacement vectors from a sufficient number of independent movement events at finer temporal resolutions, and the large sample size provides material for proper estimates of functional form and parameters of respective space-use models. Referring back to the empirically derived home-range ghost paradox in Eq. 1, how does the Markov-compliant memory model behave in more detail at the statistical-mechanical level, which applies for real data like a series of spatial fixes along an animal's path? May this approach, when coarse-grained to a MemRW model to comply with actual fix sampling intervals (the q-aspect), solve the statistical oddities of home ranges with respect to non-trivial area expansion and a fractal dispersion of fixes?

First, consider a homogeneous environment in coarse-scale terms. Homogeneity is typically invoked in a model construction to highlight some basic intrinsic properties of the system, while heterogeneity is added later to make the model more realistic with respect to specific environmental conditions. In the case of environmental homogeneity, all visited food patches are—in the generic model that highlights principles—typically similar. They are thus subject to the same degree of resource exhaustion upon visit, same regrowth rate after being left alone, and so on. Consequently, they have a similar revisit probability later on—in the long run. Sometimes this revisit will happen swiftly after re-establishment of food quality, if the moving animal happens to be nearby (the bias vector, calculated from terms with subscript in range from j to n in Eq. 6, may be pointing strongly towards this patch). At other times the animal happens to be farther away, and the revisit probability is consequently smaller. However—in overall terms, which may be considered synonymous with "in statistical-mechanical terms" when both q and sample size are large, a given memory-dependent return event may be expected to go to any previously visited location (a property expressing a homogeneous environment), given that the model animal has *long-term, spatial memory*.

To get a better grasp of this important property, consider the random scatter of fixes in the middle pane of Figure 17. Then add a "food patch" at a random location within this arena. The distance from this location to a randomly collected fix of the animal will be expected—in a large set of replicate selections of patch and fix location—to be constant. So the probability that the animal at a random point in time returns to the randomly picked food patch will also be constant—in the long run, in a homogeneous environment, and at the coarsened scale of system observation (large q).

To conclude this first important property of the coarse-grained system, implementation of spatial memory-invoked returns to previously visited patches may (in the generic model) be random when the system is simulated at the statistical-mechanical level. The returns may also, for the sake of model simplicity, go to exact historic locations along a path. However, this is not a prerequisite. I have tried simulations (unpublished results) where the return targets are chosen in the vicinity of previous locations, but this has not influenced the general model properties. The property of picking site fidelity targets randomly implies a homogeneous environment. For example, if one in one hundred fixes tends to be unexpectedly close to one or another previous fix location in area section A, and one in one hundred fixes tends to be similarly close in area section B, these two sections may be regarded as similar with respect to strength of memory-dependent site fidelity. If this property is found over a large set of area sections, the total area may be considered homogeneous at the given level of analysis in space and time (in later chapters I will specify and simulate environmental heterogeneity).

Second, local bias towards specific visited locations from any given point in the arena at any given time will tend to cancel out in the long run. Hence, without risking loss of vital mechanistic micro-scale influence on macro-scale properties, the bias effect may be ignored at the statistical-mechanical level—given that patches are randomly distributed. Therefore, the vector field representing bias in a calculus-based model (partial differential equations) does not need to be explicitly implemented. It will turn up in the space-use pattern from large q and fix sample size as an emergent property from site fidelity! Model simplification may be applied also for the directional persistence provided by the correlated random walk term of the movement at micro-scales: owing to large q, correlated random walk could be replaced by a Brownian motion, where next-step direction is uniformly distributed over $0\text{-}2\pi$ radians, meaning equally probable step direction for all directions. It is thus a matter of personal preference whether one uses biased correlated random walk or a biased Brownian motion when simulating MemRW at the statistical-mechanical level. However, this conjecture will be tested by simulations below. In contrast, at a micro-scale representation of this system, it makes sense to apply correlated random walk for the search component of movement, owing to the physical principle of least action and the premise of small q (behavioural scale). In a micro-scale and behavioural context, it is easy to visualize that an animal that changes direction every small time interval (for example, every second) will not spend its search energy optimally relative to an individual that keeps the direction until something that requires its attention appears along its path.

3.4 Statistical-mechanical simulation of MemRW

3.4.1 Model simplifications and simulation methods

The two statistical-mechanical system properties referred to above with respect to transforming the mechanistic model in Eq. 6 to the MemRW model regard random target locations for memory-dependent returns and ignorance of local bias towards specific targets. The premises were a homogeneous environment (a statistically uniform dispersion of food patches and their respective qualities) and an infinite memory horizon. That the process is observed at the statistical-mechanical level implies that it becomes representative of a large range of actual execution rules—simple or more elaborated specifications of the respective terms in Eq. 6—at the micro-scale level. This is an important aspect of coarse-graining a system from the mechanistic level—where all the terms in Eq. 6 are specified with respective functions and parameters—and the statistical-mechanical level. Many combinations of values from the respective search terms in Eq. 6 at a given point in time may lead to a similar resultant vector for the bias at that point, and similarly for the terms representing spatial memory. Thus, the statistical-mechanical level describes general system aspects—including emergent properties like strength of site fidelity and expected size and stability of a home range—at the expense of mechanistic details of the movement process. In other words, we may keep focus on the properties of the specific statistical-mechanical universality class under scrutiny (in this section MemRW represents the universality class), including its specific observable variables and parameters that may be applied for ecological inference.

In more extended (non-generic) system designs, there may be a need to specify more than two statistical-mechanical properties. For example, in our simulations of MemRW in Gautestad et al. (2013), we explored two additional aspects: degree of directional correlation in the correlated random walk component and magnitude of memory horizon (infinite or time-limited spatial memory). The degree of directional correlation from one step of movement to the next was based on a wrapped Cauchy distribution (Pasternak *et al.* 2009) using persistence parameter values 0 (classical Brownian motion; no persistence), 0.50 (some persistence), and 0.95 (strong persistence).

Implementing a memory horizon means that the simulated animal remembers only recently visited patches (within a trailing memory window), and consequently it never returns—unless by random encounter—to previously visited patches further back in time. The memory horizon was given three conditions: infinite (uniform return step probability to each of all historic locations in a given series), a trailing window of 500 time increments (returns to latest 500 locations only), and five increments. In long-lasting periods for fix collection (very large $T/\Delta t$ ratio, where Δt represents q), such a memory horizon should be expected to lead to a transition towards memory-less space use at the statistical-mechanical scales. In short: the memory horizon becomes very small relative to the temporal extent of the observed system, and properties of a classical and thus memory-less Brownian motion should appear despite memory effects at fine temporal resolutions.

For each of 3x3 combinations of directional persistence and memory horizon, five replicate series of 10,000,000 steps were performed. Critically, the temporal unit scale for increments (quantified by q in the model description above) was assumed to be large enough to implicitly embed many mechanistic movement events at micro-scales along the path. In this case, movement as a series of steps at the unit scale of simulation was implicitly defined to represent a statistical-mechanical representation of the path. Thus, the causality behind shifting the position from one location to the next at this temporal resolution is information-less. In a statistical-mechanical context, it is only the large collection of such locations and displacements that reveal something about the underlying process. To further underscore the statistical-mechanical nature of the path, each series was sampled as one "observed fix" at Δt=1000 unit steps (from step 1,000,000 onwards to discard initial transient effects). This resulted in a final fix series of length 9,000 steps.

Frequency of return steps was set to 1:100 for all series. In other words, on average every 100th time increment, a return event to a randomly picked previously visited location on the path was performed (the MemRW model description above gave the background for picking targets randomly). Since Δt was set to ten times larger than the average interval for a return event, the sampled series should be non-autocorrelated (Swihart and Slade 1985), and return events will tend to function as a "scrambling force." At temporal scale larger than the average return interval, two consecutive fixes are expected not to be closer in space than two randomly picked fixes. However, if area use expands in a home-range ghostlike manner as formulated in Eq. 1, there may in the latter case be some influence from sampling from an expectedly larger home-range extent. This aspect will be returned to in a later section.

Additionally, twenty replicate series of MemRW were also run under a higher-frequency sampling $\Delta t=20$, allowing for testing paths in the autocorrelated domain since the return interval was set to a larger scale $\Delta t=100$. In addition to comparing simulations of MemRW with the power law expansion of home-range area in Eq. 1 (expected: power exponent of ca 0.5), the spatial scatter of fixes was also tested for scale-free properties (fractal dimension). The method applied was the simple "box counting" of incidence, *I*; number of virtual grid cells containing one or more fixes, over range of spatial scales for the given fix magnitude N=9,000. The fractal dimension, D, was then estimated from the slopes of log-transformed *I* as a function of log-transformed grid cell size. For an approximately space-filling pattern, one would expect $1.5<D<=2$. For an aggregated pattern, self-similarly distributed over a range of grid resolutions, one should expect $D<1.5$, as in our analyses of sheep movement and other data sets. For example, in analysis of fifty-two individuals of red deer *Cervus elaphus*, we found $0.9<D<1.2$ (Gautestad and Mysterud 2012, Gautestad *et al.* 2013).

3.4.2 Simulation results

The simulated paths at unit time increments showed the expected pattern over the given ranges of directional persistence and memory horizon (Figure 19). Once more I underscore that—from a statistical-mechanical perspective—one should ignore the details of the respective paths in Figure 19. They are just representing a sample of a larger set of replicates. Even the successive turns and step lengths along each path should be ignored. Many variants of mechanistic details (movement rule variants) may lead to a similar statistical-mechanical meso- and macro-scale representation of the process defined as MemRW, which—in the spirit of system condensation—has been simplified to capture two essential statistical-mechanical properties of the referred mechanistic implementation of spatial memory.

With reference to comparison with Eq. 1, area from the incidence method verified a power law expansion for MemRW with exponent between 0.20 and 0.26, but only under condition of infinite memory horizon (Figure 20, column 1). This exponential range of magnitude 0.06 over nine variants of directional persistence is very narrow, despite quite different visual appearance of simulated paths (Figure 19). Thus, at the statistical-mechanical level of analysis, mechanistic details with respect to strength of directional persistence seem to have little influence on the power exponent for observed area expansion when memory is infinite. However, while MemRW seems to comply with a non-trivial power law expansion, the exponent is only half the empirically derived "home-range ghost" range ≈0.5 in Eq. 1.

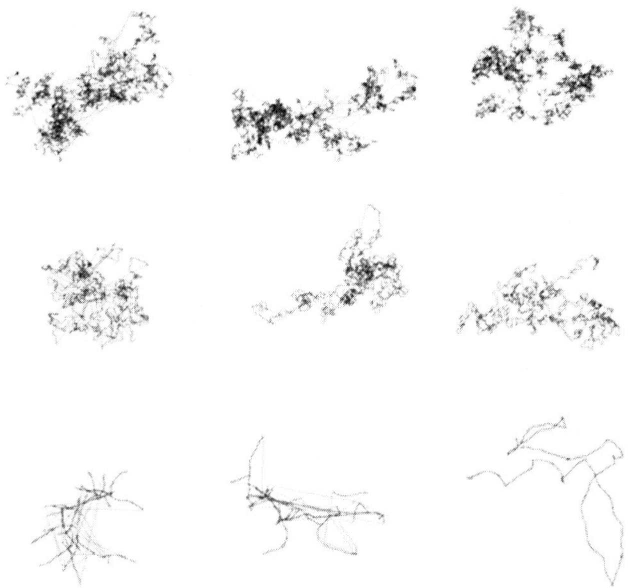

Figure 19. Simulation examples at unit time scale, illustrating MemRW with different degree of memory horizon (from left column towards right) and directional persistence (from top row towards bottom row). Memory horizons: infinite, 500, and 5; directional persistence: 0 (within arena size 4,000 spatial units), 0.50 (arena 10,000 units), and 0.95 (arena 40,000 units). The 4,000 steps in each series are connected successively by inter-fix lines. Return steps are marked as grey lines. The first step following a return step (at frequency 1:100 on average) has a random, non-persistent direction relative to the incoming step direction.

For the varying memory horizon series (column 2 and 3 in Figure 19-20), the exponent for space-use expansion was close to 1, with actual slopes in the range from 0.77 to 0.81. A slope equal to 1, where area is expanding proportionally with N, is expected for classical movement models without site fidelity (Gautestad and Mysterud 2013). Again, the result shows power exponent resilience (small range of slope, 0.04) over a broad range of constrained memory conditions. Neither in this case did the degree of directional persistence of simulated steps influence the results with respect to the power exponent of area expansion. In other words, correlated random walk behaved almost as Brownian motion (no directional persistence), in statistical-mechanical terms.

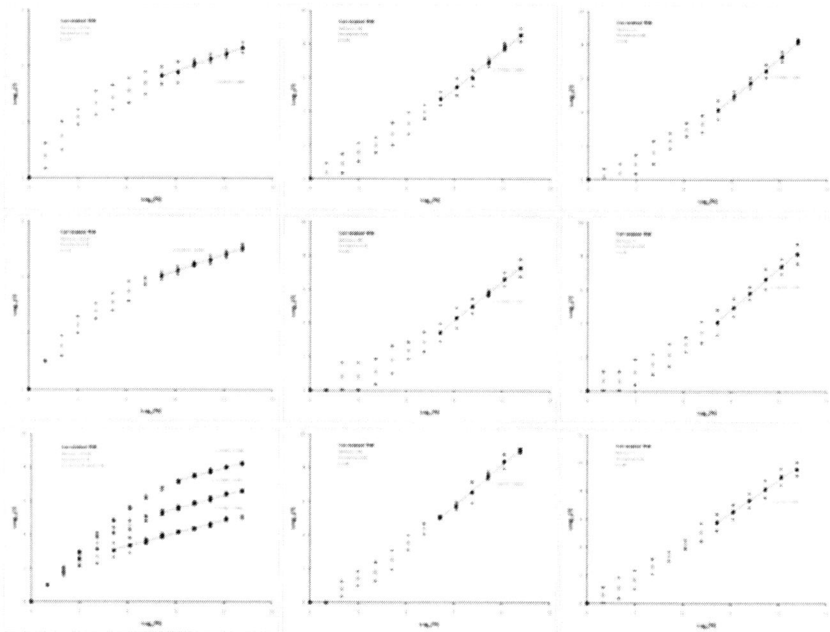

Figure 20. Log-transformed incidence (home-range area) as a function of log-transformed sample size of fixes, N; based on MemRW data in Figure 19 but sampled at frequency 1:1000. +/- 1 Standard deviation about the mean from five replicate series is shown as dashes. The small-N range was excluded from the regression range due to the small-sample effect [where log(*I*) is approximately proportional to log(N)] or applying very small grid cells relative to the underlying movement process. A grid resolution k=1:40 relative to arena size was used for all combinations of non-infinite memory (column 2 and 3), and k=10 was used for infinite memory (column 1). For one of the latter conditions, infinite memory and step directional persistence 0.95, the slopes are shown for three variants of resolution, k=1:10, k=1:20, and k=1:40. The combined graph (lower left) illustrates how the dilution effect becomes most prevalent at the highest spatial resolution k=1:40.

The fractal analysis of MemRW revealed D in the range between 1.56 and 1.65 (Figure 21). D equals the negative of the slope (negative slope gives positive D). The finest-resolution range (small *k*) was excluded due to the dilution effect (Gautestad and Mysterud 2012): when the sample size of fixes becomes too small to reveal the statistical fractal at fine resolutions the slope trivially tends asymptotically towards zero with continued decrease of k [see Gautestad and Mysterud (2012)]. The coarsest-scale range $-3 < \log_2(k) < 0$ was also excluded, due to a combination of integer artifact (large grid cells) and so-called "space-fill effect": at very coarse resolutions all grid cells tend to contain at least one fix and thus appear space-filling, as expected from a classical D=2 pattern from a dense object (Gautestad and Mysterud 2012). Also for the estimates of D, the range of exponents was narrow: ca 0.09. Contrary to the area expansion exponent, the result was *similar* for infinite and constrained memory horizon. Thus, studying D in isolation will not be sufficient to distinguish a memory-less process like correlated random walk and a similar process with spatial memory influence, MemRW. The degree of aggregation—expressed by D—was similar over all nine variants of directional persistence and memory horizon. A stable D over a wide range of grid resolutions implies that the spatial aggregation was self-similar over a substantial range of scales. However, as for the analysis of area expansion as a function of fix sample size, the results from the fractal analysis deviated from empirical results. MemRW fix patterns were not as strongly aggregated as in the sheep and other data, where D was found closer to 1. The MemRW fractal dimension is thus of similar magnitude as for classical Brownian motion, which will be illustrated later in additional simulations.

3.4.3 Conclusion

To summarize, the MemRW simulation showed a too small rate of home-range area expansion relative to empirical data (power exponent ca 0.25 in simulations relative to 0.5 in empirical studies), and thus a closer compliance to an asymptotic model. With arithmetic rather than log-transformed axes, the power function with this small exponent will appear to swiftly level off with increasing N. MemRW also deviated with respect to fractal dimension for the dispersion of fixes (D close to 1.5, rather than close to 1 as found in fix patterns from real animals). However, the scale-free aspect of MemRW was satisfied in both respects. Thus, MemRW seems to be a step in the right direction with respect to resolving the two paradoxes non-trivial N-dependence and non-trivial fractal dimension of fix dispersion. Question then remains: what is missing to obtain closer compliance with empirical data?

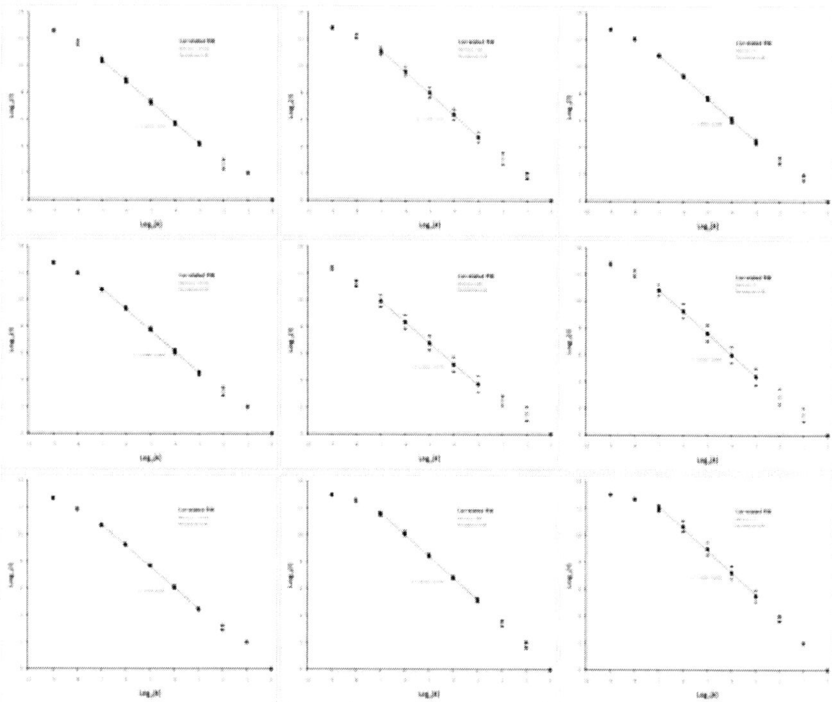

Figure 21. "Box counting" (number of grid cells with one or more fixes, as a function of grid resolution) of a MemRW process over a range of spatial resolutions was applied to estimate the fractal dimension, D, for the 3x3 combinations of memory and directional persistence as shown in Figure 19, but sampled at frequency 1:1000 (i.e., ten times smaller than the return step frequency) to ensure non-autocorrelated series for analysis. +/- 1 Standard deviation about the mean from five replicate series is shown as dashes. The range of spatial resolution, *k*, over which D was estimated, is shown by log-log linear regression between small and large magnitudes of *k*. D is estimated as the negative of the slope (becoming positive when slope is negative).

MemRW is a statistical-mechanical representation of memory-influenced space use; derived from a sub-variant of a mechanistic and scale-specific universality class; biased correlated random walk. Correlated random walk, which becomes Brownian motion when sampling the path at sufficiently large intervals, is in physical terms linked to the Markov premise. Hence, spatial memory is in the MemRW model utilized in a moment-to-moment manner, meaning that only one goal (the bias aspect;

drifting towards a spatial target) is active at any moment in time. As a consequence of this, the process is also definable at the micro-scale level (the behavioural, mechanistic resolution) where the actual movement behaviour is directly reflected in the jaggedness of the path. The process is scale-specific owing to application of correlated random walk (or Brownian motion).

While this Markov-compliant model architecture with full coherence between mechanistic model principles and classical mechanics seems to replicate memory-influenced movement to some extent, is it not fully compatible with at least our own empirical results on animal space use (lower fractal dimension and stronger sample size dependence on area demarcation). It may be necessary to take a bolder move—stepping beyond mechanistic modelling, and thus also leaving the Markovian premise behind, to find solutions to the home-range ghost and other paradoxes to be described in later chapters.

A potential direction in this respect was already hinted at in section *1.2 Pinpointing a feasible reason for ignorance*, by referring to the hypothesis that animals (with cognitive capacity to utilize a memory map) may relate to their environment in a multi-scaled, hierarchical manner. Hopefully the foregoing detailed review of basic principles and properties of mechanistic models has substantiated the proposal that memory implementation is difficult to achieve under the constraint of the Markovian principle. Thus, not only should a memory map be accounted for, but the sequential kind of processing that characterizes a mechanistic model may need a qualitatively different extension. Prepare yourself for yet another statistical-mechanical concept: parallel processing!

3.5 The next step

Invoking a similar argument as I did when I defended the very simplistic and basic illustration of memory-less movement in the foregoing chapter, in the present chapter I begin to approach site fidelity and how it will require and extension of basic statistical mechanics. Many sophisticated approaches have been presented over the years to model site fidelity in the context of home range theory, but in my view they all fail because they basically rest on the principle of a Markovian, mechanistic kind of processing. Thus, in the upcoming chapter the reader should be prepared for a fundamental step away from Markov.

4. Movement Physics Beyond Markov Compliance

4.1 System complexity

In the previous chapter the theoretical system descriptions have all been compliant with classical mechanics and thus premises for classical statistical mechanics. At one point in the presentation, I called this system architecture simplistic. A realistic system representation may be complicated, for example, with a large number of terms in Eq. 3-6 and/or a very heterogeneous environment for the execution of the movement rules or population dynamics. In quantitative terms, the opposite of simplicity is complicatedness. However, there is another, qualitatively differing aspect to consider: system *complexity*.

> *I would give my life for the simplicity on the other side of complexity.*
> (Unknown origin)

The term "complexity" is difficult in its own right. For example, on Wikipedia you may find the following attempt to define it:

> *Complex systems present problems both in mathematical modelling and philosophical foundations. The study of complex systems represents a new approach to science that investigates how relationships between parts give rise to the collective behaviours of a system and how the system interacts and forms relationships with its environment. ... The key problems of complex systems are difficulties with their formal modelling and simulation. From such a perspective, in different research contexts complex systems are defined on the basis of their different attributes. Since all complex systems have many interconnected components, the science of networks and network theory are important aspects of the study of complex systems. A consensus regarding a single universal definition of complex system does not yet exist.*
> www.Wikipedia.org, May 2014

A good entry point to the study of complex systems is the website of the Santa Fe institute (www.santafe.edu), which was founded about thirty years ago for the sole purpose of exploring this aspect of science. The Santa Fe community also provides free online lectures on complexity (www.complexityexplorer.org). Even a brief review of the intro pages of the Santa Fe website should make it clear that (a) complexity is fundamentally different from complicatedness and (b) there may still be a long way to go before the pieces for a new theory of complexity fit together and become as coherent and solved as potentially yet another compartment under the *science of simplicity*. The latter can, in my view, be summarized as the science of mechanistic and Markov-compliant systems. What if this approach is futile with respect to understanding biophysical complexity in the context of animal space use? Since the present chapter is called "Movement physics beyond Markov compliance," it should be clear where we are heading.

Again for the sake of confronting the strangeness of the initial home-range results presented in the first chapter of this book, let us this time consider the paradox of observer-dependent area expansion from a classical deterministic-mechanistic angle. Consider the following (very) hypothetical scenario, again starting with the mechanistic billiard ball example in Figure 8. Consider starting at position A with the same movement speed and direction. Instead of assuming the obvious, this time you actually perform the extra eight relocations at time intervals $\Delta t = T/9$, just to make sure that calculus derived by Newton and others is correct. Surprise! A period T later, instead of finding the ball at location **B**, you see that it has progressed beyond this expected position. Confused after additional experiments with similar results, you have to conclude that movement speed has increased as a result of your observation intensity on the path. The smaller the Δt and the larger the sample size N, the larger the deviance. By increasing number relocations N during T, you are apparently distorting the trivial expectation, and the path seems to increase as if you had instead increased the magnitude of the observation period T for a given Δt. You apparently have found that the observed path length is increasing whether you increase N by increasing observation frequency to $T/\Delta t$ by reducing Δt, or you alternatively increase the observation period T.

Of course, this example—where a deterministic-mechanistic model prediction is mysteriously falsified—is just a thought experiment. In the real world you will *for sure* find that adding the extra observations on the ball's path does not distort the expected location **B** after time T. This conclusion may be considered sufficiently verified empirically as well as theoretically and thus safely considered unquestionable for all scales larger than the quantum mechanical domain! The reason is simple, given the following three premises (among a larger set of system conditions, which we again ignore for simplicity): (a) the ball's movement complies with the mechanistic framework outlined by Figure 8 and Figure 9; (b) the ball is thus moving in accordance to a Markov-compliant process; and (c) the actual path is continuous and step-by step unidirectional at fine resolutions, meaning that there is no expectancy of "surprise" jaggedness of the path at these fine resolutions relative to Δt. At micro-resolutions the ball complies with the principle of least action. In other words, the path is differentiable in Newtonian terms. A larger temporal resolution (finer increments) means a closer compliance with a smooth and unidirectional path segment. Even the jagged path in Figure 9 will be locally stretched out to a unidirectional one at sufficiently resolved micro-scales (applying higher-frequency path sampling). At these micro-scales we find the system's characteristic scale, where "action" (in physical terms) is executed in a moment-to-moment manner, which ensures local equilibrium with respect to responding to local conditions at a given point in time in accordance to principles already described. Thus, an autocorrelated path as shown in Figure 9 is in the long term expected to show increased area demarcation with increased sample size, if N is proportional with T. Similarly, area will also increase if N is constant but Δt and T are increased to the same extent. This is trivial. From what we have looked into so far, it should also be trivial from the *non*-autocorrelated fix samples in the context of Figure 7 and Figure 17, based on the principles described in the foregoing chapter, that increasing N within a given T (implying smaller Δt for larger N) should give no area increase beyond the repeatedly stressed small-N artifact.

Still, the empirical data from sheep, bear, and other species shows expanding home-range demarcation as if the size was in a strange manner dependent on observer intensity rather than sampling time as such. Thus, one should at this point be equally perplexed by this home-range ghost paradox as one would have been in the hypothetical example just described, where a deterministic-mechanistic model prediction is mysteriously falsified by reducing Δt and observing an extended path. However, rest assured: I will soon present an extended system description where the paradox may be solved without invoking metaphysics.

As already stated, a mechanistic framework and Markov compliance are two sides of the same coin. I will describe this in more detail and simultaneously lay the groundwork to return to the initial paradox in the sheep data, as expressed by Eq. 1. Alternative system extensions, where the laws of mechanistic and Markovian descriptions are embedded as a sub-section of a larger framework, will then be needed.

4.2 The memory aspect and resilience to model details

Moving beyond the mechanistic paradigm would imply two modifications of the movement process, but with substantial implications for modelling the statistical-mechanical aspects of animals' space use: infinite spatial memory utilization and scale-free temporal memory utilization. "Infinite memory" in model terms implies that the system—within the defined spatial and temporal scale extent—shows behaviour that would indicate mathematical compliance with infinite memory if the system borders were extended to infinite scales. If a model for infinite memory is found to fit observed animal movement within the area covered and the given period of path sampling, the observer would still safely assume that an extension of these scale borders would eventually reveal a transition where compliance with infinite memory would break down. This "infinite" is only locally and temporally applicable.

To understand spatial memory in the context of modelling animal movement (whether memory is infinite or with defined demarcations), consider the arguments from the preceding chapter, where mechanistic details of a Markovian-compliant approach involving spatial memory were transformed to a statistical-mechanical version, MemRW. It was argued that a wide range of mechanistic rule compositions (generally specified to be compliant with the mathematical form of Eq. 6, one set for search terms and one set for site fidelity-influencing terms) could be condensed into meso-scale algorithm for goal-oriented returns to historic locations: occasional positional shift from a current location to a historic location, selected randomly among these locations. The MemRW model architecture was explicitly defined as statistical-mechanical (coarse-grained) under premise of Markov-compliance and a scale-specific movement process (correlated random walk). This made the system potentially compliant with a "smooth" and thus locally differentiable vector field model variant, based on calculus-based mathematics (dX/dt equations in simplified terms) rather than incrementally executed mechanistic movement rules.

Interesting system properties emerged at the statistical-mechanical level in our MemRW simulations, like the fact that variations in strength of directional persistence in simulations had little influence on the power exponent for area expansion. However, if infinite memory was replaced by a memory horizon, another consistent pattern emerged: dispersal-like (Brownian motion-like) area expansion from the given protocol (incidence from non-empty grid cell counting). However, both the power exponent of area expansion and the fractal dimension of the spatial scatter of fixes were not in compliance with our empirical results from various sets of position data.

In all the nine simulated variants of MemRW (Figure 19), a temporal equilibrium was assumed in the neighbourhood of the animal's current spatial position; successive steps at unit time increments were assumed to be totally executed—given the assumed Markov-compliant behavioural rules that were active for the set of movement-influencing events during that particular increment—before considering the step result in the upcoming increment. The Markov assumption was physically implicit from the choice of a biased correlated random walk. In other words, this is simply a requirement from the premise of a mechanistic model at "hidden" micro-scales for the set of $1<=i<=j-1$ non-memory terms in Eq. 6. However, before I relax these constraints, I briefly turn towards other variants of memory implementation in movement models.

4.3 Site fidelity under continuous time random walk (CTRW)

MemRW in the foregoing section refers to a statistical-mechanical representation of a mixed mechanistic model to implement spatial memory, biased correlated random walk. However, there is another approach that has been proposed to implement long-term memory effects, albeit not in a spatially explicit manner: "Continuous Time Random Walk" (CTRW). The step length distribution and a waiting time distribution ("resting" between steps) describe mutually independent random variables. This independence between jump lengths and waiting time to perform the next step makes the difference between CTRW and ordinary random walk (including Brownian motion).

CTRW is a memory-influenced mathematical model and non-Markovian owing to very long or infinite dependence of its previous path, but in the sense of temporal memory only (what has happened in the past relative to now). CTRW does not describe a system with spatial memory, where current dynamics are explicitly influenced both by current and older parts of the animal's path, not spatially explicit memory (where things happened relative to present location). The probability of "still waiting" versus "moving on" is recalculated on a moment-to-moment basis, for example, by setting a counter to a value drawn from a (for example, power law) distribution and then "counting down" towards zero from one time increment to the next. When the current waiting time has elapsed, the animal moves (in accordance to an independent distribution for step length) and a new counter value from the waiting time distribution is set. From this perspective CTRW may even be said to satisfy a Markovian process with slowly decaying effect from historic events. Consider first, what is a waiting time? In mathematical terms it is the time from arrested movement to starting moving again. However, in practice for animal movement, it regards how long an animal stays in a small section of space (a patch) before it leaves this section. The animal may do a lot of intra-patch movement ("rumbling and tumbling"), but this does not count until a sufficiently large displacement is recorded. "Sufficient" is defined subjectively by the researcher (that many researchers may agree on the same definition does not make the rule objective!). In my own theoretical development on the MRW model, I recently described the challenge to define this "finest resolution" for a movement path in practical terms—for example, from GPS fixes that are collected at high frequency—in *The American Naturalist* (Gautestad 2013b).

A long waiting time in CTRW terms implies in practice that—at the defined spatial resolution (grid cell size) for the walk at the given temporal resolution (inter-fix interval)—movement may be hidden at finer scales. I will now press a bit deeper into the argument stating Markov compliance for CTRW from the perspective of statistical mechanics.

A key aspect of a Markovian process may be expressed as

$$P(x_n, t_n | x_{n-1}, t_{n-1}; \ldots ; x_1, t_1) = P(x_n, t_n | x_{n-1}, t_{n-1}) \quad (7)$$

The stochastically defined next-step dynamic at the animal's present location x_n at time t_n is on one hand independent of its entire past and depends on the foregoing location x_{n-1}: the dynamics may be condensed to a set of movement-influencing terms within a mechanistic framework. Such sets of condensed terms were described in Eq. 3-4. From this angle, Markov-compliance is satisfied in Eq. 7. On the other hand, the entire past from t=1,2,…,n is simultaneously influencing the present dynamics $P(x_n, t_n)$. Conceptually this potential for a paradox is resolved by assuming a process where historic influence is slowly decaying. Since CTRW's long-term temporal memory is assumed to be continuously reflected in the system's current behavior by a random number generator function for successive step calculations (moving or staying), and there is no spatially explicit memory involved, CTRW is compliant with Eq. 7.

However, in a strictly mathematical sense, the process is correctly termed non-Markovian. The physical complications from putting the process into a given environment with specific scales of observation are then not considered. Thus one may argue that CTRW is Markovian in a statistical-mechanical sense only if the actual object is connected to the fine-scaled grid cell size that represents the difference between staying and moving. The probability of staying is then representing the slowly decaying influence from history, translated to a stochastic process for the staying time. The location x in Eq. 7 is mathematically a zero-dimensional *point* in space representing the location to stay put for the time being, not a more or less microscopic virtual cell (or cube) with spatial extent where some rumbling and tumbling may take place at finer spatial resolutions during a given interval of time relative to the respective scale grains for system observation. From this mathematical perspective where x is a point in space, CTRW is non-Markovian.

Markovian or not, CTRW has clearly shown its merit as representative for important aspects of complex movement. For example, Bartumeus et al. (2010) show in a clever analysis involving CTRW how both distributions—step lengths and waiting times—may be obtained from GPS data from seabirds. They formulate the waiting time distribution as expression of site fidelity [see also Giuggioli and Bartumeus (2011), where mathematical aspects of the spatial and temporal properties are further enhanced]. In this model stronger site fidelity implies a smaller median staying time in superior foraging patches (Bartumeus *et al.* 2010). Thus, a "patch" is in this respect the same as what I referred to above for the spatial extent of the position x in Eq. 6.

In terms of diffusion, the CTRW is typically expressing the waiting time by a power exponent. As repeatedly stated, a power law implies a scale-free process. Some simulated paths are sampled at different temporal resolutions. Scale-free movement from Lévy walk, which is related to CTRW under specific conditions, is shown in Appendix A2. In this case it quantifies how the mean square spatial displacement is expected to grow as a function of time. Consider how the area of a circle—with the path's starting point at the circle's origin and the moving object at the circle itself—is growing under weak and strong site fidelity. An exponent of one reflects normal diffusion, which means that the area of a circle with the animal somewhere at the circumference is expected to grow proportionally with time. An exponent less than one implies strong site fidelity, expressed as scale-free "sub-diffusion" [in terms of Benoit Mandelbrot, a pattern emerging from an anti-persistent path (Mandelbrot 1983)]. An exponent larger than one implies an opposite kind of space use: compliance with a so-called Lévy walk. This is referred to as "super-diffusion" (persistence in the path), owing to the occasional emergence of very long displacements among mostly small steps. With respect to the seabird GPS data (Bartumeus *et al.* 2010), sub-diffusion and Brownian motion-like exponential distribution of step lengths was found in superior foraging areas given by presence of fishery discards, while super-diffusion with Lévy walk-like distribution of steps was found elsewhere. The modal shift from Lévy walk towards Brownian motion in the step distribution is accompanied by a scale-free waiting time function for inter-step pauses and respective spatial and temporal parameter intervals for the Lévy walk and Brownian motion conditions.

Thus, CTRW deviates qualitatively from the MemRW already presented. Site fidelity under CTRW implies a continuum involving temporal memory (slowly decaying influence from historic events) in order to include a potential for very long waiting time events at a given location. However, execution of this behaviour (still waiting or leaving) is strictly local and sequential, with a slowly decaying influence from intrinsic memory. Spatial memory is not involved. The mathematics of CTRW—when parameter-tuned towards Lévy walk-like path execution—may thus be related to the fractal reorientation clock conjecture (Bartumeus and Levin 2008) referred to in *0.3 Optimal foraging in stochastic movement models*.

This kind of memory implementation in a space-use model—despite the *temporally* scale-free distribution of waiting times under CTRW, which consequently implies infinite memory—is not *spatially* explicit. There is no aspect representing goal-oriented returns to previously known sites. Another way to stress this distinction is to *consider how memory of the past is influencing the future*. In the context of movement, on one hand, the future may be influenced by the past but not beyond the current, immediate behaviour. For example, deciding to initiate a long or short step in a direction may be determined by the current situation and a longer or shorter interval of influence from the past, like in CTRW (slowly decaying memory effect on current staying time and directional persistence). On the other hand, memory may also influence the future in a more subtle manner. For example, a decision to return to a previously visited location involves a memory process where both the past and the future (a temporally coarser-grained goal) are influencing the actual path. In short, the goal may be stretching beyond the present moment in time, owing to belonging to a coarser temporal scale.

To summarize, MemRW implements *spatial* memory effects within a mechanistic, Markovian framework (scale-specific processing). CTRW implements scale-free *temporal* memory effects but no spatial memory. In MemRW site fidelity is spatially explicit: the model animal is attracted to specific historic locations, which leads to the emergence of a home range. In CTRW site fidelity is expressed as the probability of staying put or moving on, influencing the local diffusion rate in a temporally scale-free manner, but with no emergence of a home range. Both MemRW and CTRW are based on the presence of mechanistic system properties at micro-resolutions (MemRW only by assumption, since it is a statistical-mechanical representation of the path), which allows for calculus-based model design at these fine scales.

In the following, the spatial and temporal memory dimensions are combined into a common framework, multi-scaled random walk (MRW). Like MemRW, MRW is also expressing site fidelity from long-term memory, by returning occasionally to previous locations. In contrast to MemRW, the local dynamics are not in equilibrium with local conditions, including memory influence (mechanistically conceptualized as a vector field in the context of the MemRW model). Contrary to both CTRW and MemRW, the MRW approach will *require* a statistical-mechanical description for both the dynamics and the statistics. Consequently, a mechanistic frame of reference is—according to the hypothesis for MRW dynamics, to be presented next—not sufficient.

4.4 The next step

Both the concept of complexity and the principle of CTRW in the present chapter are cognitively challenging, and may scare some readers from moving on to the next steps of development. However, there is light ahead in the tunnel. One key aspect of the gradually emerging extended framework for animal space use is termed parallel processing. Hopefully, it will ignite a Eureka moment for you, as it did for me. It may represent a solution to one of the main theoretical challenges in the biophysics of memory-enhanced space use of animals.

5. PARALLEL PROCESSING: THE CONCEPT

5.1 A heuristic introduction to multi-scaled random walk

MRW was formulated back in 1991 (Gautestad and Mysterud 1993), but in a conceptual and prosaic manner to classify movement under influence of long-term memory as a contrast to traditional Brownian motion. The spatially scale-free aspect of space use was illustrated as a superposition of virtual grids on a set of fixes, where the next-level grid resolution had 2^k times larger cells (linear size) than the foregoing level (Figure 22). Thus, the grids satisfy power law scaling. After log-transformation with base 2, the respective levels k=1, 2, 3 … have 2^{-k} number of grid cells within a given spatial extent. These levels of spatial resolution were then conjectured to represent various degrees of cognitive coarse-graining—parallel processing—of an individual's perception of its environment, with information stored in its memory map.

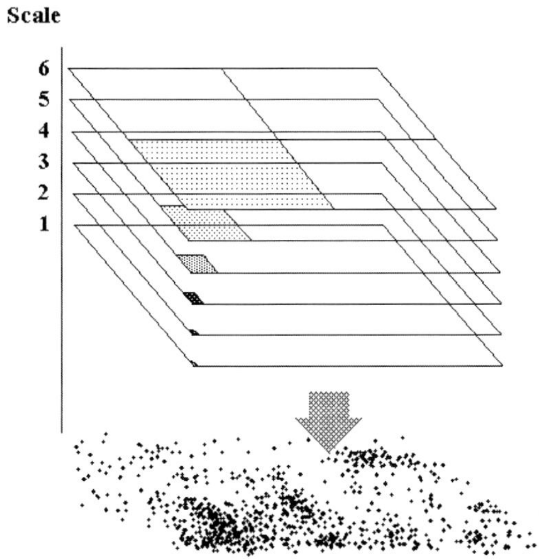

Figure 22. Representation of the scale-free spatial space use postulate based on spatial memory capacity of non-classical type; parallel processing.

This kind of cognitive capacity to "zoom in and out" over a range of spatial resolutions owing to processing of spatial memory, according to the MRW model conjecture, implies that the neighbourhood will have different size at different scales. It also implies that a decision to move to a neighbouring patch at a coarse spatial level will—in average terms—imply a coarser time resolution for execution. Additionally, a section of a movement path during execution of a coarse-grained "move" will show a larger displacement than a finer-grained move, owing to (typically) a stronger degree of directional persistence during the coarser move. Execution of n goals to target a patch at a coarse scale relative to a given position of the animal will generally involve a longer time increment to fulfil than n goals at finer resolutions. According to the MRW conjecture, a memory map involves spatio-temporal coarse-graining of information that has accumulated during visits to other locations relative to its present location. Thus, while memory in a mechanistic framework traditionally regards instances along a time axis, a memory map *sensu* MRW regards explicit cognitive processing involving zooming capacity of both time and space axes. Figure 22 goes even further: it postulates that this kind of cognitive processing in a statistical-mechanical sense is mathematically scale-free (within trivial scale ranges), owing to the power law structure of the scaling axis. It also postulates that—in addition to utilizing past experience to optimize movement behaviour in the present—this information is also utilized to determine future goals that may be active at any given point in time but not necessarily actively executed at this particular moment.

This conjectured property clearly creates new paradoxes from the perspective of Markovian mechanics: *causality is no longer explicable in terms of responses to local states and events at a particular instant in time.* Memory of the past is influencing the future in a more complex manner than in the MemRW model. With reference to the discussion of CTRW above, the position x at time t in Eq. 7 has in the MRW framework been given spatial extent. Further, with reference to MemRW, local dynamics are in MRW, not in equilibrium with local conditions. The latter described influence from the past through what I described as a memory field for strength of bias towards previous locations. Parallel processing implies that goals are simultaneously active at several temporal resolutions (in parallel), meaning that MemRW's concept of a memory-based vector field cannot apply. A vector only makes sense in a Markovian, mechanistic frame of reference owing to local differentiability of the field. The next step in a MRW cannot be explained deterministically as a superposition of a finite (but large) number of independent components, as required by a

Markov-based system. Thus, MRW describes a non-mechanistic system while still being dynamic. As an intuitive image, consider a movement path that is full of surprising moves when studied at a temporally fine-grained resolution. The surprise factor represents the sudden and unpredictable execution of latent goals at coarser temporal resolutions. These "longer term" goals are explicable from respective coarser temporal scales but not at the fine-grained resolution, where the behaviour is attempted to be decomposed into specific movement rules according to mechanistic model principles. Execution of strategic goals typically comes as a surprise, given the local conditions at the given point in time and within the current field of perception. In other words, mechanistic causality becomes blurred.

Causality in mechanistic terms requires that a zero-dimensional position x at a small increment in time may represent the dynamics, for example, by averaging some system attribute (position at time t in the present context) over a small neighbourhood of x. This zone of approximately local equilibrium with respect to the change of state (acceleration of deceleration of movement speed) allows for the system descriptions in Eq. 3-7. MRW is based on a conjecture that deviates from the premises of Eq. 3-7. Let us now dip the toe deeper into this murky water, and show how this path towards non-mechanistic dynamics may potentially open for novel model designs for complex systems.

It should be clear from the preceding chapters describing mechanistic processes that a Brownian motion-compliant movement—a memory-less process—would involve one fine-grained grid resolution only, representing the process' characteristic scale (where x in Eq. 7 may be represented by a point in space). In statistical-mechanical terms, this would equal the mean free path under the given conditions, as described in a previous chapter. The MemRW movement class may capture spatial memory effects, but in a mechanistic (Markovian) and thus temporally scale-specific manner with local causation maintained. In other words, the next move could be calculated on the basis of local conditions within the field of perception in combination with (mathematically, in superposition with) the memory "field" of vectors pointing to respective past locations, leading to an algorithm-based, mechanistic calculation of what to do in the next instance of time. Thus, the scale-free postulate conceptually illustrated by Figure 22 does not apply to MemRW. The other mechanistic approach towards memory implementation that was described above, CTRW, may reflect the temporal memory effects in the form of a scale-free waiting time, which may limit diffusion but without spatial memory

(Zaburdaev *et al.* 2014). The present approach regards a third direction. MRW seeks to integrate the two dimensions—spatial and temporal memory—into a common framework. In addition, a postulate about execution of goals at different temporal resolutions means that the future (where the goals to reach specific spatial targets reside) is also influencing the present behaviour, not only the past. The surprise factor described above is a manifestation of future goals on present behaviour.

MRW regards a statistical-mechanical model representation of memory-influenced, *complex* space use, since the basic idea of Figure 22 is incompatible with a mechanistic kind of dynamics. In summary, while a mechanistic system may embed memory-influence on current dynamics (non-spatially as in CTRW and Lévy walk, or spatially explicit as in MemRW), a scale-free influence from the past on future behaviour requires—according to the parallel processing postulate—a non-mechanistic kind of system dynamics. In simple terms, setting future goals that may or may not influence the actual movement behaviour close to the present moment will require a kind of non-Markovian dynamics that extends the mechanistic MemRW process.

Allowing for a digression at this point, the *concept of complexity* may often be accompanied by a painful cry from scientists who are cognitively boxed within the mechanistic frame of reference with respect to model formulations. The science of complexity embeds the study of scale-free processes in systems with many degrees of freedom (for example, an animal population or a large set of GPS fixes reflecting space use by a single individual). In such systems there may often appear what are called emergent properties: "bottom-up" and "top-down" flows of influence on a defined entity are interacting in a still little understood manner. The active research field of complex systems is typically attempting to force spatio-temporal scale-free processes into what I somewhat provocatively call the system narrowness of the mechanistic paradigm. Hence, perhaps this framework is an *erroneous starting premise*? Perhaps the mechanistic-Markovian paradigm needs a supplement through an extended kind of statistical mechanics? This would—as already proposed—require some serious out-of-the-box thinking. Most basically, it would potentially imply that mechanistic is no longer synonymous with dynamical! The latter may include more than the former, since a mechanistic frame of reference is a special system condition at one of the endpoints on a continuum between Markovian and non-Markovian processing in statistical-mechanical terms.

This extended space-time framework for systems dynamics has been called "parallel processing" from the onset of our theoretical development. The concept of parallel processing back then in 1991 to about 1998 was by and large verbally defined and accompanied by graphical illustrations like Figure 22. In the following I first summarize the published descriptions of parallel processing as a preparation to later chapters where this concept is mathematically sharpened with respect to its statistical-mechanical representation.

5.2 Initial descriptions of the parallel-processing conjecture

In my doctoral thesis in 1998, I described parallel processing basically as follows. Consider putting an individual or a set of individuals (a spatial subpopulation) on a square (cell) of one of the grids. As seen from "above" (i.e., orthogonally), the individual or the population are both a part of a square at a grid that has larger squares, and an even smaller part of a square that is even higher in the pile of grids. An event that seems to be a long distance apart in space (and time) from a small-cell point of view is perhaps at the same square, or just at a neighbour square, as observed from a higher-level grid with larger squares. Larger squares represent (in average terms) increasingly "strategic" cognitive overview of the environment. Consider that an individual takes a colour in accordance to which scale level it primarily related its movement to during a given time interval. Let us say that the animal is blue for the intervals when it has moved rather tactically to the neighbourhood at a fine grid scale, and red during intervals when it moves more directionally persistent towards more distant locations (the neighbourhood at a larger grid scale). A snapshot of the colour distribution over a population is, from this hypothesis, expected to show a hyperbolic distribution ($1/2^k$ as many individuals relating to its environment at the kth larger scale level at a given point in time, using logarithms with base 2 for the scale levels).

The previous paragraph gives an excerpt from my 1998 doctoral thesis describing MRW and parallel processing in a rather hand-waving manner. For example, no protocol was advised with respect to distinguishing "blue" from "red" movement modes. Whether the modes were a superposition of fine- and coarse-grained scale-specific processes or a more complex structure where more strategic goals are influencing finer-grained goals in a cross-scale manner was only loosely indicated, but implicitly assumed (otherwise the model would resemble composite Brownian motion; see details in a later section).

In the paper on the bear data shown in Figure 4, parallel processing was further elaborated on (Gautestad *et al.* 1998):

> *... MRW is based on the assumption that information at coarse resolutions in the long run influences successive displacement steps to the same magnitude as information at finer resolutions (note: this description was accompanied by Figure 19 above). This means that the product of scale (measured as, for example, km^2) and number of deterministic movements on this scale can be expected to be constant from one scale to another (up to the scale of the home-range extent). In the long run, many short-distance movements should thus be intermingled with fewer displacements on coarser scales. If so, optimization of habitat use within a home range is uniform over scales up to the size of the home range. In other words, MRW is based on an assumption that the rules for patch choice should be applied by the animal to (ideally) all patches at all spatial resolutions (scales) within its home range, and not on fine scales alone. The number of patches within a given demarcated home range is larger for finer-resolved patch resolutions. Thus, more local movements to and from neighbour patches can be expected on a fine spatial scale relative to (fewer but larger) displacements on a coarse scale when an animal seeks to equalize its optimizing effort over a range of scales. MRW represents a generalization of the classic and scale-specific random walk model to a scale-free random walk model. To summarize, while a classical random walk process involves tactical and sequential movements due to events at the scale of the field of perception or finer, MRW includes more strategic influences due to environmental conditions as they appear at resolutions between the field of perception and the home-range extent. In practice, we will sometimes see large and relatively directed movements which appear inexplicable from current local conditions. However, from a coarser perspective in space and time, the movement can be explained, demonstrating the complexities of parallel processing. The MRW involves two assumptions which are not valid for the classic random walk under home-range constraints on free dispersal. First, optimization of habitat use is expected to take place with equal effort on all scales from the field of perception to the home-range extent, and second, the field of perception is assumed to be small relative to the extent of the home range. ....*

Again, the foregoing paragraph contributes some clarification of the MRW and parallel-processing concept, but still it is far from an applicable model in statistical-mechanical terms. In a paper in *The American Naturalist* in 2005 (Gautestad and Mysterud 2005), the parallel-processing concept was again described, and this time with particular reference to statistical consequences from potentially self-reinforcing use of specific patches:

> *... Homing behaviour violates the basic assumption of independent successive revisits to given habitat elements inside a home range, with serious consequences for, for example, the applicability of the standard low order Markovian-based protocol ... for testing habitat preference at the individual level (note: this protocol was specified as comparison of local density of fixes and local habitat class; often referred to as use/availability analysis). Generalized linear model-based resource selection models offer other approaches to estimating habitat preference. For example, return frequency for each patch or distance to most recently visited patch could serve as covariates in a logistic regression model of resource selection. However, even if such spatially less explicit protocols are applied, it is logical to expect spurious results if the animal tends to revisit strategically a subset of a priori equally available patches of same size and type in a self-reinforcing manner. It follows from Markov-based mechanics that the given set of patches should be visited with equal probability, but the staying time in each patch would depend on patch quality. A low-order Markov-based assumption of equal revisiting probability then raises the risk of mixing the statistic signals from (re)visiting frequency and intensity of patch use. ...*
>
> *... if complex MRW-like processes in the population are driving intrinsic dynamics in the direction of positive feedback and fractal distribution of individuals, challenges like the paradox of density inflation (note: density inflation will emerge as a consequence of self-reinforcing revisits to specific locations) will have to be solved by developing and applying a model framework from which this kind of complexity can be understood from first principles. ... Our main take-home message is that a framework expansion from sequential to parallel processing may be needed in order to start building ecological theory from dynamic assumptions more realistic than the current linear, scale-specific, or "quasi-multi-scaled" approximations. Our three suggestive tests for MRW-based kinetics are just a starting point on a new direction of research (note: these protocols will be described in following chapters). ...*

In a paper in *Ecological Complexity* (Gautestad and Mysterud 2006), the parallel-processing concept was further elaborated on, and this time with explicit reference to fractal pattern of fixes as an emergent property. In the following excerpt from this paper, it is also referred to as Lévy walk as a representation of scale-free search:

> *The MRW model describes an animal that is mixing short-term, local movements ("tactical habitat utilization") with less frequent but more far-ranging commuting ("strategic displacements") in a manner that self-*

organizes into a scale-free, or fractal, weighing of tactical vs. strategic movement events in the long term. In this respect MRW is fundamentally different from a classic random walk process. The latter is scale-specific, since successive steps in a spatial random walk path obey a low-order Markovian function (Turchin 1998): local and short-term influences at the characteristic time scale for the process determine in principle both the direction and length of the next spatial displacement. This kind of purely tactical space use can be characterized as a sequential, or serial, process of space use. On the other hand, MRW can be conceptualized as a movement process where the next step is also influenced by directed, strategic displacements due to decision making that is based on information processing and functional responses involving coarser spatial and temporal scales in parallel with the immediate tactical scale. For example, an animal decides that it should tend to move towards a previously visited location far beyond its field of perception at its present location; not necessarily at the moment of the decision, but within a given coarser time period. This means that this strategic displacement may take place over a relatively long time period in comparison to the successive, shorter-term tactical steps during this time. During the strategic time interval, the integration of all the successive, intermediate tactical steps has brought the individual to its strategic destination. Thus, with reference to Fig. 2 (note: Figure 23 here), an animal performing MRW will occasionally move strategically towards a long-distance destination, while still responding tactically to successive local events during the drift towards the destination. This involves spatially explicit memory effects on movements, since the long-distance destination is given by reference to historic events, i.e., based on previous visits to—and experiences from—that particular distant location. Thus, the constraining (drift-like) influence of strategic movements on finer-scaled movements extends the displacement dynamics from a purely scale-specific, sequential Markovian process to a multi-scaled (and potential scale-free) parallel process, where the coarse-scaled displacement processes define constraints on finer-scaled parts of the movement.

Strategic movement based on spatial memory effects (memory map utilization) brings up a potential for a positive feedback from animals tending, for example, to make long-distance revisits to familiar patches, at the expense of less utilized patches, in a self-reinforcing and scale-free manner. …

The text above states that MRW "self-organizes into a scale-free, or fractal, weighing of tactical vs. strategic movement events in the long term." In this respect observe a little detail in Figure 23's lower right pane, which concretizes this statement. The area of each of the "blocks" in the diagram is of similar size; from high and narrow to short and wide. This creates a more precise description of the spatio-temporal postulate on memory influence relative to the initial presentation in Figure 22. Mathematically, the integration of space use over a given scaling interval of grid resolutions (the x-axis of a block in the diagram shows grid cell size, given as L^2) leads to a similar solution at all scales (positions on the x axis). In more simple terms, number of grid cells that have been traversed during a given time interval multiplied by the size of grid cells at the chosen scale of reference is—on average—constant. This creates the hyperbolic form of the distribution in Figure 23. In short: in the long term, an animal that is processing memory-based information in accordance to MRW will show a large number of small steps (on average equal to the square root of the cell sizes in Figure 23), fewer medium-size steps, and very few large steps. If the block area of each scale interval is similar, the

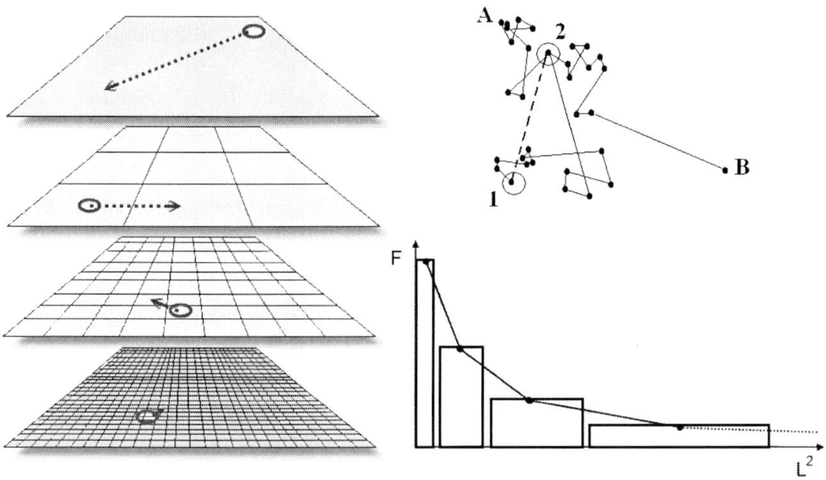

Figure 23. A multi-scaled random walk (MRW) is conceptualized as multi-level series of steps of variable length L on a "sandwich" of lattices where the next higher scale level is characterized by a *k* times larger grid cell size. When all steps are projected down to ground level, a mixture of frequent, short steps and more infrequent, larger steps from a start position A to an end position B will emerge, as illustrated in the inset to the right. The step from 1 to 2 represents an infrequent strategic "return step" to a previous location. In the long run, the dominance of Lévy flight steps over strategic return steps will maintain a power law distribution of steps to neighbour grid cells at various scale levels $k = L^2$, as illustrated in the histogram to the right.

given individual has utilized (it has "related to") its environment with equal statistical weight to each of the grid resolutions in Figure 22 and in the main pane of Figure 23. The space use is here referred to from a statistical-mechanical perspective (the properties of the distribution), and not a "behavioural mode" perspective. This system property—equalized utilization of information at different scales, which was first referred to in mathematical terms in the 2005 *American Naturalist* presentation (Gautestad and Mysterud 2005)—will be returned to frequently below.

5.3 Parallel processing in later publications

In a 2010 paper (Gautestad and Mysterud 2010a), parallel processing got a more stringent conceptual description by contrasting it with the mechanistic/Markovian model architecture for memory influence. The latter was termed MemRW at the statistical-mechanical level in foregoing chapters. The following paragraphs contain an excerpt of this specification:

> *...models have recently been proposed where memory of conditions in patches outside the individual's current position has been implemented in the movement rules (Dalziel et al. 2008, Fryxell et al. 2008, Getz and Saltz 2008, Mueller and Fagan 2008, Nathan et al. 2008, Schick et al. 2008, van Moorter et al. 2009). In this manner the low-order Markovian-structured model framework is expanding the spatial scale for environmental influence on successive step decisions from the local neighbourhood (field of perception) to a broader "radar field" with long-distance conditions recalled from a memory map. The processing is still scale-specific in the time domain, based on successive low-order Markov-compliant next-step decisions. To make this approach feasible in practice where local conditions at each time step have to "compete" with hundreds of similar sets of distant conditions recalled from the memory map, these models are defining distant conditions in space and time to have less influence than local and current conditions. Otherwise, history (reference memory) inputs would in practice block for more tactical space-use optimization based on short-term working memory and direct input from senses. However, real animals utilizing long-term memory frequently make sudden long-distance returns to other parts of their home range, often based on experiences far back in time. These "strategic moves" may seem paradoxical from a tactical low-order Markovian perspective. For example, the current patch may still contain ample resources at the time of departure (Fryxell et al. 2008, Wolf et al. 2009). On the other hand, the move may make sense in a coarser scale perspective. This kind of behaviour violates the Markovian model design. Consequently, what is needed for improved realism is a broadened statistical-mechanical framework that explicitly includes the*

effect from goal-oriented and memory-dependent (i.e., long-term), strategic kind of space use in parallel with the more tactical behaviour. MRW embeds the potential for scale-free space utilization beyond the "mean free path" scale, as conceptually indicated in Fig. 2 (note: in the paper; reproduced as Figure 24 here).

With reference to Fig. 2 (*note: Figure 24*), a denser distribution of resource patches implies a smaller spatial and temporal characteristic scale for low-order Markovian-compliant movement interrupts. Consequently, whether an underlying directional persistence in a given Markov-compliant path is due to a correlated random walk or a Lévy walk step from the long tail part of the distribution, the mean free path represents in statistical terms an environmentally defined scale cut-off (spatial truncation) of any such process-intrinsic directional persistence. At coarser observational lags than the mean free path, both types of motion become Brownian motion-like with a Gaussian distribution of step lengths and uniformly distributed random step directions due to the randomization effect from interrupts. Only at finer observational lags than the mean free path can the distinction between a correlated random walk and a Lévy walk (independent of the aspect of memory effect) be distinguished (Bartumeus and Levin 2008, Reynolds and Rhodes 2009). Since the mean free path in a low-order Markov context is normally much smaller than the home-range scale, a model for memory-influenced intra-home range movements may, in a Markov context for simplicity, use random walk or correlated random walk rather than Lévy walk. The distinction at coarser scales than the mean free path will disappear.

But in real data, the Lévy pattern seems to extend towards larger scales than the mean free path. Consequently, a cognitive capacity for a memory map utilization, whereby directional persistence may survive the mean free path constraint under the general Lévy walk / Brownian motion continuum, is a necessary model expansion in order to understand how a scale-free space use may take place over the full scale range from the individual's field of perception and the magnitude of the home range.

As already mentioned, in order to implement cognitive map-influenced movements into a Lagrangian statistical mechanical home-range model (*note: the term Lagrangian basically means that one is studying the distribution of step lengths rather than the spatially explicit dispersion of fixes*), it has been suggested that trajectories of opportunistic movement sequences (displacements not explicitly directed towards a previous location), represented by a Brownian motion or a Lévy walk, are interrupted at a small average frequency by directed and memory-dependent returns to

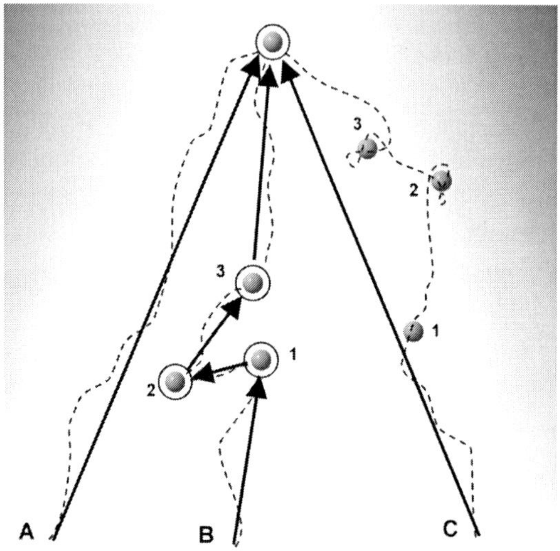

Figure 24. Conceptual illustration of the qualitative difference between spatial displacements of foraging individuals with short-term memory only (paths A and B) and a cognitive map-influenced multi-scaled random walk (MRW; path C). The paths A and B could, for example, represent a correlated random walk with strong directional persistence or a specific displacement characterized by the long tail part of a Lévy walk distribution of step lengths. Starting from different positions the three individuals happen to reach a common end position. Individuals A and B have no initial goal to reach this location, while C has a long-term goal either to reach this specific patch (i.e., a directional return step, based on prior visits to this location) or the location's neighbourhood in coarse-scale terms. Both conditions require a capacity for cognitive map utilization. The respective fine-grained paths are indicated as dashed lines, while the successive realized step vectors at a coarser unit temporal scale (observational time lag) are simplified as straight-line arrows. Resource patches that happen to enter the respective individuals' field of perception—imagined to be very small relative to the total distance covered—are marked as bullets. A series of intra-patch micro-movements prior to leaving the patch are assumed but not shown. Successive steps in a memory-less process (paths A and B) are driven by the output from a combination of the individual's current internal state and multi-dimensional inputs from the environment within the current spatial limit of its sensory organs (e.g., sensed resource patches, distance to resources, perceived dangers, etc.). These time-specific inputs are then considered weighted in accordance to the respective defined terms' strength in the movement function and superimposed to generate a resultant vector for the next displacement in the series. Consequently, path A shows a relatively uninterrupted and unidirectional displacement towards the final location, since a lack of environmental "interrupts" during the walk makes the resultant vector from one time step to the next primarily dominated by the directional persistence term inherent in the correlated random walk. In contrast, the path B individual successively hits a series of resource patches (marked by numbers), and at each of these three events, a new resultant vector is calculated for the displacement out of the patch (recalculations are marked by circles). Since successive displacement vectors are determined by local and temporal conditions, the next vector in the series becomes independent of the previous one. In contrast to this classic low-order Markovian condition, path C indicates that the individual when leaving a resource patch has maintained the strategic (longer term) goal to reach the final location. Thus, the resultant vector at points 1-3 are not fully determined by conditions at these respective locations and points in times. In other words, spatial memory allows for long-term strategy (a goal to reach the final location) as a parallel process to short-term tactics (resource patch utilization, etc. along the path). This behaviour contrasts with the serial (sequential) processing in paths A and B with a mathematically well-defined local causality structure for the displacement vector adjustments during the total time interval. This framework characterizes linear statistical mechanics, including Brownian motion and diffusion.

previously visited locations (Gautestad and Mysterud 2005, 2006, Börger *et al.* 2008, Dalziel *et al.* 2008, Gonzalez *et al.* 2008). These approaches deviate with respect to process rate structure: single-scaled (i.e., a low-order Markovian compliant walk) or multi-scaled (as in MRW).

One approach regards correlated random walk displacements where only return steps are memory-influenced, and exploratory steps are defined as Markovian-compliant random walk or Lévy walk (Börger *et al.* 2008, Dalziel *et al.* 2008). Truncation of directional persistence takes place at the mean free path scale as explained above. In the second type of memory implementation, the MRW model, both the exploratory steps (Lévy walk by default) and the return steps maintain directional persistence beyond the mean free path scale. This is implicit in the conceptual description of MRW as successive steps to neighbouring grid cells with linear scale size k, in a superposition arrangement of grids where the next and coarser-grained grid has k^2 larger cell areas, and the frequency of next level steps also fall off as a power law (Gautestad and Mysterud 1993, 1995, Gautestad *et al.* 1998). Thus, exploratory steps and return steps are both memory-dependent and goal-oriented, but the former regards displacement to a finer- or coarser-grained neighbourhood (i.e., a less specific location) rather than to a specific target location. This difference in memory effect implementation represents not only an opportunity for distinguishing the two classes by fractal analyses of space-use data, but also a more complex movement than provided by the classical Markov architecture. …

The concept of Lévy walk has been referred to several times above. In a 2011 paper in the *Journal of Theoretical Ecology* (Gautestad 2011), this concept is defined as follows:

> *… Fractal-generating space use without explicit reference to a causal link to memory has also been studied, in particular as a consequence of Lévy walk (Viswanathan et al. 1999, Hagen et al. 2001, Sims et al. 2008, Webb et al. 2009, Humphries et al. 2010, Seuront 2010). Lévy walk is a kind of random walk with a power law-compliant tail in the distribution of step lengths. However—like random walk and Brownian motion—it represents a memory-less (i.e., low-order Markovian compliant) process, where the current displacement is a function of conditions within a narrow trailing time window (Gautestad and Mysterud 2005). A stochastic process like animal space use has the Markov property if the conditional probability distribution of future states of the process, given the present state and the past states, depends only upon the present state; that is, the past is irrelevant because it doesn't matter how the current state was obtained. MRW, Brownian motion, and Lévy walk each represent a statistical-mechanical level of system abstraction and analysis at a basic level, and*

> *the fractal dimension D represents a feasible parameter to distinguish between these variants of space* use (Gautestad and Mysterud 2010a). …

In the last paragraph above, I refer to Lévy walk as a memory-less and thus a Markovian-compliant process. Lévy walk, which is still a little-understood process in general statistical-mechanical terms, is spatially memory-less but is temporally scale-free. The latter may, in my (modified) view, imply that a specific expression of Lévy walk should be considered compliant with parallel processing in the time domain, but still not in the space domain.

I will return to more detailed descriptions of Lévy walk later, but I include the definition above as an initial, simple reference to "scale-free search." In other words, a typical series of fixes collected from a Lévy walk path will show a distribution of step lengths that satisfies many small steps, a few medium-range steps, and a very few very large steps, since coarse-scale goals in the time domain are generally positively correlated with coarse-scale displacements in the space domain. Further on the parallel-processing concept in the *Journal of Theoretical Ecology*:

> *… ecological theory has to cope with the complexity from "memory-influenced particles," i.e., the property of potentially long-distance and long-term strategic (goal-oriented) homing directly towards previously visited locations—return steps. Since these target locations occasionally may be relatively far away, a kind of directional persistence is needed during the displacement. For example, the behaviour must allow for overriding unfavourable local conditions during the return step, and also include the continuation of a strategic move even when this is occasionally interrupted by a series of local direction-influencing events. This property of combining short-term tactics while maintaining the long-term strategy based on memory may imply a parallel processing kind of space use, with more complex movement mechanics at the statistical level than a classic random walk or Brownian motion (Gaussian or Poisson step length distribution), correlated random walk (directional persistence), or Lévy walk (power law distributed step lengths) can express. One modelling approach, based on Multi-scaled random walk and a parallel-processing postulate … seeks to formulate this tactics/strategy complexity in an explicit manner and within the framework of non-equilibrium statistical mechanics (Gautestad and Mysterud 2010a). … The MRW model explicitly shifts focus from the mechanistic to the statistical mechanical level a priori, i.e., the execution scale in simulations is defined to be coarser than the assumed mechanistic scale. It … is … defined as memory-influenced space use, which is scale-free both in space and time (Gautestad and Mysterud 2005, 2006, 2010a). The latter implies a postulate of a parallel processing kind of space use, where short-term goals are executed with*

faster process rates than more strategic goals. For example, a "within the next day" goal may run in parallel with more frequent "within the next hour" kinds of goals. Due to the hierarchical structure of process rates (Allen and Starr 1982, O'Neill et al. 1986), strategic goals may constrain shorter-term tactical goals (Gautestad and Mysterud 2010a). ... a collection of MRW step lengths typically shows Lévy walk-like properties ("fat," power law tail of the distribution of step lengths). From the composite process of power law distributed displacements and more directed returns targeting previously visited locations, a multi-scaled home range will emerge. The space use becomes strongly fractal-like in a statistical sense. (Fig. 1) (note: this is Figure 25 here). ...

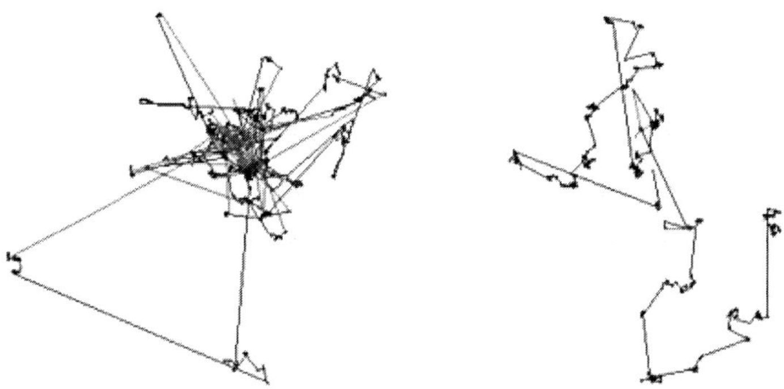

Figure 25. Two simulation examples of 4,000 steps each, from MRW sampled at unit time interval. Successive Lévy walk-like displacements are marked by black lines, and occasional returns to previous locations are marked by grey lines. The left-hand path shows an animal returning to a previous location at a frequency of 1:100 steps on average, and where all locations have equal return probability ("infinite memory"). The individual on the right-hand side returns at same frequency, but in a more time-constrained manner, to locations within a trailing time window covering the last 20 steps (short-term cognitive map utilization). By removing the connected lines (as in the figures to follow), a complex space-use pattern appears from the aggregated set of fixes even under the given condition of a homogeneous environment.

The last sentence points to the important fact that scale-free space use like Lévy walk will generate a fractal (self-similar) dispersion of fix collections from its path. Thus, we are now closing in on one of the space-use paradoxes described in this book: the fractal dimension of real animal fixes was found to be smaller (implying stronger scale-free aggregation) than expected from a classical model for movement.

Since I have now given you a glimpse of the Lévy walk concept, and *de facto* have also allowed you a preview of an actual MRW simulation in Figure 25, where Lévy walk and memory-dependent returns are combined, I want to include an extra excerpt from the 2010 paper in *Ecological Complexity* (Gautestad and Mysterud 2010a), where parallel processing is explained more stringently and in more detail than the descriptions above:

> *... For a MRW the scale-free nature of exploratory steps extends beyond the scale of the mean free path. This MRW-characteristic property is a consequence of parallel processing with a hierarchical structure of process rates.*

Conceptually, consider a walk where memory map-based strategic step decisions are taken in a scale-free and ratio-dependent manner. This implies a parallel-processing type of cognitive map influence on displacements, where the process rate continuum, in accordance to hierarchy theory (Allen and Starr 1982, O'Neill *et al.* 1986), can be made discrete in model terms by defining temporal scale ratio in a power law fashion: strategic movement-influencing decisions can be simplified to take place over a set of discrete characteristic time resolutions which scales as Δt^n, where the positive integer Δt represents the chosen logarithmic base for unit time, and n=0,1,2… For example, if choosing base unit Δt =10, strategic decisions at scale n+2 are 1/10 as frequent on average as scale n+1 events and 1/100 as frequent as scale n events. An important property of parallel processing is coarser-scale decisions becoming a constraining factor for the more high-frequency displacements running their course in parallel at finer time scales. In other words, when an individual has made up its mind for a relatively long-term directional move, finer scale step events may take place during this coarse scale move but are not overriding the strategic decision (a side effect from the finer process rate).

The coarser-level displacements may take place relatively suddenly or more drift-like from the perspective of a finer scale level in the total process. As an example of the former, when you drive to the University in the morning, you may have to stop to fill the tank of your car or you may be disturbed by other short-term events that may even make you consider alternative routes. However, these interrupts (*sensu* Fig. 2; *note: this is Figure 24 here*) are normally not sufficient to override your destination goal for the next hour. You are able to keep long-term focus thanks to parallel processing. If your brain functioned in a purely Markovian fashion (all decisions taking place at a given, relatively high-frequency time scale), this kind of scale decoupling and constraint structure would not be feasible. Either you would have to engage a very strong filter for incoming signals (no eye on the tank needle) or you would in effect perform a biased random walk with a renewed directional persistence at every time step event. This would, like the strong filter case, lead to a relatively smooth and directional drift-like motion towards the goal. In contrast to this, a parallel-processing structure also allows for relatively sudden and long-distance displacements, in addition to performing steps with short-term disadvantage for a longer term advantage. A moose may decide to move to the neighbouring valley within the time frame of a day, at an indeterminate point in time as seen from a resolution of minutes and hours. In fact, if this displacement takes place rather suddenly, it may appear unexplainable from the perspective of fine-scaled events and conditions like having to commute over a low resource and potentially dangerous matrix habitat. Only from the perspective of the time scale of a day does the displacement appear as deterministic and advantageous in fitness terms.

In more physical terms, as an example of parallel processing, consider a particle performing a classic Brownian motion at unit temporal scale $\Delta t^0=1$ from micro-scale dynamical interactions with the environment. Further, consider giving the particle a memory map (with constantly renewed information from the particle's successive positions), which is continuously "utilized" in a virtual manner at three temporal scales Δt^1, Δt^2 and Δt^3 (e.g., time ratios 10, 100, and 1,000). Let these three processes run as three independent random walks, called particle "Blue," "Green," and "Red," on the virtual memory map. At every 10th step on average, Blue instructs the real particle to jump from its current physical position to the location defined by Blue's analogous position at the memory map. Similarly, but at lower frequencies 1/100 and 1/1000, the real particle gets step directives from these levels. At every such "strategic displacement" event, Blue, Green, and Red perform a relative mental repositioning to location (x,y) = (0,0) where the origin represents the physical particle's new location. Then the three virtual particles and the physical particle continue their respective dynamics until the next interruption. Displacement instructions from Red will on average imply a much larger step length than a unit-scale Brownian motion step—on average $\sqrt{1000}$ as large (i.e., covering a 1,000 times larger grid cell area in the grid superposition analogy) since a random walk particle's root mean square is expected to expand at this rate. Similarly, the 10 times more frequent Green instructions than Red instructions imply a $\sqrt{100}$ times larger step and a Blue instruction $\sqrt{10}$ times larger. Consequently, the Brownian motion particle gets low-frequency displacement boosts that in overall terms make the path's distribution of step lengths change from a classic Poisson (exponential decay in the tail part, characteristic of Brownian motion and random walk) to a power law decay in the tail part, *with Lévy walk characteristics but without the classic mean free path truncation.* We carefully avoid stretching this "virtual particle" analogy into a biological interpretation of actual memory map mechanisms at this stage. We just underscore that this thought construction may visualize important aspects of MRW versus a low-order Markovian-compliant space-use process. MRW represents a parallel-processing model construction, which (from the fractal geometry approach and the space-time scaling function to be described in this paper) can be tested against other statistical-mechanical classes of space use models. …

5.4 Why invoke the parallel-processing conjecture?

So far I have presented two paradoxes from empirical data—the home-range ghost (Eq. 1) and the fractal dispersion of fixes ($D \approx 1$). Later a third issue will be described, the so-called "Z-paradox." Based on the data we have studied so far, there seems to be an apparent generality of these statistical-mechanical properties of space use among vertebrates where—as a rule of thumb—site fidelity in the form of a home range is involved. I have now provided a walk-through of some basic classical system premises, for the sake of supporting the claim that there are in fact real paradoxes to be solved. If there are no paradoxes, one expects the challenges to be solved under the classical paradigm. A model for memory implementation—MemRW, which was compliant with the classical premises of mechanistic modelling—has also been presented. Unfortunately, these results did not solve the challenges. Thus, additional arguments for a deeper drilling to find solutions were given in the presentation of conditions for Markov compliance and definition of complex versus merely complicated systems. Finally the concept of parallel processing was presented. Parallel-processing influence on a specific path may be summarized as in Figure 26. Over the following chapters I will show how the application of this concept as a conjecture for a statistical-mechanical model has resulted in theory and simulations that generate space use which reflects all three paradoxes. Thus, under this conjecture the specific system properties appear paradoxical only from the perspective of classical premises.

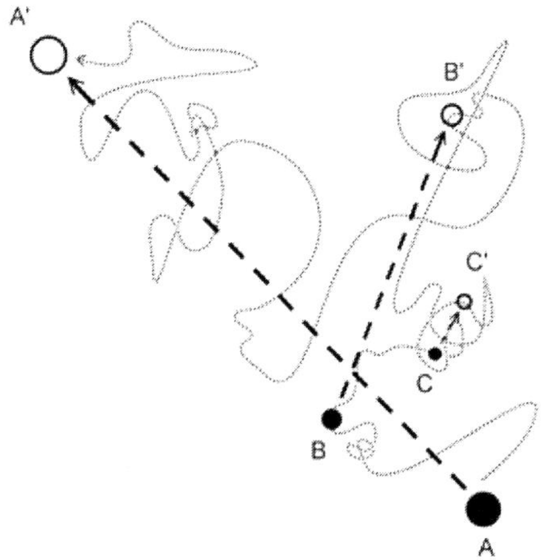

Figure 26. Memory-influenced movement under the parallel-processing conjecture is conceptualized by three hypothetical goals along a spatial path (dotted line). The goals are postulated to be executed at three separate temporal scales: long term (arrow towards target A', to be reached within time t_a', decided at location A at time t_a), medium term (B and B'), and short term (C and C'). The difference in time intervals for the three targets implies different process rates, and consequently an option to execute several goals at finer temporal scales for each goal at coarser scales. A premise is that the finer-scale goals do not conflict with the coarser-time goals. For example, several medium-term goals may be executed during the time interval from t_a to t_a' (only one is exemplified; from B to B') and even more are possible during even smaller intervals (for example, the example from C to C', as long as these do not conflict with the longer-term goals B' and A'). For simplicity longer-term goals are illustrated to reflect longer-distance targets. Even if this normally is the case on average (in statistical-mechanical terms), it is not a necessary condition.

Basic-level (generic) simulations should be applied to test compliance with basic-level statistical properties. From sets of spatial GPS relocations as representatives of space use, a statistical-mechanical theory should be invoked. This theory should then explain the emergence of both aspects $z \approx 0.5$ and $D \approx 1$ in a satisfactory manner. Other aspects that need a similar compliance will be described and handled over the following sections. The target for this development is thereby clear-cut: presenting a *generic* model that is on one hand able to replicate *basic* statistical-mechanical properties of real space-use data and on the other hand has an accompanying theory that is able to exterminate the paradoxes when these are studied from the perspective of the classical framework.

Before this challenge is taken care of, there is little point in adding new aspects to the generic model for the sake of improved realism relative to specific biological and ecological conditions. Hence, the main goal for the present book is to keep focus on the basic statistical aspects of animal space use, and study the extended framework's statistical-mechanical properties. Defining framework *extension* rather than framework alternative is important. Framework extension implies a quest to build bridges between the existing framework (classical statistical mechanics and Markov-compliant dynamics) and the novel aspects. Framework alternative means a theory that floats in the middle of a large ocean, with no continent emerging on the horizon. Thus, theoreticians on firm ground should neither be expected to observe the theory out there, nor bother (!), before some bridges are presented to them. Flashing some light now and then from the little island is not sufficient.

Bridges towards firm land should consist of mathematical and statistical descriptions—including simulation experiments—that show how $z \approx 0.5$ and $D \approx 1$ may change towards their "classical" counterparts $z \approx 0$ and $1.5 < D <= 2$ by changing or fine-tuning specific model parameters and terms. When the bridges are sufficiently developed and explored in simulations, a set of *test protocols* for real-life confrontations is needed. The extended theory, like all theory, needs to be able to produce hypotheses with accompanying assumptions, with the classical system conditions providing the respective null hypotheses. At that point, if hypotheses from the extended theory are repeatedly verified from tests on real data, the alternative hypothesis may over time mature from an idea and a concept to a proper theory in scientific terms. This new framework will then contain the old framework with some new compartments added. Depending on conditions (for example, if memory influence is verified or not), the most compliant compartment should be invoked to model a given real-world system. In short, classical conditions should invoke classical tools, while non-classical conditions should be invoked where this is more realistic (Gautestad *et al.* 2013, Gautestad and Mysterud 2013).

With respect to the parallel-processing concept, bridges towards the mainland began to emerge after the statistical-mechanical perspective was introduced more explicitly in the modelling work over the last ten years or so (Gautestad and Mysterud 2005, 2006, 2010a, b). At present there have in fact emerged some testable hypotheses with classical premises as the respective null models, and some tests have begun to emerge (Gautestad *et al.* 2013, Gautestad and Mysterud 2013). In the following I present some of the pieces in the jigsaw puzzle to which the parallel-processing postulate has given birth.

In this quest for broader system coherence, it is crucial to document statistical-mechanical coherence between two levels of abstraction: one should be able to study either a population of individuals occupying the same area or study space use by a single individual under a common framework. In a step-by-step manner a theory may then emerge. The existing theory in this context embeds nothing less than the mechanistic, Markovian framework already described, with all its facets and including its classical statistical-mechanics representation. At the population level the classical framework typically regards diffusion law compliance and allowance to use local density of individuals as proxy for intensity of space use (habitat selection). At the individual level the coherent theory regards Brownian motion (with variants) and a sequence of fixes as a proxy for intensity of space use. A sequence of developments towards an extended theory with inter-level coherence will be presented in the following sections. It includes both already published elements and novel aspects presented here for the first time.

5.5 The next step

Until now the individual level space use has been the main focus, and the important concept of parallel processing has been presented as a way forward to implement memory effects. In the upcoming chapter I turn towards implementing this principle in modelling of populations. Again, a few additional physical concepts are introduced, in order to bridge behavioural ecology to extended principles of population redistribution.

6. MODELLING PARALLEL PROCESSING

6.1 Inter-level model coherence and ergodicity

From the prosaic descriptions of parallel processing in the previous chapter, you should hopefully get an intuitive glimpse into the basic idea of this approach and an understanding of the arguments for a statistical rather than a mechanistic approach. However, to bring this concept and its movement representation MRW further in the direction of a framework with testable hypotheses, some mathematical stringency is needed. Following this, simulation protocols are needed to test whether the statistical properties of the generated patterns are in compliance with patterns in real data.

In (classical) statistical mechanics a population of equally behaving and independently moving individuals within a given environment may represent a sequence of GPS fixes from a single individual, given that an important *premise of ergodicity* is satisfied. In simple terms, ergodicity means that a mean location of a set of particles (for example, a snapshot-based average position of a number of individuals within a given spatial extent) equals the average position of a set of locations that are collected over a time extent from a single particle in this set of particles (for example, a sample of GPS fixes of an individual).

Ergodicity is a statistical-mechanical concept related to the mean field approximation that provides the cornerstone for standard population dynamical models (Ovaskainen and Cornell 2006). Thus, if population dynamics are simulated in discrete steps like $F(P_{t+1}) = F(P_t)$ where the right-hand side is the function that describes the evolution of number of individuals P within a given spatial extent from time t to t+1, a randomly chosen individual in this local population is assumed to have reached "full mixing" (having potentially been everywhere within the given area extent from t to t+1). If this has not been the case, an important premise of ergodicity compliance—and thus mean field compliance—for common population-dynamical equations is broken.

If ergodicity is not satisfied for the population as a whole, one may (under the premise of a classically behaving system) redefine a sufficiently small area size for the population so that a given individual can be realistically assumed to have satisfied a potential for "having been everywhere" at this more localized scale and at the given time resolution. For example, if you plot a moose's position, it will, with 100% probability, be located somewhere within a circle of size ten kilometres at the next location ten minutes later. However, the probability for the next location is not equal over the area of this size, so the condition for ergodicity is not satisfied. But if you consider the moose to be anywhere within a circle of (for example) ten meters after ten minutes—given that you (for some reason) know or may reasonably *assume* that it has not left this smaller circle yet—ergodicity may be satisfied at this local scale.

Alternatively, one may consider the original arena size but instead invoke a larger time interval for fix collection to allow for fulfilment of the full mixing premise. The moose may be anywhere within the larger circle of ten kilometres if a fix interval of ten hours is considered. In this manner, a potential for ergodicity may be fulfilled even for the larger area. You—the observant reader—may now immediately see a connection between the ergodicity condition and the condition for non-autocorrelated fixes. The latter is—under classical system conditions—basically defined from the same requirements!

6.2 Population kinetics: premises for the classical approach

I will first elaborate on the intrinsic assumptions for modelling space use at the population level, before switching the coin back to the individual level. I have already described the conditions that have to be satisfied for traditional, spatially *implicit* population models of type dP/dt (continuous time dynamics) or P_{t+1}/P_t (discrete time). In a nutshell, a premise of full mixing at the given space-time system demarcation is required to satisfy ergodicity and sufficient mean field approximation. In systems where the ergodicity premise is broken, population dynamics may also be modelled either by continuous time or discrete time, but space needs to be defined explicitly for finer resolutions of the system's spatial extent. A discrete model with explicit space involves a coupled map lattice model (Kaneko 1993), where a virtual grid for local population demarcations—the "lattice"—and a time increment are defined.

If the lattice is assumed to represent local populations in a metapopulation context (Hanski and Gilpin 1997), meaning that there is a relatively sharp transition zone with respect to the spatial scale of high and low migration rate, the model becomes more complicated. Here, just consider a lattice in the form of a virtual grid superimposed to represent system description at a finer spatial resolution within a larger extent. In standard spatially explicit population dynamical modelling, the population is in discrete space-time models assumed to be intra-cell ergodic at the scale of lattice cells and choice of time increments, but non-ergodic at coarser resolutions under the same increment. Thus, the grid resolution is typically (or should be) defined close to the spatial scale where a transition between ergodic and non-ergodic dynamics is found. A continuous rather than time-discrete space-time model under non-ergodicity means application of partial differential equations, where the population is assumed to be "locally well mixed" in small sections of space within small increments of time. A continuous model that is formulated by ordinary differential equations always has to assume that the system is ergodic at the scale of the system extent for the actual population.

In the extremely simplistic (parsimonious) model to follow, a discrete space-time scenario is simulated under classical premises of local (grid cell) ergodicity and Markov compliance by the constituent individuals in the population. Figures 27-32 show a typical simulation output under what I refer to "classical" conditions and using a homogeneous environment for illustrative purposes. The time period for the simulation is assumed to be short enough to allow for ignorance of birth and death rates. It will be trivial in follow-up model extensions to include also these rates. The movement speed is assumed to be constant on average at micro-scales. Again, this assumption may be relaxed in more extensive model variants. The speed is measured along a fine-resolved and smooth movement path by an individual, assumed to be seen at a substantially finer scale than the actual grid cell resolution. While the speed is constant on average, it may of course vary from resting to sprinting at these micro-resolutions. However, while averaging over many micro-time increments, this variation will not influence the overall pattern in statistical-mechanical terms (the diffusion rate, or average inter-cell migration rate) at the chosen grid resolution and time increment.

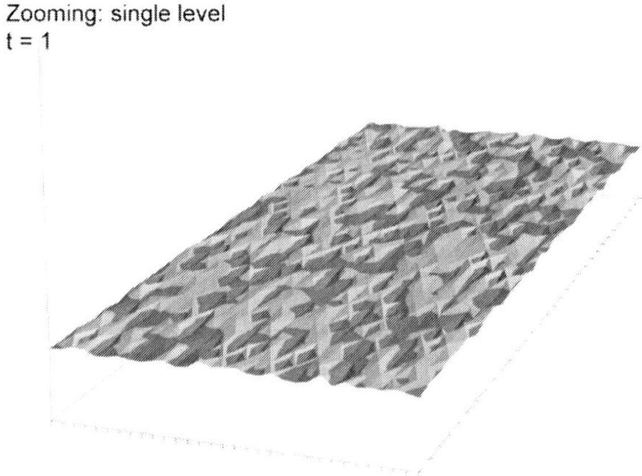

Figure 27. Output from the Zoomer model under condition of scale-specific conspecific aggregation in a homogeneous environment.

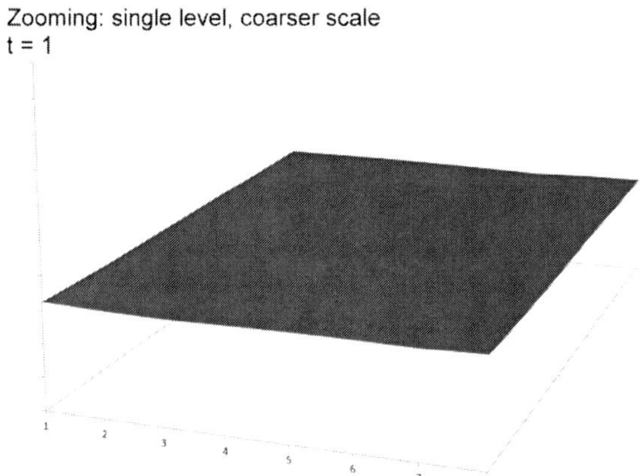

Figure 28. The population dispersion in Figure 27 coarse-grained to a larger virtual grid size. Fine-grained fluctuations are practically averaged out.

As unit time interval Δt for the discrete dynamics to be modelled in the following, consider Δt to be sufficiently large to allow an animal under unidirectional motion with this average speed to be hypothetically able to reach the defined system border ("arena") regardless of current position within this cell. If a path is jagged rather than expressing unidirectional motion (*i.e.*, the typical condition), the individual will traverse just a small part of the arena during Δt. In this respect I refer back to Figure 9. For simplicity I will generally refer to discrete time intervals as t, t+1, t+2, etc.; implying that the interval between successive iterations of the model has an implicit duration Δt. A close consideration of the duality of a population as a density variable over space and time on one hand and a collection of individuals on the other hand implies that the description might be better described as *population kinetics* than population dynamics. The latter is normally focusing on inter-generation time scales, while in the following we are aiming to study the abundance fluctuations from local movement; i.e., at much finer temporal resolutions. In a strict sense, kinetics and dynamics are interchangeable concepts, but I prefer to use kinetics for temporally fine-resolved dynamics. For the sake of avoiding a drift towards zero local density under classical movement terms, let us assume that the environment is closed, like a lake for a fish population (with no migration-providing creeks leading in and out) or an island for a terrestrial population. The necessity of this assumption was described in *0.5 Reflection of individual space use at coarse scales*.

Further, consider that the environment is homogeneous except for a "self-inflicting" (intrinsic) variation in local population density. Extending the model to describe a heterogeneous environment will be trivial, but is avoided at this point to highlight the novel aspects as clearly as possible. The intrinsic variation in abundance emerges from a rule that specifies a slight attraction towards conspecific individuals. In other words, after a while one should expect local congregation of individuals and thus a locally heterogeneous population density. Inclusion of conspecific attraction is important for population dynamical modelling under a broad range of conditions, as described in *0.5 Reflection of individual space use at coarse scales*.

For increased realism we also add a rule that allows an individual to occasionally move to a location in the neighbourhood due to extended search during Δt. As already described in Chapter 2 and Chapter 3, classical conditions of diffusion imply expected compliance with a Gaussian probability kernel for a given individual's position after a new time increment. If unit grid cell size is set equal to the variance of the expected displacement kernel (squared standard deviation) at the given Δt, most single-step displacements will be intra-cellular. Inter-cell movements will mainly go to closest neighbour cells, and jumps to second-closest cells can be ignored for simplicity.

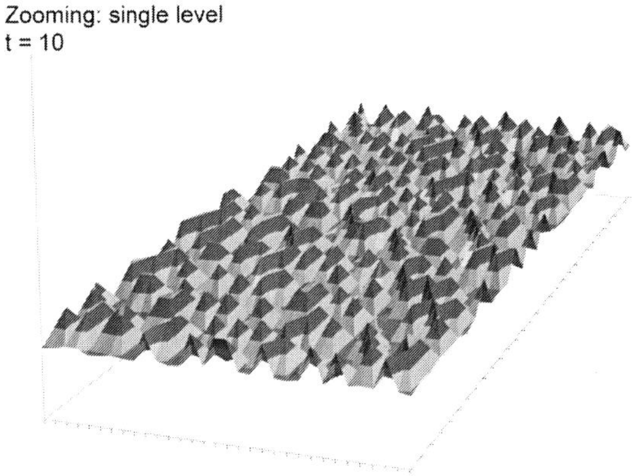

Figure 29. Zoomer model output after 10 iterations, under condition of scale-specific (classic) conspecific aggregation and redistribution.

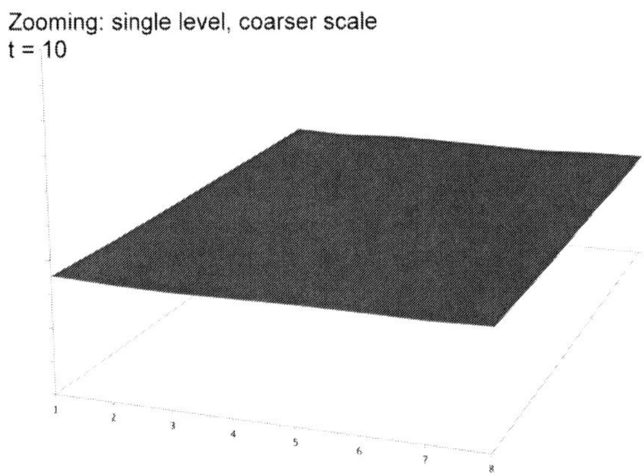

Figure 30. The output in Figure 29 coarse-grained to a larger spatial scale.

To improve model realism further, a third rule is included, whereby a large percentage or all individuals are assumed to be moving away from a given location with a high degree of directional persistence (correlated random walk) if local density surpasses a critical level during a given time increment. These individuals are then redistributing themselves randomly within the arena during this increment, as a simplified implementation of such events. The larger area for the redistribution follows from the shift to a correlated random walk or approximately linear moves during these events, but movement speed is still assumed constant at the standard magnitude defined for the system. The exhausted cell is assumed to be restored to standard resource level within one increment, whereby new immigration from surrounding cells may occur.

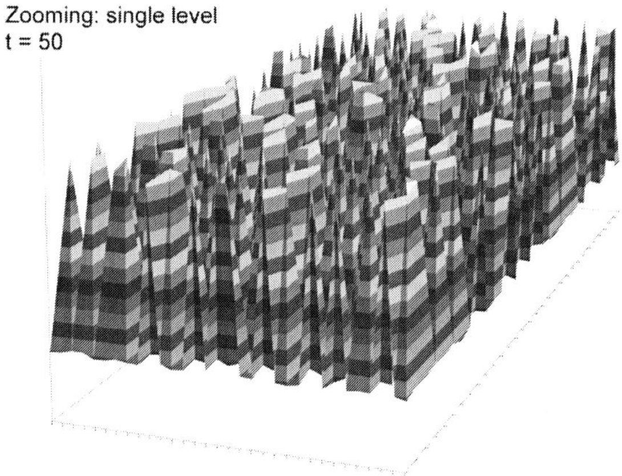

Figure 31. Model output under classical conditions at t=50, showing the effect from local crowding and successive redistribution (see main text).

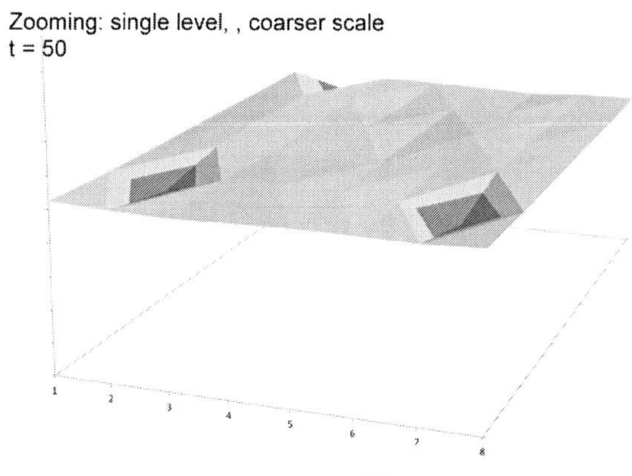

Figure 32. Coarse-grained presentation of the population dispersion in Figure 31. At three locations the population abundance at t=50 is temporarily and locally depressed due to a recent crowding and re-distribution event. These instances are better visible at this coarse-grained resolution. However, in general, the distribution is quite homogenous due to the condition of an environmentally homogeneous environment.

In this scenario let a sudden event occur, whereby all individuals are scrambled around in their environment. This "catastrophe" is introduced for the purpose of studying the dynamics towards equilibrium. Thus, the purpose is to study the population kinetics under the given rules and conditions transforming the dispersion of individuals M over time following the event, from the perspective of M_{kj} over a range of k and j, where k equals spatial scale and j is the spatial location of a virtual grid cell at scale k within the given arena. This situation—illustrated as time at t=1 where local population density is M_{kj}=1000 individuals on average over the set of j cells at unit grid resolution k=1—is shown in Figure 27. In Figure 28 the same dispersion is displayed from a coarser spatial scale. How do the individuals reorganize themselves—from the perspective of population dynamics—under this classic framework?

6.3 Classic population kinetics

According to classic population kinetics with Markov-compliance, one should expect the population to comply with a diffusion-like movement, for example, as described in the context of Figures 13-15, but strongly influenced by the conspecific attraction rule at fine scales (grid cell at level 1). Ergodicity is assumed for this resolution and time increment, which we define to be of magnitude of the individuals' average perceptual field. There will be a locally varying mean free path for resource encounters owing to the heterogeneous environment at finer (hidden) resolutions than unit scale. However, at unit scale and larger, this micro-scale heterogeneity is assumed to be averaged out except for the defined influence from conspecific attraction and local avoidance of crowding (temporary cell extinction). The individuals are frequently bouncing into conspecifics at these micro-scales, and are consequently—from the premise of classical mechanics—following a random direction upon departure from this position. Thus, from the perspective of grid cells at level 1 and at time resolutions of Δt and larger, a single individual's path is expected to comply with Brownian motion (except during crowding events, when and where Brownian motion temporarily switches to correlated random walk or unidirectional motion).

Individuals have no long-term relationship with conspecifics, owing to lack of long-term memory. Thus, subsequent encounters between a specific pair of individuals are random. As ecologists we do of course all oppose this premise for the population, but it is necessary to invoke it here for two reasons: (1) it is required from the premise of Markov-compliant dynamics and (2) indirectly it also pinpoints a potential lack of realism of classical population models under a broad range of biological-ecological conditions where spatial memory is involved. So, bear with me at this point. The Markov premise will be relaxed in the following scenarios.

Even if individuals in the present scenario prefer to stay in a better-than-average patch and thus are following the marginal value theorem (Charnov 1976), lack of spatial memory means that the spatial positions for successive encounters of conspecifics are rapidly forgotten. As repeatedly specified, Brownian motion is a memory-less process. However, the individual may carry with it an impression of an "average habitat quality" without breaking this Markovian premise, in compliance with the marginal value theorem.

Owing to lack of spatial memory, from one instant in time to the next, there is a larger than zero (albeit small) probability that a given individual with a resource patch within its field of perception will lose contact with this patch. This will also happen if it deliberately—owing to local food exhaustion—starts a unidirectional move or correlated random walk to search for a new encounter of resources or conspecifics. In the first case, this may lead its path out of the present grid cell and over to the next during Δt. In the case of local depletion, the animal will be redistributed somewhere inside the arena during this interval, with identical probability pr. grid cell. Consider standard movement to be characterized by a diffusion rate at unit scale (k=1) of magnitude 5%. A constant rate is a feasible simplification since we have defined the environment to be homogeneous on average at scale k>=1.

Under these rules and conditions, the dynamics are simulated for fifty time increments, starting with the randomization (scrambling) event. Figure 29 shows the situation after ten iterations. The local density of individuals has swiftly been more heterogeneous owing to conspecific attraction and some local crowding events, but less so at a coarser spatial resolution (Figure 30). At a later time t=50, the local aggregation has increased further (Figure 31), to the extent that it is visible also at coarser scale (Figure 32).

6.4 Population kinetics under the parallel processing premise

6.4.1 Structure of scale-free space use

Under the parallel-processing premise, population kinetics becomes qualitatively different from the foregoing classical scenario. Again we are starting with population kinetics under the same environmental conditions, dispersal rate from level $k=1$, including a tendency for conspecific attraction, and reaction to local crowding above a critical magnitude of local abundance. Other influences and responses are, in this simplistic case, assumed to be hidden at finer micro-scales relative to the chosen unit scales for space and time. For this scenario, now re-simulated under parallel-processing conditions, let us again assume $k=1,2,...,m$ (virtual) grid layers on the area for the population. The next grid level $k+1$ has a cell size, u_{k+1}, four times as large (the side length is twice as large) as the cell size at scale k. For six scales this involves u=1 (at level $k=1$), u=4, u=16, u=64, u=256, and u=1024 (level $k=6$ with 32x32 cells). These layers of grids represent successive, discrete scales (representing slices from a continuum) in our population, organized in a hierarchical manner; similar to what was described in Figures 22-23. It is easier to explain the parallel-processing mechanisms with discrete scale levels, but one can assume that the true dynamics are taking place over a continuous range of scales. Each successive scale level is given the same *relative* size, since its cells are 2^2 times larger than the level below, regardless of scale. In topological terms, we say that the virtual grid layer structure then becomes *scale-free.*

The key difference from the classical scenario in Figures 27-32 is that the population under parallel processing is dynamically responding to conditions over the entire range of these scales, not only to the conditions at unit scale (k=1).

We can again count number of individuals, M_{kj} ($j=1,2,...$), in different cells (j) at a given level and at spatially coarsened levels (larger k), as we did in the previous section for classical conditions. The individuals are now relating to their environment in a scale-free manner; owing to spatial memory, they are able to allow conspecific attraction to transcend beyond the unit scale that characterized the scale-specific process in the previous section. They are also able to perform a scale-free search, implying a shift from Poisson-compliant to *de facto* compliance with a power law-like distribution of step lengths.

Now we start the dynamics for the six levels in the present configuration, representing a scale range from 2^{2*0} to 2^{2*5} (1 to 1,024) cells pr. level. Since *k*=1 represents a resolution where u is defined to be of magnitude the field of perception, individuals must invoke their memory map to compare local conditions at any scale beyond $2^{2*0}=1$. In other words, tactical displacements below *k*=1 are effectively averaged out. We are thus observing the dynamics of respective M_{kj} at a statistical-mechanical level for all scale resolutions from *k*=1 to *k*=6.

6.4.2 Distribution of space use over a scale range

From the initially randomized population after repeating the scrambling event as specified for the scale-specific dynamics, consider choosing 5% of the individuals from each cell at scale 1 randomly, and call them "zoomers" during the upcoming Δt. The zoomers represent the part of the population which, during a given small interval in time, leaves their present cell at our unit cell size u at scale *k*=1, and then redistributes themselves according to the average conditions of conspecifics at a coarser scale than the field of perception. Owing to introduction of spatial memory and the parallel-processing postulate, the zoomers may target neighbour cells at different levels $1<k<(m-1)$, where m represents the spatial limit for parallel processing at time resolution Δt. Thus, multi-cell displacements from the perspective of *k*=1 may be either coarse-scale exploratory steps or return events to specific locations outside the animal's field of perception. This behaviour thus deviates from the classical movement assumption, where multi-cell displacements are always exploratory (moving away from local crowding, without specific strategic targets). For simplicity, consider zoomers to be individuals that, during Δt, execute targeted returns to specific locations within a surrounding area at scale *k*, which is "reached" as Zoomer to this level during the actual time increment.

The remaining individuals not performing zooming during Δt confine their movement to their present cell during this interval (inter-cell population kinetics are thus not described at scales $k<1$). However, every individual gets a new and equal chance to become a Zoomer in the next interval. Owing to the premise of a parallel-processing limit of extent equal arena size, zoomers at the highest level (*k*=6) are redistributing themselves at the same or lower levels.

In general, the actual percentage of zoomers from *k*=1 is positively correlated with the chosen time increment: larger increments imply larger percentage of zoomers at these coarser temporal scales. Thus, there is no specific interpretation of a Zoomer rate of 5% at time scale Δt, prior to linking this rate to a specific unit scale in space and time for a real population system. These scales should be estimated from analysis of real population or individual-level data, based on the model outline presented here. However, since this estimation of context-specific characteristic scales depends on the parallel-processing postulate, the property should first be verified for the actual data. This book provides several methods for this test.

In a random manner, tag each Zoomer with a random number between two and six with weights 1, 1/4, 1/16, 1/64, and 1/256, respectively, representing scales 2^{0*2}, 2^{1*2},..., respectively. This weight describes the non-uniform probability of being tagged with a number from two to six, where the number six represents m, the upper limit for *k* in this scenario. For example, the probability of being tagged with the value three is 1/4 the probability of being tagged with the value two. The actual number describes the highest scale level *k* which this particular Zoomer is expected to refer to with respect to target execution during Δt. With reference to Figure 23, a "level 3" Zoomer is an individual that in this interval reaches level three in spatial grid terms. As already specified, this spatial reorientation requires spatio-temporal memory utilization. In order to optimize habitat and site choice under a *postulate of uniform statistical-mechanical weight* over space and time (as a generic model for parallel processing, soon to be described in detail), I assume no preferred scale for optimization for 1<=*k*<=m. An individual is expected to zoom 1/4 as often to level *k*+1 as it happens to zoom to level *k*. A mathematically more stringent specification of the equal weight over a scale range was given by the presently added comments to Figure 23 above. A statistical-mechanical deduction in mathematical terms of this scale-free redistribution (based on kinetic energy dispersion) is described in a following section. In the present example, 1% in the population (1/5, or 20%, of the 5% zoomers) is reaching each of the five levels during one time increment. Averaged over all individuals the zoomers to level *k* become uniformly spread over the *k*-1 levels from this parallel-processing postulate. In other words, each level is equally important with respect to respective zoomer's decision about what to do next. Scale-free memory utilization facilitates this parallel processing property.

This scale-free inter-level distribution of zoomers has an interesting population-level property. Each individual has a small probability of reaching a high scale level because of a lower weight for a larger *k* in the tagging procedure for zoomers. However, at the population level this is compensated for in the long term—i.e., in statistical-mechanical terms—by the larger number of individuals that contribute to zooming to a given level at larger *k*. For example, the probability that a Zoomer reaches scale level four is 1/16 the probability of reaching level two. At level four there are on average sixteen times as many individuals within a given cell at this scale owing to its larger size. These individuals contribute to the aggregated probability for number of zoomers reaching level four. Therefore, (a) the number of zoomers at each level is constant when counted over the total arena, and (b) the ratio of total number of zoomers pr. cell at level *k*-1 relative to level *k* is constant on average over all cells. These constancies follow from a premise of equal environmental utilization at different scales. By necessity for a scale-free processing up to level m, individuals are—as was specified in the introduction to the model description—able to reach (in a target-oriented manner) all area locations from any position during a given time increment in the simulation. So, (*k*)-level zoomers are under the given premise of a constant movement speed implicitly assumed to show a more unidirectional path than (*k*-*i*)-level zoomers at levels lower than *k*. The strongest degree of directionality is thus in average terms performed by the highest-level zoomers, at level m.

An upcoming description of the MRW model explains in more detail how the zooming process should be interpreted in statistical-mechanical terms. A single zooming event performed by a given individual to a specific location within a virtual grid cell at scale *k* is not sufficient to provide meaningful information. Only the *ensemble* behaviour of the total set of zoomers during Δt is of interest. According to the parallel-processing postulate, when this is translated to the population level, one should—when a single cell at respective scales is considered—expect $1/2^2=1/4$ as many zoomers to level $k+1$ relative to level k for $1<k<m$. This ratio is the same as the expectation from the *time* average for a single individual performing MRW with step length distribution showing power exponent $\beta=2$ (the actual equation will be given a few pages down). Since ergodicity is assumed to be satisfied within the given area extent of magnitude m, the ensemble average (population of zoomers to level *k*) during a single interval may be expected to show the same inter-scale dispersion as the time average from many displacements of a single individual, collected over many intervals. The rationale for $\beta=2$, explained in an upcoming section as equal habitat utilization by a single individual at respective spatial resolutions, is a statistical-mechanical expression for such equal weight. Simultaneously, it provides the complementary rationale for Zoomer dynamics at the population level. Inter-level coherence is thus

maintained in statistical-mechanical terms between individual-level MRW and population-level Zoomer dynamics.

For the ensemble of (k)-level zoomers to a given cell at scale level k+1 embedding the given set of four departure cells at level k, we can in a statistical manner assume that the individuals distribute themselves randomly (remember; statistical mechanics!) within the cell at scale k where they—from respective memory map considerations at this resolution—find the most attractive conditions. According to the present rules under condition of a homogeneous external environment, the sub-cell with the highest experienced number of conspecific encounters (according to memory map processing) is the preferred one. This proximate mechanism of zooming in to the highest local density of attractive elements increases the aggregation of individuals among cells at scale k, relative to the randomized distribution we started out with.

In the model I assume that the zoomers to level m—reaching the highest level of their memory map utilization during Δt—prefer the same grid cell among the four neighbour cells at scale m-1 owing to their familiarity with the environmental conditions here in coarse terms, which has shown that this cell tended to have higher population abundance relative to the other cells at this level. In short, in an open environment m defines the average extent of the memory map of respective individuals and in the present scenario also equals the extent of the available arena.

In a similar manner as for (m)-level zoomers, which distribute themselves at scale m-1, let the (m-i)-level zoomers (i=2, 3, … , k+1) behave in the same locally aggregative manner with respect to their levels of zooming. At finer spatial resolutions (smaller k), more parallel executions of zooming and local redistribution are performed during a given Δt, owing to a larger number of grid cells at finer resolutions. For the next time increment, repeat the same Zoomer rule all over again. For each iterative turn in this simulation, the distribution should be expected to become more and more aggregated over all scales between k=1 and k=m-1.

In contrast to classical conditions above (see Figures 27-32), an aggregative tendency of individuals at levels k>1 will be intrinsically driven by a process that is *multi-scaled*. It is also scale-free owing to the power law scaling of the virtual, multi-scaled grid structure. A slower process rate for individual inter-cell movement (inter-cell movement rate) at successively larger spatial resolutions is compensated for at the population level by larger population abundance (larger M) at coarser scales. Thus, dispersion rate is of similar magnitude: 1% pr. level in this example.

As for classical conditions I define a critical number of individuals, M_{max}=1000, a cell at scale *k*=1 may hold (the local carrying capacity). It is logical to define the functional and numeric response to upper density threshold to resolutions within this unit scale, since this is where individuals first are responding to crowding (for example, competition for limited resources or space). Owing to local crowding effects, either 80% or all individuals in a cell at Level *k*=1 with $M_{1j} > M_{max}$ become zoomers in the next time interval. In other words, the sub-population in this cell goes temporally extinct or close to extinct. I could have chosen a less dramatic rule, for example, letting 50% rather than 80-100% of the individuals become zoomers upon passing the critical level of crowding, but the resulting overall pattern of interest here would not be qualitatively affected. To verify this I run a scenario without any critical level for local abundance. If local "extinctions" represent v% of all cells at scale 1 during one time interval, the total number of zoomers in this interval becomes 5%+(v%)*R, where R equals the rate of crowding dispersal (R=0 or 0.8 in the present simulations). For simplicity, full restoration of local resources is assumed—as under classical conditions—to take place during the next time increment.

6.4.3 Emergence of a scale-free population dispersion

Now, let the show begin! We again start with a scrambled condition that equals the pattern shown in Figures 27-28. The spatial density of individuals after a while is, however, very different from the classical scenario. At unit scale the population has become strongly aggregated as time is progressing, and this multi-scaled aggregation is clearly visible also upon coarse-graining (Figure 33). The reason for the successively increasing aggregation that transcends a large range of scales is a *self-reinforcing* intrinsic process (conspecific attraction) where an initially subtle difference in local conditions at fine scales leads to zoomers tending to "land" within these localities at the expense of random displacement. While classical conditions led to local aggregation only, parallel processing leads to a broader definition of "local"; i.e., transcending a scale range.

Figure 33 includes results from two conditions: the upper row shows results from the "no crowding effect" rule, while the lower row shows the effect from defining a critical maximum abundance at scale level *k*=1 (see details above). If local aggregation passes this level, 80% of individuals in this grid cell are randomly dispersed at next iteration. The most obvious difference is a stronger multi-modality of population dispersion, with no grid cells with M>1000 (the critical level for dispersal) at level k=1. The crowding rule creates frequent local perturbations of local abundance, but may be considered more realistic in ecological terms relative to absence of such a rule.

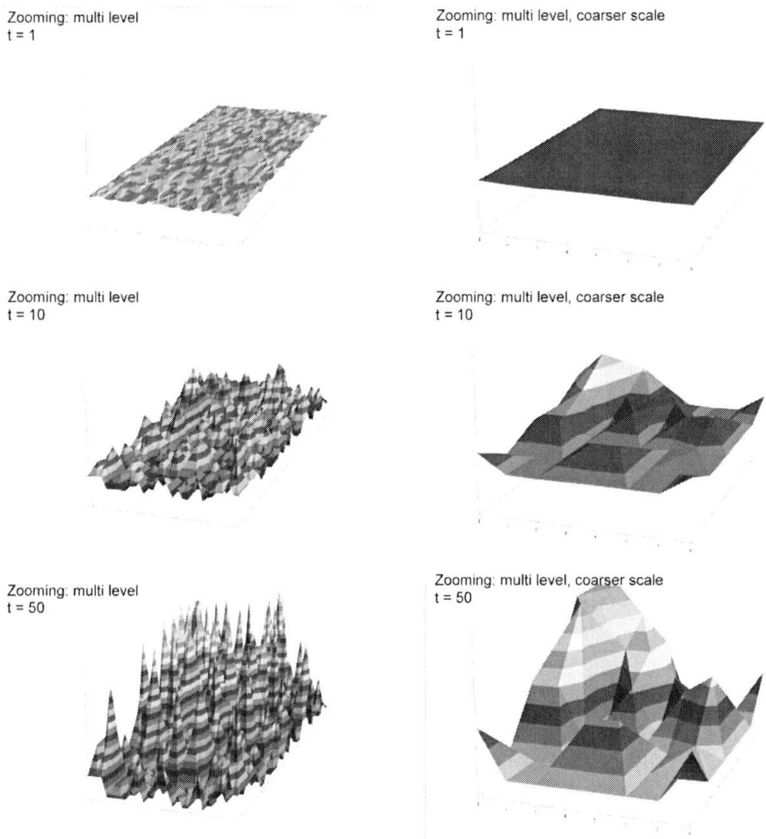

Figure 33. Additional output examples from the Zoomer model, this time under parallel processing conditions. See main text for details.

6.4.4 Mathematical formalism of parallel processing

The main points of the extensive verbal model description in the previous section can be condensed into a single equation. The only dynamic aspect of the environment is local population abundance, where "local" is level-specific. With a tendency for conspecific attraction as the main driver of population kinetics in this simplified scenario, the actual model for the parallel-processing dynamics of M_{kj} individuals within respective virtual grid cells of size u at spatial level k can be summarized as

$$M_{kj}(t+1) = M_{kj}(t) - rM_{kj}(t) + \sum_{k=k}^{m-1} \mathbf{E}Z_{k+1}(t)/2^{2(k-1)} \mid 1 <= k <= m-1,$$

$$\begin{aligned} \mathbf{E} &= 1 \mid u_{kj}(t) \in \{u_{kj}(t) : Z_{k+1}(t) > 0\} \\ \mathbf{E} &= 0 \mid u_{kj}(t) \in \{u_{kj}(t) : Z_{k+1}(t) = 0\} \end{aligned} \qquad (8)$$

The sigma term in Eq. 8 is the immigration ("zooming-in") rate to a specific cell at scale k, summed over contributions from all scales from k+1 to m. The parameter r is the accumulated emigration (zooming-out) rate over the range of memory-influenced dynamics. Defining k=1 as the unit scale, 5% in the present scenario with six levels gives 1% zooming rate pr. level from k=2 to k=m. To contrast this generic form of parallel processing with classical statistical mechanics (scale-specific and Markov-compliant redistribution of individuals), consider the latter as Eq. 8 under the special condition k=1 and m=k+1.

Observe that unit level k=1 is the finest scale represented by Eq. 8. This level is set relative to the coarsest scale for memory influence, m, in addition to by necessity being equal to or larger than the animals' field of perception. At finer resolutions the principles of statistical mechanics of a spatially explicit population system are easily violated, since deterministic responses to specific local events (environmental input) are averaged away to a lesser degree. Thus, unit scale k=1 for a real system may be defined as the finest spatio-temporal resolution that ensures a statistical-mechanical representation of migration (zoomers). If one chooses to study the dynamics from a larger unit level relative to the same top level m, the range m-k' is now reduced relative to the previous level k=1. This is expressed mathematically as

$$k' = k + s \qquad \mid s = 1, 2, 3, \ldots m-1 \qquad (9)$$

In Eq. 9, $k' > k$ implies that the system is studied at a coarser level than this transition level. Zoomers at finer scales than level $k' > k$ do not leave cell j

at modified level $k'\equiv1$ relative to original level k, but contribute to reduce the net emigration (out-zoomers) from this cell at level k' relative to level k. Therefore, for $k'>k=1$, the emigration rate r, representing out-zoomers, will be reduced to

$$r' M_{kj}(t)= rM_{kj}(t)*[(m-k'-1)/(m-k-1)] \qquad | \; k' > k;\; k\equiv1 \qquad (10)$$

For example, if $k' = k+1$ and $m = 6$, $r'M_{kj}(t)$ is reduced by a factor of $(m-k'-1)/(m-k-1) = 4/5$, which equals 20% reduction of migration rate at this coarser scale. Additional zoomers at finer resolutions than k' redistribute themselves within respective cells at these "hidden" scales and thus do not contribute to zooming out.

As for classical conditions the actual magnitude of r is obviously a function of the magnitude of the time increment implicit in an iteration of Eq. 8. A larger interval Δt implies larger r. While Eq. 8 regards the statistical-mechanical properties of parallel processing in its simplest possible formulation, more terms and factors would have to be added to reflect improved ecological compliance with specific systems. One such additional element is explored here: local cell extinction owing to temporary resource depletion. In the example above, this regarded level $k=1$ the finest grid scale $2^0=1$ for cells where $M_{kj} > M_{max}$ and where local emigration thus was set to 100% ($r\equiv1$). However, for simplicity Eq. 8 is without fine-scaled extinction specified.

$Z_{k+1}(t)$ represents the "incoming" $(k+1)$-zoomers to cell Mkj during the interval from time t to t+1. The canonical parameter **E** equals one for all j at scale k where $M_{kj}(t)$ receives immigrants from the higher level $k+1$ during one time interval, and all else is zero. According to the present rule for zooming in, a given cell j at level k will receive zoomers from one or more higher levels if this coarser cell that is embedding cell u_{kj} at that interval in time has higher than average M relative to neighbour cells at respective levels. Neighbourhood is thus represented by a parallel superposition of grid cells at different magnitudes of k above its own level $k=1$. A neighbourhood at larger k (implying a larger cell, embedding cell u_{kj}) satisfying higher M than average relative to this coarser cell's neighbourhood leads to additional zoomers percolating down to cell kj at level $k=1$. However, these zoomers are distributed homogeneously among $2^{2(k-1)}$ cells at level $k=1$ in the neighbourhood of this cell. The expected number of individuals actually settling in cell u_{kj} at level $k=1$ equals $Z_{k+1}(t)/2^{2(k-1)}$ from levels where **E**=1 during this time increment.

6.4.5 Parallel-processing properties

Six aspects stand out from the parallel-processing population model:

- It represents non-Markovian dynamics, owing to the summation term with m>1 in Eq. 8. A given animal relates to its environment at different scales *simultaneously*. In short, an individual's spatial "neighbourhood" transcends its local field of perception.
- The process is non-mechanistic in classical terms, since local causation at a given instant in time is "fuzzy," owing to spatially multi-scaled dynamics. In short, short-term behaviour (the Markov domain) may simultaneously be constrained by medium- and long-term influence, with respective differences in process rates (coarser time influence implies smaller rate).
- The *net rate of change* of local population density pr. cell and pr. unit time is decreasing as we study the dynamics from coarser spatial resolutions. This is a consequence of the combination of (a) longer term parallel influence on movement-influencing factors (for example, cognitive processing and execution of goals) and (b) longer term influence may allow for wider range-directed displacements. For example, returning to a distant target typically requires a longer term goal the longer the distance. Considering a spatial snapshot of many individuals' positions (the population dispersion at a given point in time) will then—according to the parallel-processing postulate—reflect a mixture where many individuals are executing short-term responses, some are in the process of executing some longer term goals, and even fewer happen at that point to be steered primarily by even longer term goals. While cell-specific migration rate falls off with increasing level k, owing to larger M at coarser levels, the migration (zoomer) rate is constant from the perspective of the entire arena with spatial extent m.
- The spatial scale levels are postulated to have a log-linear (scale-free) representation, reflecting an animal's postulated cognitive perception of its environment, with respect to memory-dependent processing of information.
- The population dynamics are coherent with individual-level MRW dynamics; thus, Eq. 8 may—under a premise of ergodicity—represent "ensemble average" as opposed to "time average" if M independently moving individuals are replaced by M relocation fixes of a single individual.

- Eq. 8 represents a statistical-mechanical, non-equilibrium description of complex population dynamics under influence of spatio-temporally scale-free memory utilization.

All these bulleted points require deeper explanations. These will come in later chapters. At this stage I just wanted to provide a glimpse into parallel processing, by contrasting it with population dynamics under classical conditions. At this stage the model conditions have been extremely generic: maximally simplified and condensed far into the domain of a caricature of a real system. However, this is the strength of models: start with the simplest situation, and build in more realism in a stepwise manner as soon as the basic properties have been described and tested from various angles.

One of the critical steps regards testing coherence between parallel processing as it is simulated from the perspective of statistical mechanics, both at the population and the individual level, is to document compliance with ergodic principle. This compliance has to be met under specific scale demarcations over space and time for a given system. As a preview of this aspect—while waiting for individual-level simulation of parallel processing in later chapters—Figure 34 gives a conceptual impression of the population kinetics for eighty-five individuals both under classical (scale-specific, Markovian, and memory-less) conditions and parallel-processing conditions. It is assumed to show a snapshot of expected individual displacements for a small and randomly collected number of individuals zooming out during a given time interval from a given cell shortly after the "scrambling event" (close to a homogeneous dispersion of individuals). Thus, local aggregations have still not emerged. While Markov-compliant movement leads to all eighty-five individuals tending to displace themselves to their respective neighbour cells, eighty-five zoomers under parallel-processing conditions tend to distribute themselves over a larger scale range in a scale-free manner.

In the classical situation—reflected in Figure 29-30 in population terms and upper pane in Figure 34 in individual terms—respective individual displacements at Level 1 are relatively short and basically limited to steps to neighbour cells in accordance to the mean free path considerations already outlined. In the parallel-processing situation, ignoring sampling error and with reference to the starting conditions leading to Figure 33 after fifty increments, the lower pane of Figure 34 shows compliance with Eq. 8-10 with an expectancy of four zoomers at level m-1 for each Zoomer at level m, sixteen zoomers at level m-2 relative to level m, and sixty-four

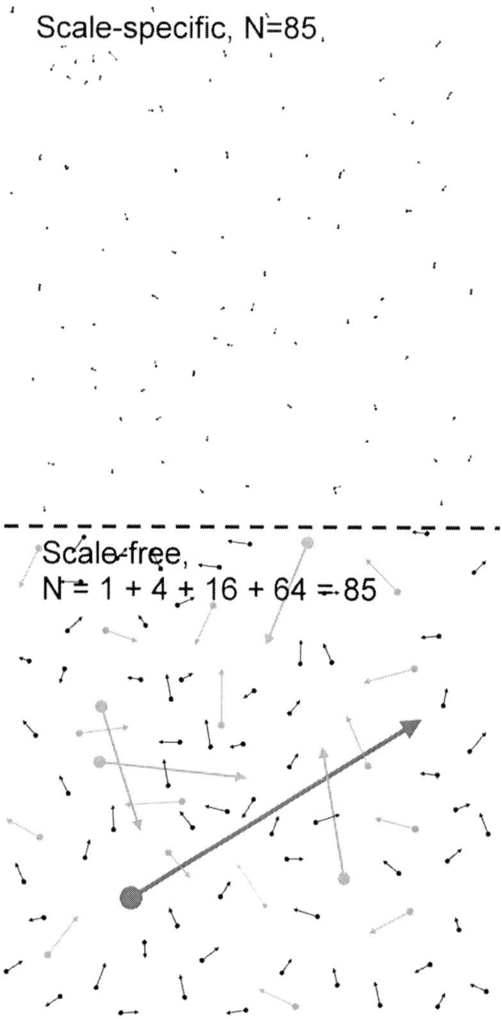

Figure 34. Illustration of population-level representation of parallel processing, where respective 85 individuals at a given small interval in time are represented by a displacement vector from start to end of the interval. In the scale-specific case (upper part), displacement vectors are of similar size, complying with a negative exponential distribution of displacement lengths. In the 5-level scale-free space-use variant, some individuals show larger displacements (lower part) in compliance with a power law distribution.

zoomers at level m-3; all relative to one Zoomer to level m. The 128 zoomers to level m-4 and 256 zoomers to level m-5 are not shown, to avoid clutter. Thus, the inter-level dispersion of zoomers follows the equation $Z_{k+1} = Z_k/\beta$ with β=2 for log-scaling of levels based on logarithmic base 2.

Level (m-5)-zoomers are, in present model terms, the level represented in the upper pane. Under condition of scale-free space use, only 85/256 (ca 33%) of all dispersing individuals to neighbour cells during a single increment are shown in the upper pane. Under condition of scale-specific space use, all zoomers move to neighbour cells. The comparable total number of zoomers from and to level 5-1 (level 1) is not 85/256, but (85 + 128 + 256) = 469. Thus, for a scale-specific space use, the upper pane shows only 85/469 = 18% of the total.

As already described, to understand how zoomers to a higher level may be as abundant as zoomers to lower levels (under condition of equal weight of space use at different levels), one should acknowledge that the number of potential zoomers at level k+1 at any given moment is 2^{2k-1}=4 times as abundant as zoomers at level k, owing to four times larger cell size and four times as large M_{k+1} relative to M_k. In other words, the zooming rate at respective levels regards the given percentage of zoomers relative to all individuals M at respective levels, where M is increasing with increasing level. Thus, smaller probability pr. individual to zoom to a higher level is compensated for by a larger number of individuals pr cell at larger level.

If the ratio of zoomers in the parallel-processing condition was different from the specified uniform distribution over the m levels—zoomers at level k+1 departing from expectation of 2^{-2} as many zoomers relative to those zooming to level k on average—there would not be equal weight between processing of spatial memory over a range of scales. At level k+1 there are one-fourth as many cells to relate to and four times as many individuals pr. cell on average. The total set of level-specific population density of respective Zoomer classes $\{M_k\}$, which for the given system extent is expressed as M_k/u_k, is under the parallel-processing condition homogeneous over the actual range of scales if individuals belonging to respective levels during a small time interval are distributed in accordance to cell size, which equals 2^{2k-1}. Further, the system's process rate is special: the classical and parallel-processing architectures differ with respect to how processing is distributed over the scales. In other words, they express qualitatively different dynamics. Under classical conditions processing takes place at fine scales only (between level 1 and level 2 in the present simulations, with cell-to-cell migration rate 5% at level 1). At coarser levels, close to nothing happens during a given time increment (m-1=1 for this minimum scale range where m=2). In other words, when m=2 r in Eq. 8 can be considered zero—and thus ignored—at coarser scales than k=1. In this case r regards the dispersal rate between neighbour cells at level k=1 only (with small and ignorable error). Under parallel processing where typically m>>1, there is—under generic conditions—equal magnitude of processing at all levels from 1 to m. This is specified by Eq. 10, which states that higher levels are characterized by relatively smaller r. Fever individuals are dispersing themselves at coarser scales during a given increment (thus, smaller process rate for the population as a whole at this level), but they are under generic conditions compensating for this by displacing themselves over a larger distance on average.

6.4.6 Dispersion of kinetic energy over a scale range

In statistical-mechanical terms, parallel processing under the generic condition is equalizing kinetic energy pr. unit area over the range of scales 1, 2, …m. Kinetic energy E is expressed as $E= \frac{1}{2} M'v^2$, where M' represents an object's mass and v represents velocity ($d/\Delta t$, where d is distance moved). If…

(a) we set M'=M for simplicity (allowing mass to scale proportionally with number of individuals in a cell),
(b) we observe that individual zooming from level k+q to level k+q+1 is one-fourth as abundant relative to zooming from level k+q-1 to level k+q, but this is compensated for by four times as large M in cells at next higher level (same zooming rate in terms of percentage of M),
(c) cell diameter is proportional to net movement distance at a given level,
(d) movement speed is constant on average,
(e) larger net displacement when zooming to larger levels implies a stronger degree of directional persistence during these events,
(f) cell diameter (or side length) is proportional to the square root of their area, and
(g) $u_{k+1} = 4u_k$

… we have for the parallel processing condition:

$$E_{k+1}/u_{k+1}= E_k/u_k \quad | \; u_1 < u_k < u_m, \beta=2 \qquad (11a)$$
$$E_{k+1}/M_{k+1}= E_k/M_k \; | \; M_1 < M_k < M_m, \beta=2 \qquad (11b)$$

Four times as large kinetic energy pr. cell at next higher level is balanced by normalizing to unit cell size or considering a single individual.

For parallel processing the net displacement during an increment is on average twice as large for (k+1) Zoomer as for (k)-zoomers, owing to point (c) in the list above. Under classical conditions, expected net displacement at level k+1 is only proportional to the square root of displacement at level k (owing to the classical diffusion law). Thus, dissipation of kinetic energy is quickly deteriorating as we coarse-grain the classical system towards larger cell sizes, unless much extra energy is constantly added in a specific manner to compensate. Such external "forcing" would require the externally added energy to be distributed in a scale-free manner to satisfy a distribution as under the parallel processing. In other words, *parallel processing implies, owing to distributing kinetic energy (directed movement) in a scale-free manner, a potential for animals to utilize their environment over a range of scales more effectively, in comparison to a classical Markov-compliant kind of space use.*

The "temperature" (a so-called intensive variable of a system) is constant and homogeneously dispersed over the range of scales, owing to a constant movement speed of the individuals. However, the kinetic energy (an extensive variable) regards net displacement at various scales. Larger displacement (at a larger level) means same kinetic energy expenditure pr. individual, owing to constant movement speed on average but a more unidirectional kind of movement at larger levels. Fewer individuals zooming from level 1 to successively higher levels are compensating for by the nested hierarchy of virtual grid cells: the number of nested level 1-cells and thus M_k are larger at levels k>1, and E_k/u_k (or E_k/M_k) is constant between respective k.

In classical statistical mechanics k=1 may represent the so-called "heat bath" where energy is dissipated from potential energy, via kinetic energy (movement) to heat. In parallel processing the heat bath is expected to be spread uniformly over all scales from k=1 to k=m, given that Eq. 11—the conjecture which states equal weight of space use over a scale range under generic conditions. However, at coarser resolutions than m (in an open environment), the system will behave classically; actually pr. definition, since m is the spatial extent of parallel-processing compliance at the given time increment. If the actual time increment (unit time) is increased, m may also increase and thus contribute to stretching the parallel-processing range toward coarser scales. The ultimate limit, m_{max}, depends on many factors, for example the time it typically takes for an individual in the actual environment to traverse its home range.

The relationship with respect to equalized kinetic energy expenditure over a scale range under condition of parallel processing can be read also from Figure 23, which showed the expected scale-free distribution of step lengths for a given individual in a population. If the exponent β for a power law-compliant step length distribution satisfies β=2, the area in each of the rectangles that are marked in Figure 23—where the x-axis shows L^2 rather than L—are of similar size. Each rectangle symbolizes the integral of displacement sizes (summed space use) within respective log-scaling bins for grid cell area. The latter is proportional to L^2. Thus, since Figure 23 regards statistical mechanics of a specific individual, there is coherence between ensemble average (the population) and the time average (fix sample of a given individual). This parallel-processing coherence between the population and the individual level will be explored in more detail in later sections.

6.4.7 Redefining statistical mechanics

In previous sections, the concept of parallel processing, accompanied by the MRW model, was defined as representation of a non-mechanistic system while still being dynamic. Thus, two universal classes of system dynamics were defined: the classical mechanistic frame of reference (Markov compliant dynamics) and the non-mechanistic counterpart (parallel processing). In a statement in the preface, I argued that a proper implementation of long-term memory effects required this system extension for the sake of biophysical consistency. I also argued that a statistical-mechanical representation of animal space use was necessary for proper modelling (dynamically or statistically) of this kind of complexity. Modelling such systems by a mechanistic design will have to struggle with "surprise factors": events that are inexplicable at micro-scale but explicable from a coarser time scale. In other words, execution of respective strategic goals at coarse time resolutions will by necessity appear in a given time series of behaviour as unexpected events from the perspective of a temporally fine-resolved system dynamics. These statements do in fact require yet another specification, involving the concept "statistical mechanics" itself. Before clarifying this, let us briefly recapitulate the nature of parallel processing and the concept of surprise events.

Consider an individual that happened to pass location A at three separate points in time. Let us assume that local environmental conditions and the individual's internal state are identical at all three events. Following the first event, the animal moved towards west during the following time increment. Let's assume that this behaviour nicely confirmed the prediction provided by a detailed mechanistic model for this individual and its environmental conditions. During the second visit the animal again went towards west in the following time step, increasing our belief in the model even further. However, consider that the third visit surprisingly was followed by a step towards east. Perhaps this was towards a non-preferred habitat patch, relative to the preferred habitat in the western direction. From a mechanistic perspective, the step towards the east is interpreted as statistical noise. In simple terms, it represents an inexplicable outlier among otherwise model-confirming events. However, consider that in all three instances the animal also had an identical strategic goal running in parallel with the temporally finer-grained tactical responses. One of these could be a return to a resting patch somewhere towards east, within a relatively wide time frame (for example, within an hour or two). When actually executed, this goal required in this example passing matrix habitat between the present location and the target location.

From a temporally fine-grained perspective of minute-by-minute dynamics, the coarser-grained goal had to be considered non-deterministic, invoking statistical noise onto otherwise model-compliant behaviour. The goal's execution will appear non-explicable from present conditions. Thus, it is non-mechanistic in a manner that was explained in previous sections. Imagine execution of strategic goals as invoking surprise events on finer temporal resolutions than the time scale for the actual goal. In the example above, there might have been a latent chance for the goal to be actually executed at each of the three visits to location A, but only one of these events experienced the effect from this goal to move towards the resting patch during the following time increment (for example, one minute).

Hopefully this additional hypothetical example of parallel processing contributes to underscore my claim that two kinds of dynamics are needed: mechanistic and non-mechanistic. In previous sections I explained and exemplified in a similar manner why a statistical-mechanical system representation—coarse-graining for short—was necessary to capture the essential dynamical properties of parallel processing.

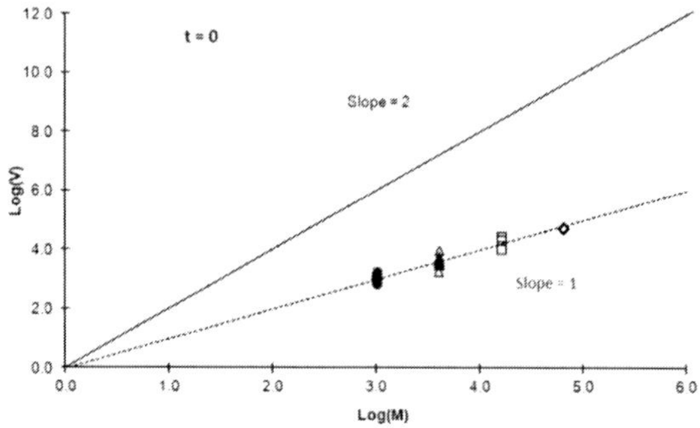

Figure 35. Spatial variance (V) of local population abundance (M) under Zoomer model condition of scale-specific space use. Initial random dispersion (t=0) in a homogeneous environment shows V proportional with M (observe log-transformed axes).

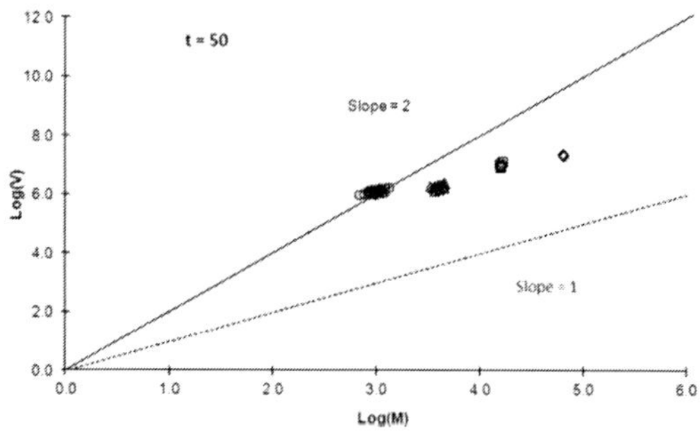

Figure 36. At t=5 the dispersion is more heterogeneous, but still showing V proportional with M but with larger intercept with the y-axis.

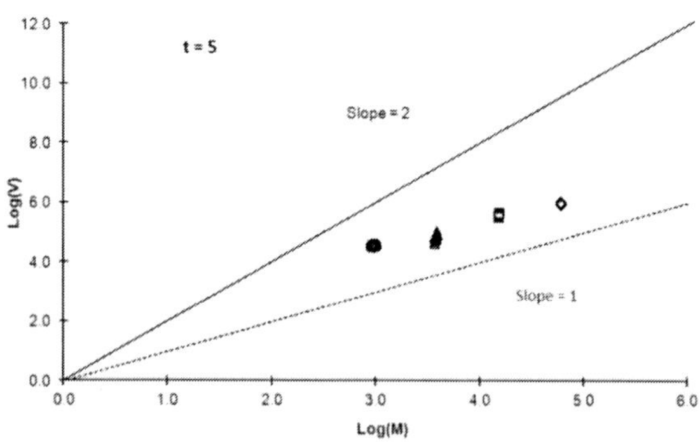

Figure 37. Additional time progression, to t= 50.

However, the term "statistical-mechanical" contains the non-compatible element "mechanical"! Thus, formally the meso- and macro-scale representation of a non-mechanistic system should be termed "statistical-non-mechanical." However, this is both confusing and linguistically convoluted, so for simplicity I will have to—at least for the time being—keep referring to non-mechanistic dynamics at macro scale as statistical-mechanical, implicitly stating that this framework embeds two qualitatively different kinds of statistical mechanics. It is tempting to invoke the extended term "statistical dynamics," but I leave that for later. Therefore, in this book wherever I refer to statistical mechanics, it should be clear from the context whether I refer to the Markovian/mechanistic kind or the parallel-processing/non-mechanistic kind of system representation.

6.5 Statistical properties of a parallel-processing system

6.5.1 Scale-specific versus scale-free spatial variance

One additional property that distinguishes classical (Markovian) population dynamics from the parallel-processing counterpart will now be described in detail from a pattern-observing (statistical) perspective, rather than from the process perspective. The strong aggregation pattern in Figure 33 was referred to as more "clumped" both at fine and coarser scales. How can this be concretized in statistical terms? *The variance relative to the mean* is a standard approach to describe such heterogeneity, and may be categorized as a statistical-mechanical system aspect if one considers the process behind this statistical relationship.

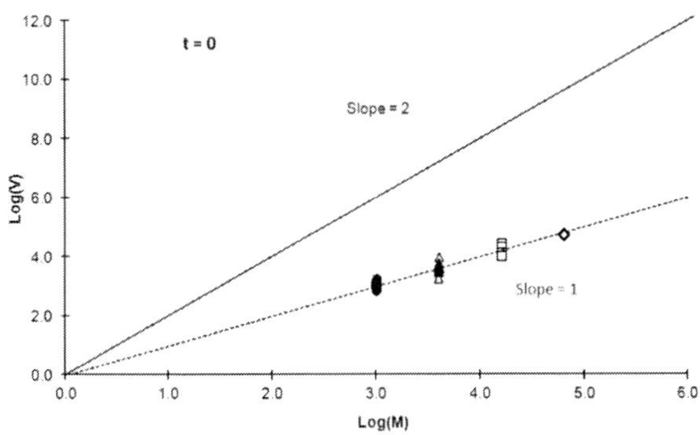

Figure 38. Spatial variance (V) of local population abundance (M) under Zoomer model condition of scale-free space use. Initial random dispersion (t=0) in a homogeneous environment shows V proportional with M (observe log-transformed axes).

Variance, V, in a sample of population measurements, is measured from variation of M*kj* at one particular scale (resolution u), whether we include all or only a subset of the cells at this scale in our series. A larger sample of M would give a better estimate of V of the true M. How the variance from parallel-processing conditions varies over spatial levels *k* compared with classical conditions is shown in Figures 33-40.

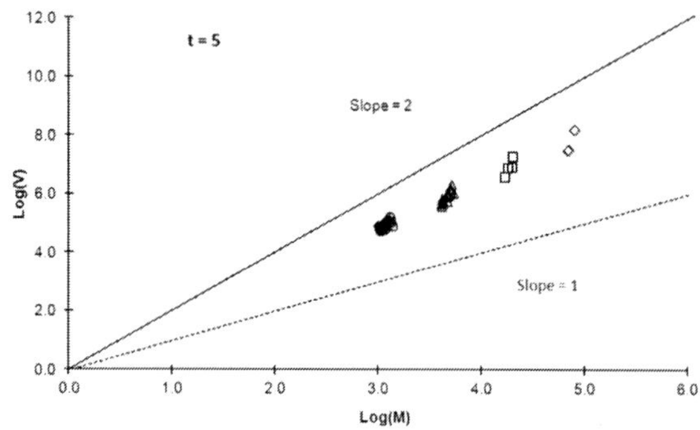

Figure 39. Scale-free space use, t= 5.

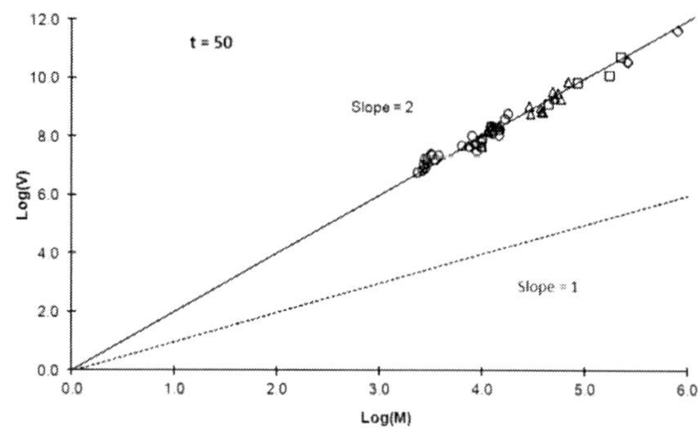

Figure 40a. Scale-free space use, t = 50. The population dispersion now shows V proportional with M^2.

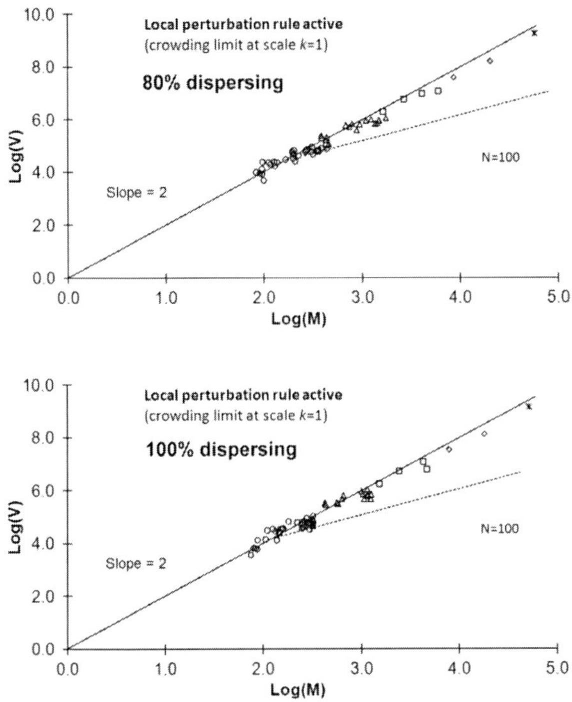

Figure 40b. Scale-free population dispersion under more complex movement rules. See main text.

Figure 35 shows the standard expectation from Markovian (scale-specific) dynamics of individual space use the moment after spatial scrambling. Log-transformed variance is changing proportionally with log-transformed average population density at the respective spatial resolutions (grid cell size). Since the slope z of log(M,V) plots satisfies z =1, the dynamics are Markovian in statistical-mechanical representations. The local dynamics in respective grid cells are inter-independent at the finest resolutions and thus also at coarser resolutions where neighbouring cells are aggregated. In Figure 36—only five time increments later—the variance has increased in the system as a whole owing to the tendency for local aggregation of individuals. However, the log-linearity with z =1 is approximately maintained. A similar state (z =1) is seen in Figure 37, reflecting population heterogeneity at t=50. In short, a stronger spatial aggregation as a function of a scale-specific space-use process maintains the well-known property where variance at coarser resolutions is a simple sum of the variance in respective sample units (grid cells) at finer resolutions.

Figures 38-40a show the similar situations at t=0, 5, and 50 for the evolution of spatial aggregation under parallel-processing conditions, with the local crowding dispersal rule deactivated (upper row in Figure 33). In this scenario, the more intense kind of clumping corresponds to a log-linear slope z=2. The system is approaching z=2 in an asymptotic manner, meaning that large numbers of iterations, t>>50, will show a similar pattern. Figure 40b shows that activating the local crowding rule with 80% or 100% dispersal for cells reaching M>1000, which leads to frequent self-inflicted perturbations on abundance, still shows z=2 for M over a range of k. The null model, z=1, is inset for comparison. This result, where z=2 is maintained even under such strong level of perturbations, is quite remarkable, given the apparently very different spatio-temporal dynamics with and without this rule (Figure 33). Thus, the intrinsically driven and scale-free movement process is, in the context of the process' statistical-mechanical V-M relationship, *very resilient* with respect to influence from self-inflicted perturbations. The only difference observed was that the iterations had to be run somewhat longer (from N=50 to N=100 in these examples) to reach the asymptote log(V)=2*log(M).

This result implies that variance is proportional with the square of local number of individuals pr. grid cell at the actual scale:

$$V = aM^b \quad (12a)$$

$$\text{Log}(V) = \log(a) + b*\log(M) \quad (12b)$$

To summarize, while classical, scale-specific (Markovian) dynamics from level k=2 and larger led to $b \approx 1$ for how the variance of M varies over a range of spatial scales, a parallel-processing-compliant system showed $b \approx 2$. Further, as can be seen by comparing the respective figures, the intercept log(a) also contains important information. While classical conditions showed a increasing with time, parallel processing showed a constant intercept a over time. Instead the exponent b (the slope in the log-log plot) increased while the scale-free pattern emerged from the initial scrambled situation.

6.5.2 South state hospitality

Let us take a breather here before turning towards yet another paradox emerging from the results in Figures 38-40. It provides some background for the phase of development of the foregoing model. In parallel with searching for a solution to the fix sample size and fractal pattern paradoxes observed in the sheep data in 1991 and in subsequent analyses of home-range demarcations, I also struggled at another system level—population kinetics—to understand animal dispersion patterns from first principles. These two levels of model abstraction, individual space use, and the process of population dispersion obviously need inter-level coherence. According to traditional approaches the respective frameworks have, from the onset of theoretical ecology, been the Brownian motion and correlated random walk models at the individual level and kinetic principle-compliant models for populations. Diffusion models have bridged the two levels. More sophisticated approaches for improved realism have also spun out from the generic theory. For example, at the individual level partial differential equations have been applied to model home range and dispersal dynamics that are not ergodic (Moorcroft and Lewis 2006, Fieberg and Börger 2012) (see previous sections). At the population level similar mathematical tools have been applied, in addition to coupled map lattice models for open or spatially structured (metapopulation) systems. To understand dispersal and the progression of invasion fronts, partial differential equations that are extended with terms for reproduction were developed sixty to seventy years ago (Skellam 1951). However, as outlined over the previous sections of this book, all these approaches have a common statistical-mechanical platform: mechanistic, scale-specific, and Markov-compliant system coherence.

While the non-Markovian parallel-processing idea for memory-influenced movement slowly began to mature over the years, it simultaneously stood clear to me that statistical mechanics is a coin with two complementary expressions. Thus, there would *by necessity* be a need for some serious drilling towards the other side of the coin: the population aspect. Otherwise an extended theory for the duality between individual and population levels of system abstraction would simply not be coherent. In short, a parallel-processing theory for individuals could never be constructed to be compatible with classical population kinetics. Thus, a painstaking climb towards a theory for the population level to complement individual-based space use slowly took place through the early nineties.

In particular, my two-month guest visit to Knoxville in 1994 sparked much progress. Professor Nils Chr. Stenseth at my own department invited me to join a lunch with Professor Stuart L. Pimm (Dept. of Ecology and Evolutionary Ecology, University of Tennessee), who visited Oslo. Within the hour Stuart invited me over to his lab. Coincidently, Professor Mike L. Pelton (who provided the black bear telemetry data referred to in the introduction) had his lab in another building of the same university, Dept. of Forestry, Wildlife and Fisheries. In an extra twist of coincidence this spring, Ivar also arrived to Knoxville, in connection with the12th Eastern Black-bear Workshop in Gatlinburg.

In Tennessee I had, over a series of weeks in February and March, the pleasure to experience almost an overdose of hospitality and inspirational discussions. In this creative environment I could gain the benefit from zigzagging between wildlife management and theoretical ecology. I enjoyed every minute, including a range of events, excursions, and presentations there unfortunately is not room to detail here. But it was an explosive experience, sparked by Nils Christian's lunch invitation to Grand Hotel in Oslo!

Before leaving Knoxville I was able to finish the first simulation framework for the population variant of modelling parallel processing. If I recall correctly, I handed Stuart the first printouts of this model the day I had to leave for the airport. Basically, the version I finished in Knoxville was the same I applied to produce the outputs in Figures 29-33. Unfortunately the accompanying theory was too fragmentarily developed at that point to allow for a paper publication of the model. It went to the shelf, but has nagged at me ever since. Thus, the present model description and results from simulating parallel processing at the population level represent the first presentation of this theory. Finally, after more than twenty years of hibernation.

As expected, the population model for parallel processing pinpointed yet another paradox under the premise of the null model represented by classical population kinetics. This *Z-paradox* adds to the two already pinpointed for the individual level: non-trivial fix sample size dependence in home-range demarcations, and fractal dispersion of fixes.

6.5.3 The Z-paradox

Lack of significant progress in developing a mechanistic model for, and the small number of interesting and testable predictions stemming from, the variance-mean regression epitomizes problems in pattern-oriented approaches to population dynamics.
Ilkka Hanski (1994), p. 19 (Hanski 1994)

Population dynamics (kinetics) in compliance with equation 9 for $b>1$ in Eq. 12 introduces an issue relative to classical statistical-mechanical premises in open (non-metapopulation) systems, which was described for the foregoing simulations. As Hanski summarized in the citation above, the commonly verified mean-variance relation with $b>1$ and typically close to two in empirical data has been a theoretical nuisance for a long time. Originally, it was pinpointed by Taylor in 1961 (Taylor 1961), and has received wide recognition (and spurred intense debate) ever since:

Over the past half-century since Taylor's initial description there have been many attempts to explain his law, yet none of these have found general acceptance.
Kendal and Jørgensen (2011), p. 1.

In my doctoral thesis back in 1998, I termed this issue *the Z-paradox*. This concept spans a wider range of systems than the original issue formulated as Taylor's power law, since the Z-paradox regards both population variance and individual space-use variance. Further, it puts particular focus on the scaling aspect of the issue, by studying expected and observed V-M at different spatial resolutions.

If we are careful to consider compliance with ergodicity over the actual scale range of system description (k=1,2, …, m), the Z-paradox should be expected to apply both to the M-V (mean and variance) regression of individuals in a population (*ensemble average* and variance at different magnitudes of k) and M-V of local dispersion of fixes from a single individual's path (*time average*). Implicitly, we must in the latter case (M representing cell-specific number of fixes) assume a series of non-autocorrelated fixes to satisfy an ergodic system. For a population, ergodicity is achieved if individuals during a given time increment have sufficient mobility potential to remove local autocorrelation, which would otherwise appear due to lack of sufficient mixing. In the following I describe the Z-paradox from the individual level perspective, to further underscore the no less than fantastic system duality between individuals and their local population representation in statistical-mechanical terms, whether the dynamics comply with a Markovian or a parallel-processing framework.

Consider a sample of GPS fixes from an animal path, accumulated during a season and under a period of site fidelity (a home range is thus emerging from the set of fixes). Let us further assume that the classical Markovian principle of serial, scale-specific movement has applied for the responses to habitat heterogeneity during this season. For example, the space-use process could be MemRW-like, as defined in *2.10 Coarse-graining the mechanistic memory model*. If we superimpose a grid on the set of fixes, mean, and variance of inter-cell number of fixes, M pr. cell of the actual size can be calculated as a function of grid scale. Some cells will have accumulated a large number of fixes, while other patches have fewer fixes. Moving to larger grid cells, M will be larger at this level.

Since we are assuming that the principle of superposition applies for analysis of this variability due to the serial process premise (see *2.6.1 Spatial coarse-graining*), we can formulate a null hypothesis for how the overall variance in this heterogeneous dispersion of fixes is expected to relate to mean number of fixes pr. cell at each spatial scale.

Let us start by superimposing a relatively fine-masked grid, defined as scale k=1, and calculate M over all *non-empty* cells and the variance (V) around this mean at this scale level. We then get our first (M,V) plot in a diagram with M on the x-axis and V on the y-axis. Larger k implies larger M. Next, we repeat the procedure by using grid cell size k=2, 4, 8, ... Since our sampling universe (for example, the defined home-range area) is constant for the inter-scale analysis, and we include the total variability of the data set in each new calculation of M and V at respective levels, we expect the ratio V/M to be constant. In other words, V = aM, in compliance with Eq. 12a with b=1. As support for this claim I refer to Benoit Mandelbrot's famous book on fractal geometry (Mandelbrot 1983, p. 367); but similar support may also be found in general textbooks of statistics. Since the sampling universe (summed area of sample units) is constant, V should increase proportionally (linearly) with M as we zoom out from fine to coarse resolutions. In short, many small units should in sum contain the same variance as the sum of larger but fewer units. With log-transformed axes for M and V, we expect a linear fit to slope = 1 (from A to B in Figure 41). The magnitude of the intercept, a, defines the degree of habitat heterogeneity (departure from habitat homogeneity). Let us consider that we started the other way round, from coarse scales towards finer scales, and from a level with larger overall spatial variance—owing to a larger a. Again our first (M,V) plot would be our starting point for a prediction about how V would change with M towards finer spatial resolutions. The null hypothesis—under condition of ergodicity and lack of environmentally invoked autocorrelation between sample units—would now be that V should decrease linearly with decreasing M towards finer scales (from point C to point D).

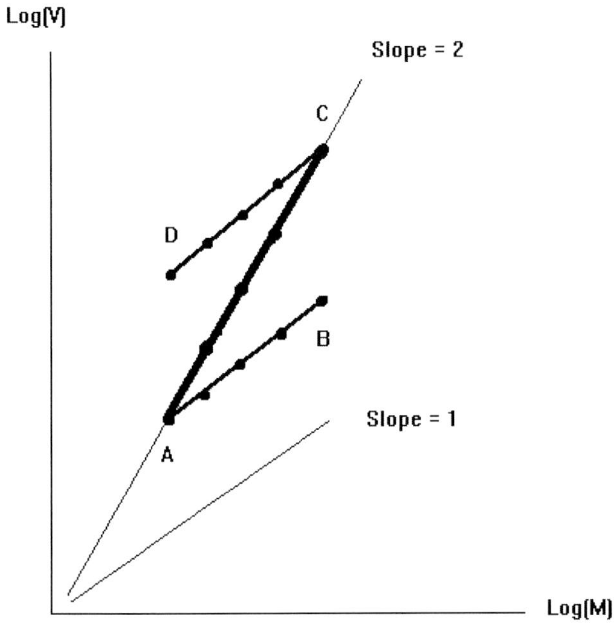

Figure 41. Illustration of the Z-paradox in a diagram of logarithmically transformed sample variance (log(V)) as a function of logarithmically transformed mean number of animal relocations (log(M)) pr. sample unit at a given grid resolution. Sufficient population mixing to satisfy ergodicity is assumed. M is changed by changing grid resolution. For a random distribution of relocations in a homogeneous environment, log(V) = log(M) (slope =1 in the diagram). In a heterogeneous environment variance for a given M may be large (for example, of magnitude described by A or D). However, by changing grain size for sample units, variance is expected to change proportionally with this size, since M changes proportionally (say, from A to B or from D to C, or vice versa). The line that is labelled "slope = 2" represents the Z-paradox, since variance is changing non-proportionally from A to C (or C to A) rather than the expected A to B (or C to D) under change of grid resolution. In other words, the line A-C is not expected under standard kind of statistical mechanics.

It is easy to verify this property (derived from the central limit theorem) in an Excel spreadsheet: Fill in a series of inter-independent numbers in a given column. Then calculate the series' mean (M_1) and variance (V_1). Larger range in the set of numbers leads to larger series variance, but series length as such does not influence M and V [beyond small-sample artifacts; see McArdle (1989)]. Next, lump together two and two numbers to produce a coarsened series, and calculate (M_2,V_2). In a similar manner, produce (M_i,V_i) at coarser resolutions. Then make a (M_i,V_i) scatter plot, and test that the regression line has a slope close to one. This test may represent a spatial dispersion of non-autocorrelated sample units: successive numbers were independent. To test the effect from autocorrelation, use the same series of numbers at scale 1, but add or subtract them successively in a random manner (50% chance for adding or subtracting). This leads to an autocorrelated series, a Brownian motion-like path in one dimension. This time the (M_i,V_i) regression will show variance changing closer to proportionally with M squared, as for population complexity. However, in this scenario, variance inflation from coarse-graining is due to the strong autocorrelation in the original series: two successive sample units (at all resolutions) will tend to have more similar-sized M relative to more distant units. Consequently, the ergodicity premise is not satisfied. The population is not sufficiently "mixed" in statistical-mechanical terms at the scale of the given time increment.

Computer simulations of a parallel-processing system as just shown comply with V changing proportionally with M^2 rather than M, despite environmental homogeneity (Figures 38-40). In Figure 41 this progression is illustrated by the line from A to C, which represents the Z-paradox. Starting from two different endpoints on the range of M, if $V = aM^2$ we do not reach the same point on the (M,V) line with slope 1, representing V changing proportionally with M. With log-transformed axes, a figure describing the letter Z appears when we add the lines for the two null model predictions. From a scale simplicity point of view (i.e., the classical paradigm), one would expect A going to B or D going to C (or vice versa when zooming to finer scales), and not A to C (or C to A). The classical and non-paradoxical results were shown in Figures 35-37, where the parallel shift of the line A-B to the line D-C with increasing intrinsic system variability represented small and larger levels of variance (difference in *a*) as a consequence of difference in spatial resolution to calculate M. However, parallel processing introduced paradoxical results from similar coarse-graining (Figures 38-40), reflected by the line A-C in Figure 41. Parallel processing of the tendency for conspecific aggregation led to intrinsically generated heterogeneity of M; both across space at a given resolution, and across resolutions. In contrast, under Markov conditions the tendency for conspecific aggregation led to local aggregation where inflation of variance was expressed through the intercept *a* rather than the power law exponent *b*.

By arranging and fine-tuning the local model habitat in some special way, it is certainly possible to get a pattern like Figures 38-40 from within the classical framework of classical population kinetics. For example, one could arrange the habitat to satisfy something like a superposition of a geometrically scaling set of sine waves or some other (rather hypothetical) construction of a scale-free environment. The environment could also be locally heterogeneous at all scales in a manner that resembles a Brownian motion, as was proposed by the Excel spreadsheet procedure above. However, this kind of spatial autocorrelation may be easily removed by differentiating a series. The model that generates the paradoxical part of the pattern in Figure 41 from *intrinsic* mechanisms—parallel-processing effects on population kinetics under condition of conspecific attraction in a homogeneous environment—is documented here for the first time.

6.5.4 The density dilution issue of classical models

With respect to understanding population dispersion at landscape scales, animals from many taxa violate the classical assumptions about Brownian motion and Lévy walk by performing memory-dependent displacements and returns in connection with site fidelity, commuting between resource patches, and optimal positioning relative to conspecifics (Ward and Saltz 1994, Ostfeld and Manson 1996, Barrett and Lowen 1998, Healy 1998, Chittka *et al.* 1999). In the population simulations above, I implemented the latter effect in a simplistic manner by defining a tendency for individuals to congregate. The simulation results illustrated how classical assumptions led to fine-grained population aggregation, while parallel processing made this heterogeneity percolate towards coarser scales. This intrinsically emerging effect on the V-M pattern was described as the Z-paradox from the perspective of classical assumptions. However, there is also another aspect of population dispersion that represents a challenge under classical models, hereby termed the *density dilution issue*.

At the population level, memory-less dispersal will, under broad range of classical model conditions, lead to population extinction, owing to lack of a mechanism for intra-specific cohesion (a tendency for conspecific attraction). Classical diffusion can easily illustrate this issue. Consider a drop of ink in a bathtub. Eventually the blue dot of aggregated ink molecules will be randomly dispersed in the entire area available—the tub. Thus, inter-molecule contact among ink molecules will gradually drift towards zero. In a population context, Allée effects will in an open environment bring the population to the brink of extinction well before the local population density reaches zero density. In the already described population model under classical conditions, illustrated by Figures 29-32, the density dilution issue was avoided by defining a closed environment. If the same initial dispersion of M individuals had been studied in an alternative scenario without area constraint, classical diffusion would in the long term lead to leakage of individuals over an increasing area, despite local rule for conspecific attraction. The net emigration happens in the border zone, where some individuals are losing contact with the population and start roaming around into emptiness.

To circumvent the density dilution issue, classical premises for population dynamical modelling simply require either (a) a closed environment where individuals are consequently self-crossing their movement paths with higher probability than in open environments; (b) net population growth (the difference between birth and immigration rate on one hand and death and emigration rate on the other hand) to be sufficiently large relative to the movement rate to keep population density behind the diffusion front above a critical level to ensure long-term survival (Skellam 1951); or (c) individuals are congregating from a common tendency to successively move towards patches with higher than average resource density, and these patches are sufficiently stable over the time scale of a generation to allow for population maintenance from intrinsic population growth.

I exclude the so-called rescue effect (Brown and Kodric-Brown 1977) from these alternatives (higher inflow than outflow of individuals), since net immigration has to come from somewhere. Thus, local rescue is just a temporary solution from the perspective of the population as a whole if the latter cannot be maintained in the long run owing to the effects under points a-c above.

The constraints with respect to achieving realistic population survival in an open environment under classical conditions are quite stringent, relative to survival of populations where individuals have a cognitive capacity to return strategically towards locations with previous encounters of conspecifics. These individuals have the option to explore the landscape over broad spatial scales, and still maintain contact with the population by occasional returns (Gautestad and Mysterud 2006). An alternative theoretical framework involving memory-influenced kinetics has been requested for many years (Conradt *et al.* 2000, Travis and French 2000). If Eq. 12 for the relationship between variance and the mean with $b>>1$ from an intrinsic process—conspecific attraction—can be verified for a population, the present parallel-processing model may provide an alternative route towards exploring population dynamics with reduced Allée effect as an emergent property.

To conclude; intra-specific cohesion from strategic, memory-dependent displacements (Figure 34), whether during the dispersal phase or in other phases of an individual's life cycle, means that each individual will occasionally revisit locations where it directly or indirectly (e.g., by scent cues) came in contact with conspecifics in the past. Thus, long-term, spatially explicit memory capacity has to be involved for non-random path crossing. Such a tendency to congregate strategically (in concert with the wide range of other movement-influencing factors not included in the simplified parallel-processing example above) has a dual effect. First, it may contribute to decreased extinction probability for a population by allowing targeted rendezvous between individuals. Second, it may resolve the density dilution issue for populations in open environments, if a tendency among purely opportunistically dispersing individuals to congregate in shared resource-rich patches is not sufficient to explain long-term population survival.

In the parallel-processing scenario called the Zoomer model, I assumed that conspecifics are parts of an expanded "resource" background of the environment. In a positive feedback manner, intra-specific cohesion will generate a non-trivial and multi-scaled spatial congregation pattern at the population level. An upper cut-off scale for this kind of multi-scaled population dispersion will then be approximately $\mu L'_{max}$, where L'_{max} is the idealized maximum displacement range for individuals in a unit time step during the broad-ranging dispersing phase, and μ is the median number of time steps between return to previously visited locations. A large μ implies that the upper scale limit for parallel processing-compliant population kinetics (termed m in the model above) may be quite coarse and thus involve landscape-level spatial resolutions. The magnitude of μ quantifies strength of site fidelity, which may be weak (large μ) but still existing even during natal dispersal (and therefore critical to avoid extinction from Allée effects). The parallel-processing link between conspecific attraction and site fidelity is also bridging population kinetics to individual-level modelling of memory-influenced space use, which will regain the focus over the following sections.

6.6 The next step

Again, by starting with classic model premises under the current paradigm, in this chapter I have gone one step further by extending the traditional coupled map lattice framework for spatio-temporal population dynamics to account for memory influence under a premise of scale-free redistribution. The Zoomer model was formally defined in Eqs 8-10, and the simulation examples illustrated how the complex population dispersion could be statistically presented by a mean-variance power law. It is particularly important to underscore that the process if intrinsically driven redistribution (conspecific attraction combined with redistribution from local overcrowding) under the Zoomer model results in multi-scaled population heterogeneity even in a homogeneous environment. The perfect storm is invoked in the theory of animal space use by considering the more realistic scenario where the environment is heterogeneous. Such a shifting mosaic of population redistribution from both intrinsic and extrinsic forces still makes the emergent pattern amendable to ecological analysis and hypothesis testing. For example, I showed how the dynamics led to a remarkably simple relationship between mean abundance and its variance. Model parameters may then be subject to ecological interpretation. In the following chapter I will make a U-turn back to the individual level, to explore how the variance-mean relationship emerges in a coherent manner under the premise of parallel processing. Time to introduce Multi-scaled Random Walk!

7. PARALLEL PROCESSING AT THE INDIVIDUAL LEVEL

7.1 Testing inter-level model coherence

If a population organizes itself according to parallel-processing principles, a lot of unexpected spatial dispersion variance emerges at coarser scales relative to classical system expectations. Simulation results confirming this property at the population level were shown in Figures 38-40. To confirm coherence with the individual level, I have run a MRW-compliant space use with ergodicity satisfied (non-autocorrelated series; the actual model details will be described in the following chapter). A result is shown in Figures 42-43. Figure 42 verifies that V is changing proportionally with M^2 also at the individual level where a snapshot of individual locations (coarse-grained to cell densities) is replaced by a sample of fixes from a given individual and studied over a range of grid resolutions. To satisfy this duality, ergodicity has to be satisfied. In the present simulation, return step frequency was 1:10 and observation frequency was 1:100. When the return interval is shorter than the fix sampling interval, successive positions of the individual become spatially scrambled. Thus, ergodicity is expected to be satisfied owing to non-autocorrelated series.

The extra variance from an exponent larger than one is equally unexpected from classical system premises at the individual level as it was at the population level. The power law structure $V = aM^2$ means that the spatial dispersion of fixes is scale-free, i.e., it satisfies a (statistical) fractal with a low magnitude of the dimension (I will return to this aspect below). In other words, the scale-free variability is maintained upon coarse-graining, rather than being "averaged away" (being proportional with change in grid size and thus with M) as one would intuitively expect from a scale-simplistic process' pattern [for example, see Figure 2 in (Wiens 1989) and Figures 27-32 above, in compliance with the expectation from *2.6 Linearization of non-linear mechanistic interactions*].

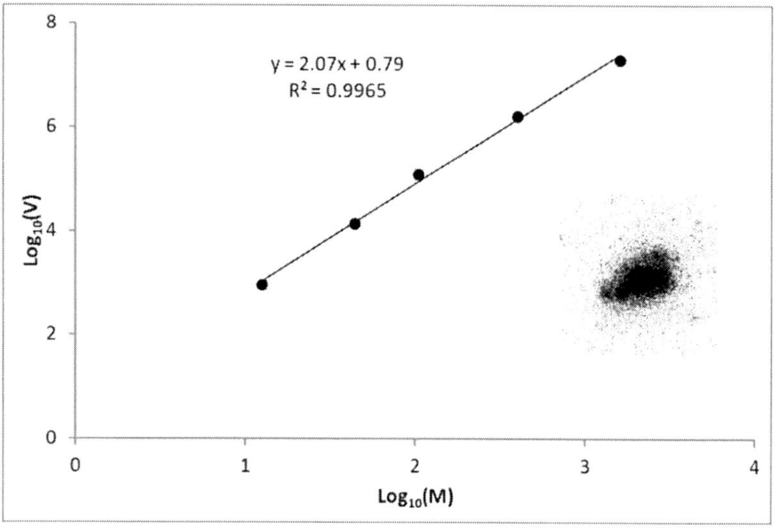

Figure 42. Variance-mean analysis of spatial scatter of relocations from the MRW model. See main text.

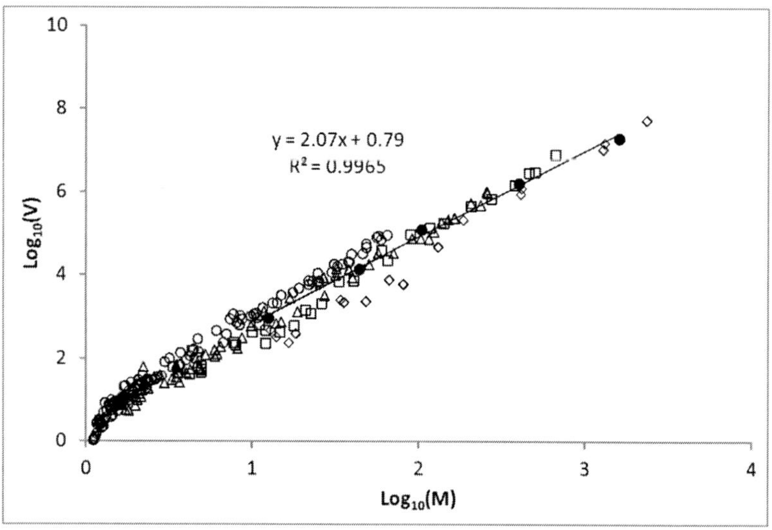

Figure 43. Variance-mean analysis where local density of relocations is split into spatial sub-sections in addition to different spatial resolutions. See main text.

While each log(M,V) plot in Figure 42 represented calculations of M and V over all available cells at respective levels, Figure 43 shows the calculations for each row separately in the respective grid resolutions. Thus, intra-level variability of M and V is shown. Again classical premises make us believe V should change in proportion to M, where M now is the average number of fixes pr. grid cell in a specific row. Now plotting M and V for respective rows, circles represent grid cells of size 1/128 (linearly) relative to the total area in the lower-right picture of fix dispersion in Figure 41. Triangles show plots at relative scale 1/64, squares show 1/32, and diamonds show 1/16. Larger-cell levels are omitted, owing to increasing statistical artifacts when number of sample units is small. To compensate for lower number of grid cells in a row versus the entire arena, each calculation of V is multiplied by a level-specific factor equal to the square root of number of sample units at a given row at the given level.

Two main results stand out from Figure 43. First, also at respective grid resolutions, the analysis confirms $M = V^2$ (excluding a statistical artifact for very low M, where number of sample units is few owing to empty cells). Second, after adjusting V with square root of sample units at respective levels, the extended sets of log(M,V) plots overlap with the result in Figure 42. The square root compensation results in an increased intercept *a* in Eq. 12a, and makes the respective intercepts comparable from one scale level to the next. This factor does not influence the slope of the log(M,V) plots at respective levels.

To conclude on Figures 42-43, if the principle of ergodicity is satisfied (non-autocorrelated fixes), space-use variability at the population level and the individual level are similar in statistical-mechanical terms with respect to inter- and intra-level mean and variance relationship. This result, which is presented here for the first time, provides strong support for the hypothesis that the Zoomer model for simulation of parallel processing at the population level is coherent with multi-scaled random walk (MRW) simulation at the individual level. The model details of the latter approach will now be described in greater detail, including MRW under non-ergodic conditions.

7.2 Simulating multi-scaled random walk (MRW)

7.2.1 The non-mechanistic premise for spatio-temporal memory

There seems to be a general consensus among experts on movement ecology under memory influence that movement should be considered a two-component process; exploratory moves (e.g., searching for food and other resources) and more strategic moves involving the utilization of a memory map to return to previously visited locations. The MRW model also contains the two facets of space use—exploratory moves and spatial memory-dependent returns—but under a premise of a statistical-mechanical execution level. Owing to the previous chapters where I have argued repeatedly for a need to replace the Markovian premise with a non-Markovian one to allow for *both* temporal and spatial memory effects, MRW and its parallel-processing postulate are based on a premise that *a mechanistic representation in the classical (Markovian) sense cannot be derived.* Thus, while simulations of MRW are dynamical, they are not mechanistic representations of movement! Therefore, the new term "non-mechanistic dynamics" is introduced.

The reason for this extremely bold statement that MRW cannot be represented by a (classical) mechanistic variant is what I consider a logical necessity from the premises that were discussed with respect to classical premises in Chapter 2 and mechanistic implementation of spatial memory in Chapter 3. In Chapter 5, I described many examples where an individual's local responses could not be derived from local causes at a given small interval in time (response time, in Markov terms). Thus, the combination of spatial and temporal memory seems to *require* a statistical-mechanical level of system abstraction. In *4.1 System complexity,* I even proposed that attempts to crack many issues in complexity theory may benefit from this approach—to accept a departure from mechanistic thinking and explore non-Markovian approaches directly at the statistical-mechanical level.

In a previous chapter I presented a statistical-mechanical representation, MemRW, of a recently proposed temporally scale-specific (Markov-compliant) and memory-enhanced model. MemRW was based on translation of mechanistic simulations in Börger et al. (2008) and Van Moorter et al. (2009) to the statistical-mechanical level. The MemRW simulations, which resulted in Figures 19-21, were basically derived from the MRW model (developed more than ten years earlier) by defining a scale-specific process for exploratory moves: correlated random walk. In contrast, in the original and still standard MRW, the exploratory step component generate steps that are scale-free, in order to comply with the parallel-processing conjecture. Conveniently, a Lévy walk (Mandelbrot 1983, Reynolds and Rhodes 2009), which has already been loosely described in several foregoing sections, satisfies the property of spatially scale-free steps. Thus, MRW regards space use where exploratory steps—representing search in conventional foraging simulations—are combined with occasional returns to previous locations. The latter represent site fidelity, which depends on spatial memory. In contrast to MemRW, the parallel-processing conjecture is replacing the Markov premise in the time domain. In this case mechanistic causality breaks down. The classical logics of causality imply a mechanistic process. This claim will be further justified in a later section.

Critically, the temporal unit scale t of a MRW sequence is assumed (as it was in the MemRW derivation) to be large enough to implicitly embed many hidden movement events at micro-scales smaller than t. In this case, movement as a dynamic series of displacements at scale t may reflect a statistical-mechanical representation of the actual process, similar to a real path sampled by GPS at intervals of—for example—ten minutes, one hour, or larger. Thus, attempting to define causality behind shifting the position from one location to the next at temporal resolution t is information-less. In a statistical-mechanical context, it is only the large collection of such locations and displacements that reveals aspects of the underlying process, as already explained in previous chapters.

7.2.2 The step-generating algorithm: scale-free movement

In a MRW simulation the first aspect of movement—exploratory movement in a parallel-processing context—is assumed to represent locations generated at unit time interval t (the temporal simulation scale) as successively independent step vectors of length L that are calculated from the following iterative process:

$$L_{\mathrm{MRW}}=a(\mathrm{RND})^{-1/(\beta-1)} \qquad (13)$$

By default, α=1 and β=2. RND is a random number between zero and one. Thus, Eq. 13 then simplifies to $L_{\mathrm{MRW}}=(\mathrm{RND})^{-1}$. A larger β implies fewer occurrences of large steps; with β=3 or larger the step generator produces steps that are more in compliance with Brownian motion. A larger β is thus compliant with fewer scale levels for parallel processing in the population model described in previous sections, and β=2 is compliant with an equal number of zoomers to each scale level, as expressed by Eq.

8. Therefore, β=2 regards the situation where individuals put equal effort—in the long run—into relating to their environment at fine, medium, and coarse scales. This was explained in connection with Figure 34.

The median step length for β=2 at temporal scale *t* is 0.5^{-1}=2 length units, since the average number from the random number generator (RND) equals 0.5. In other words, this step scale is tiny relative to (for example) a pixel size of 1000x1000 length units for the arena in Figures 42-43. A large arena is consequently needed to simulate an open environment, due to the occasional appearance of very large displacements from the step generator in Eq. 13.

Eq. 13 is the same equation that is traditionally applied to simulate a memory-less and scale-free process, Lévy walk. Simulated path examples are shown in Appendix A2. As may be understood from Figure 34, where the parallel processing in a population was conceptualized, parallel processing leads to a Lévy walk-like kind of step length distribution in statistical-mechanical terms (sufficiently large time increments). While some steps in a Lévy walk-generated distribution of step lengths are surprisingly large but limited by how far an individual may displace itself during a given interval and movement speed, most steps are tiny, so the median step length is also small.

A constant α in the path-generating algorithm above implies a given average movement speed for the simulations (Gautestad and Mysterud 2010b); when integrated to temporal scale *t*, speed is assumed constant for simplicity. Different strengths of path jaggedness at finer, hidden temporal resolutions relative to *t* may then explain difference in length of successive steps.

Since a Lévy walk path is scale-free, it “looks the same” upon sampling every x^{th} position to get a coarser-scale path (Appendix A2). “Looking the same” means that the new path will have a similar form of the step length distribution after re-scaling by a parameter (Reynolds 2008, Gautestad 2012), and the spatial dispersion of fixes also looks the same statistically after such re-sampling. Consequently, since the implicit time resolution *t* in Eq. 13 is assumed to be sufficiently large to allow for a statistical-mechanical representation of the observed path, a relatively large step at this resolution may be interpreted as a relatively unidirectional path at finer (hidden) temporal resolutions. Similarly, a small step at scale *t* implies the dominant situation with implicitly assumed high degree of local jaggedness in the hidden path details. Hence, despite a constant movement speed on average, both very short and very long steps may emerge.

The statistical-mechanical premise for the magnitude of t also makes it feasible to define successive step directions to be drawn uniformly from 0-2π radians (all directions equally likely at temporal scale t). Complicating the rules by defining directional persistence for successive steps (like in correlated random walk) will not lead to qualitatively different behaviour of the path, since almost all steps are relatively tiny and long steps are almost always followed by a sequence of tiny steps. Thus, contrary to correlated random walk, a directional persistence will have minimal influence on the path's statistical properties. To further underscore the statistical-mechanical nature of the MRW approach and give additional support to the "Occam's razor"-inspired choice to disregard directional persistence, each series was, in the context of Figures 42-43, sampled as one "observed fix" pr. $100t$.

7.2.3 Adding site fidelity to a scale-free movement path

Eq. 13 alone generates a Lévy walk-like path, with super-diffusive properties and no memory. However, the algorithm in Eq. 13 is postulated to reflect a scale-free kind of space utilization, in compliance with the population model analogue with zoomers. From this perspective, Eq. 13 satisfies an implicit premise of scale-free habitat utilization as expressed in Eq. 8-11. To allow for this similarity between a spatially memory-less Lévy walk and a parallel-processing-compliant space use under spatial memory utilization, the animal must be assumed to have familiarized itself with the actual environment. Thus, MRW must include two aspects of memory-dependent space use: scale-free search as in Eq. 13 (exploratory moves, like zooming in Eq. 8) and site fidelity to allow for exploitation of spatial memory of previously visited patches. Without occasional returns to familiar sites, spatial memory has no intrinsic fitness value.

What regards this other aspect of movement, memory utilization in the form of relating to the historic path, on average every μ^{th} time increment a step is—under default model conditions—followed by a directed return to a randomly chosen previous location in the series. Simulated path examples are shown in Appendix A3. Such return events introduce non-random path crossing—and therefore site fidelity—to the model. Without it, spatial memory has no value since the animal will recycle space use by chance only. Owing to the Lévy walk compliance for exploratory steps, this chance is very tiny and can be ignored.

The inverse of the magnitude of return interval μt (where μ is an integer larger than one) defines the general strength of this homing tendency: larger μ implies weaker site fidelity (Gautestad and Mysterud 2013), for example, during natal or seasonal dispersal. Alternatively, returning to a position within a narrow probability distribution kernel surrounding each historic location rather than landing on this location exactly would not influence the present results, according to unpublished simulation results. Thus, again I invoke Occam's razor to keep the model as simple as possible while still maintaining its generic properties.

The choice to return to a randomly picked location (or to a small neighbourhood surrounding this location) among the generated set of locations at the respective given points in time may be considered consistent with general principles of statistical mechanics under which a constant μ over the entire arena implies a homogeneous environment over the actual scale ranges of space and time. It also implies infinite spatio-temporal memory (see below). Since long and short exploratory steps from Eq. 13 are performed under the same energy expenditure (constant a, and thus a constant movement speed on average at micro-scales of the implicit path), there is no need to complicate the generic model with a distance penalty with respect to which historic location to return to. Return steps are based on the same assumption: more implicit (hidden) jaggedness while returning to a target close by, and less jaggedness when returning to distant targets. While the lines expressing return vectors in Figure 25 were unidirectional, this was just to visualise both the start location and target location for respective events.

7.2.4 Simulating space use: the generic MRW model

The parallel-processing architecture offered a solution to the commonly observed $V=aM^2$ pattern, termed the Z-paradox (Figures 35-43). The next challenge is how MRW performs in the meeting with the other paradoxes, the home-range ghost ($z\approx0.5$ in Eq. 1a) and the scale-free dispersion of fixes ($D\approx1$). First, choosing fix sampling larger than the average return interval will ensure a non-autocorrelated set of fixes (Gautestad and Mysterud 2013). Since two adjacent fixes in the sampled series will then not tend to be closer in space than two randomly selected fixes, ergodicity for scale-free space use is expected to be satisfied. In such a scale-free ergodic process, MRW simulations have verified (Gautestad and Mysterud

2010a) that space use may be quantified as number of virtual grid cells (incidence, I), containing one or more fixes in accordance to:

$$I(\mathrm{N}) = c\mathrm{N}^z \qquad (14)$$

I(N) at the choice of grid resolution is reflecting how aggregated space use is expected to change as a power law (thus, scale-free function) with respect to sample size N. Ref. (Gautestad and Mysterud 2010a, 2012). While Eq. 14 regards non-autocorrelated relocations, it may also refer to autocorrelated fixes if the sample N is increased in a time-proportional manner, rather than increasing N uniformly within a given time period. Further model elaborations below will describe how grid resolution may be optimized, and how c is a function of grid resolution and z.

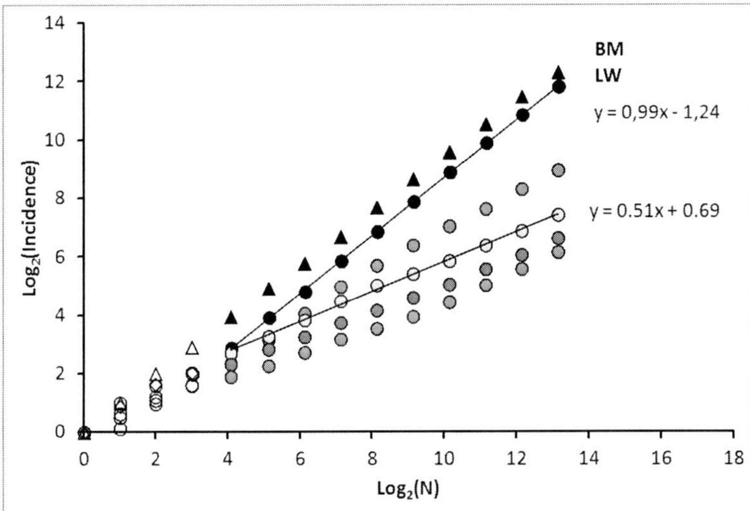

Figure 44. Space use I(N) of Lévy walk (LW), Brownian motion (BM), and MRW (lower four series). MRW is simulated under variable ratio of average return interval relative to observation interval. Both Lévy walk and Brownian motion confirm regression slope $z\approx1$ in double-log plot of I(N). The four sets of plots with the smallest slope regard MRW, where increased influence from return steps leads to $z\approx0.5$ for non-autocorrelated fixes, which is expected from β=2 and infinite memory in the MRW-generating algorithm. The transition zone from autocorrelated towards non-autocorrelated spatial relocations leads to an intermediate power exponent, approaching $z\approx0.5$ asymptotically.

Theoretical compliance with Eq. 12 and how z varies with different simulation conditions has for MRW been published repeatedly over the period from 2005 to present (Gautestad and Mysterud 2005, 2006, 2010a, b, Gautestad 2011, 2012, Gautestad and Mysterud 2012, Gautestad 2013a, Gautestad *et al.* 2013, Gautestad and Mysterud 2013). Some preliminary results based on Eq. 13 and showing compliance with Eq. 14 were presented already in my doctoral thesis in 1998. Figure 44, which is reproduced from ref. (Gautestad and Mysterud 2013), shows a comparison between MRW and simulations under classical models memory-less, scale-specific Brownian motion and spatially memory-less, scale-free Lévy walk. The latter was simulated as MRW with a short memory horizon on spatial memory influence (see next sections). In other words, Lévy walk in combination with returns only towards recent locations leads to a classical Lévy walk in statistical-mechanical terms. Figure 44 thus verifies that a memory horizon makes it impossible to differentiate this kind of memory-constrained MRW and classical Lévy walk.

7.2.5 The characteristic scale and the home-range ghost paradox

Figure 45 shows another example based on non-autocorrelated fixes, collected from ref. (Gautestad and Mysterud 2010a). Figure 45 shows the effect from varying the grid resolution. Large resolution (small grid cells) leads to a strengthened requirement for large N, owing to the *dilution effect*: a small N is simply not sufficient to reveal the scale-free pattern at fine resolutions (Gautestad and Mysterud 2012). Towards the other end of the intrinsic scale range, too large grid cells lead to all or almost all cells containing fixes already for small N. Thus, this *space-fill effect* leads to zero slope and log(c)= 0 (all fixes in a single cell) for small N, and—for larger N—an asymptotic approach towards z≈0.5 and a similarly asymptotic change of log(c) towards a value smaller than zero. By fine-tuning the scale resolution, one may therefore find the process' *characteristic scale of space use*, where log(c) = 0 and $z \approx 0.5$.

Defining observed home-range area A(N) from N fixes as I(N) leads to a spac+e-use function for sample size dependence similar to the empirically derived Eq. 1a in *1.5 The home-range ghost*. While MemRW simulations in previous sections led to the exponent z to be of magnitude 0.25, empirical data showed stronger N-dependence from z≈0.5. MRW results under generic conditions (Figures 44-45) illustrate full compliance with z≈0.5 for non-autocorrelated data. This compliance was asymptotic: if the critical ratio between return event interval and fix sampling interval is surpassed towards non-autocorrelated fixes, z is approaching 0.5 and stays at this level under additional returns pr. fix interval (Figure 44). The intercept log(c) is decreasing, though, with increasing return frequency (stronger site fidelity). Consequently, a change in strength of site fidelity is reflected in a change in the estimate of c. In this manner, the system's characteristic scale of space use (inversely proportional to c) may be applied as a quantification of strength of site fidelity. In this manner MRW both seems to have the potential to solve the home-range ghost paradox and simultaneously provides a very potent statistical-mechanical quantification of site fidelity strength.

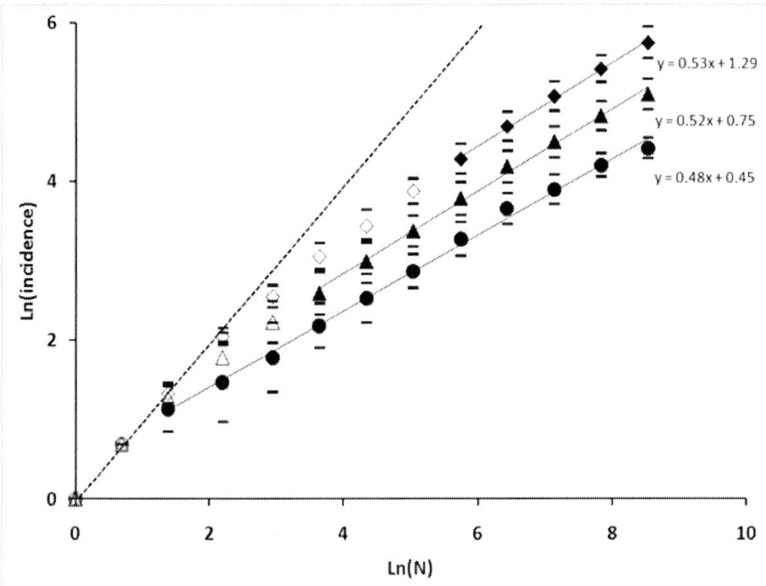

Figure 45. Ln(incidence) as a function of ln(N) from numeric MRW simulations with β=2 for the trajectories. Each series represents the average scores from five replicate runs of length N=10,000 at spatial resolutions k=1/20 (circles), k=1/40 (triangles), and k=1/80 (diamonds). Standard deviations are shown as dashes. When analysing real data in a similar manner, grid resolution should be varied to fit the scale where intercept $c\approx1$ [log(c)=0)]. This unit pixel size then equals the movement's characteristic spatial scale of space use.

7.2.6 Fractal dimension of the spatial dispersion of fixes

Of the three paradoxes presented so far, MRW in its generic form has been shown to crack both the Z-paradox and the home-range ghost paradox. It now remains—in this first approach towards MRW simulation results—to see how the spatial dispersion of fixes relates to the empirical results from sheep and red deer data showing a scale-free pattern with D≈1 (Gautestad and Mysterud 2012, Gautestad *et al.* 2013). From the statistical-mechanical relationship between D and β under condition of Eq. 13 with β=2, one should expect z≈0.5 (Gautestad and Mysterud 2010a, Gautestad and Mysterud 2013) (Viswanathan *et al.* 2000, Hagen *et al.* 2001, Gautestad and Mysterud 2006), according to the following series of relationships:

$$D_{step} = \beta - 1 \qquad | \qquad 1 < \beta <= 3 \qquad (15)$$

D_{step} is the fractal dimension of the distribution of steps from a classical Lévy walk (Mandelbrot 1983, Hagen *et al.* 2001). However, under some conditions the spatially explicit counterpart, the fractal dimension of the distribution of fixes D, is of similar magnitude as D_{step}. This also applies to MRW in the autocorrelated domain (Gautestad and Mysterud 2005, 2006) [in the non-autocorrelated domain, a series of MRW steps will show a distribution that is not scale-free; see Gautestad and Mysterud (2005)]. Figure 46, which is based on the same data as in Figure 45, confirms that the observed D≈1.

A theoretical relationship between D and z was developed in ref. (Gautestad and Mysterud 2010a). The exponent z in Eq. 14 may thus be replaced by an expression for dependence on D, which is expressing the strength of scale-free spatial aggregation of fixes (Gautestad and Mysterud 2010a):

$$D = 2(1-z) \qquad (16)$$

This leads to the following D-specified variant of Eq. 14:

$$I(N) = cN^{(1-D/2)} \qquad (17)$$

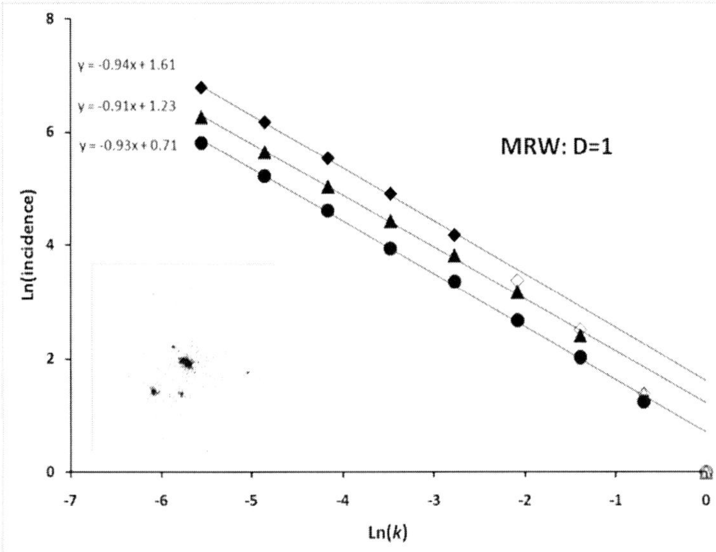

Figure 46. Box counting results, ln(incidence) vs. ln(k), from numeric MRW simulations with D_{step}=1 (β=2) for the trajectories. The series were the same that led to the results in Figure 45. Each series represents the average scores from five replicate runs of length N=1250 (circles), 2500 (triangles) and 5000 collected steps (diamonds). Standard deviations were generally smaller than the height of the respective series' icons, and so are not shown. The spatial dispersion of one of the five N=5000 sets is superimposed for illustrative purpose.

Eq. 17 with D = 1 leads to z=0.5—the home-range ghost—with a statistical-mechanical link towards the expected intrinsic emergence of fractal dimension of fix dispersion. Figure 46 shows an example [published in ref. (Gautestad and Mysterud 2010a)] of simulation results that support Eq. 15-17. Specifically, the average slope in Figure 45 (z = 0.51) leads to expectancy D = 2(1-z) = 0.98, which resonates well with D = 0.93 on average from direct calculation (Figure 46).

On the other hand, scale-free space use under β=2 in Eq. 13, but under condition of memory horizon, leads to the expectation $z\approx 1$ rather than close to 0.5 (Gautestad and Mysterud 2010a, Gautestad 2011, 2012), as was shown in Figure 44. In this case space use may appear non-complex (scale-free but not memory-influenced at the actual temporal scale) (Gautestad and Mysterud 2013). Thus, Eq. 14 and Eq. 17 regard a system with infinite spatial and temporal memory in the context of the actual scale extents for analysis.

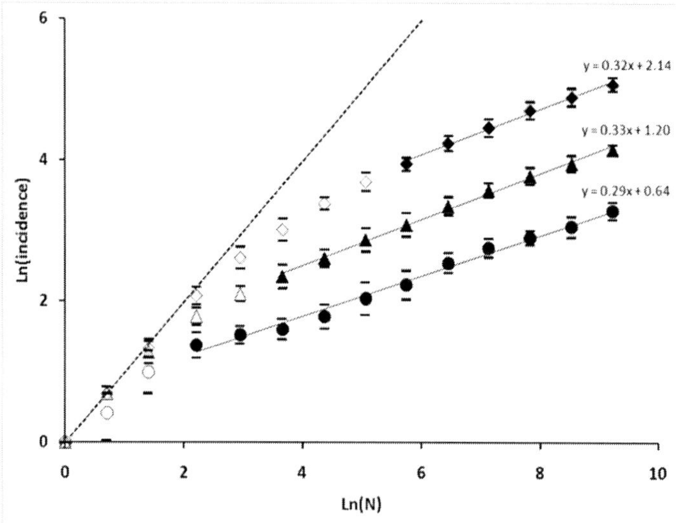

Figure 47. Ln(incidence) as a function of ln(N) from numeric MRW simulations with D_{steps}=2 for the trajectories. Each series represents the average scores from five replicate runs of length N=5,000 at spatial resolutions k=1/20 (circles), k=1/40 (triangles), and k=1/80 (diamonds). Each of the series was uniformly sub-sampled to get series of different magnitudes of N from the respective series. Standard deviations are shown as dashes.

The mechanistic counterpart of MRW has been simulated under two statistical-mechanical conditions, referred to as MemRW in previous sections. The first condition is based on replacing Eq. 13 as a step-generating algorithm with an algorithm for correlated random walk (spatial memory, but Markov-compliant time dynamics; see a later section for the actual algorithm). The results were shown in Figures 19-20, and showed $z \approx 0.25$ for infinite memory horizon and z closer to one for memory horizon. This was a similar result as for MRW under memory horizon—see Figure 44—as expected since both scale-free Lévy walk and scale-specific correlated random walk showed $z \approx 1$. In ref. (Gautestad and Mysterud 2010a) an alternative representation of MemRW was also shown, by changing the MRW "scale weight" parameter β in Eq. 13 from the default β=2 to the more Brownian motion-compliant β =3. Figure 47 shows that z=0.32 is in fact close to the Brownian motion counterpart in

Figure 20. However, while MRW with β =3 still represents a scale-free process in Lagrangian terms—but with more weight on finer-scale space use than coarser-scale resolutions relative to β =2 with equal weight over scales—MemRW with correlated random walk represents truly scale-specific exploratory steps (all weight on fine-scale space use). This may explain the intermediate z between 0.25 and 0.5 from MRW when β =3. Interestingly, even the mechanistic-compliant version of MemRW (correlated random walk-based) showed a power law expansion of incidence as a function of sample size, despite the use of a scale-specific algorithm for exploratory steps. Figure 48 shows the result from the estimates of D from the data sets with β =3 (D_{step}=2).

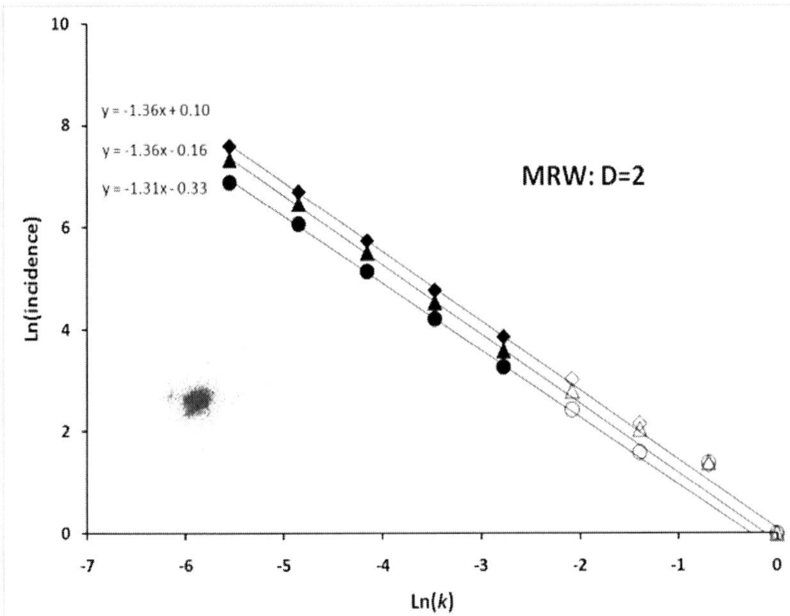

Figure 48. Box counting results, ln(incidence) vs. ln(k), from MRW with D_{steps}=2 (β=3) for the trajectories. Each series represents the average scores from five replicate runs of length N=2500 (circles), 5000 (triangles), and 10000 collected steps (diamonds). Standard deviations were generally smaller than the size of the respective series' icons, and are thus not shown. The spatial dispersion of one of the five N=10000 sets is superimposed for illustrative purpose.

Figure 48 shows the regression slope leading to D≈1.35 from studying the pattern over a range of grid resolutions. This result should, according to

Eq. 17, be compliant with the expectation z=1-D/2=0.33 from studying the pattern from varying N at a given grid resolution (Figure 47). The actual slope 0.31 on average for the latter was in fact very close to the expectation from a direct estimate of D.

In summary, MRW seems to explain also the remaining empirically derived paradox, by showing spatial fix dispersion with D≈1. Further, a bridge from generic MRW towards memory-less processes (both scale-free and scale-free) was illustrated in Figure 44 for data with memory horizon. Another transition towards classical premises was shown from z = 0.25 for MemRW with scale-specific exploratory steps in Figure 20, which is closer to the classical expectation of a home-range asymptote (z = 0). Transition between MRW and MemRW has also been illustrated by Eq. 17 and Figures 47-48, based on β = 3 in Eq. 13.

7.2.7 Site fidelity and habitat selection

Consider that c in Eq. 14 is expressing strength of site fidelity, and thus "intensity of space use" as its inverse:

$$c = I(\mathrm{N})/\mathrm{N}^z \rightarrow$$

$$1/c = \mathrm{N}^z/I(\mathrm{N}) \tag{18}$$

Consider a memory-less "space-use simplicity" scenario with z≈1 (Figure 44) and letting c' express the inverse of fix density N/I(N):

$$c' = I(\mathrm{N})/\mathrm{N}^z = I(\mathrm{N})/\mathrm{N} \rightarrow 1/c' = \mathrm{N}/I(\mathrm{N}) \qquad | z = 1 \tag{19}$$

Eq. 19 implies that statistical-mechanical conditions for classical "use versus availability" methods to estimate strength of habitat selection (Manly *et al.* 2002) may be derived. The right-hand side of Eq. 19 is expressing density of fixes pr. non-empty grid cell, under the condition that spatial memory is not influencing the dynamics at the actual scales. In other words, Eq. 19 is expressing local density as proportional with 1/c'. Consequently, *a transition from z≈0.5 towards z≈1 will determine the demarcation between complex and traditional analysis of space use.*

Under condition of complex space use, a smaller c in Eq. 14 implies stronger habitat selection strength. Thus, a "characteristic scale of space use," measured by the parameter c, may represent a proxy for the inverse of relative habitat selection strength. This characteristic scale of space use is found in Eq. 18.

7.3 Model extensions

7.3.1 Strength of site fidelity

In summary, three time scales are defined for analysis of a MRW path: the implicit statistical-mechanical compliant interval between successive displacements in simulations (t), the average return interval to a previous location ($t_{ret}=\mu t$), and the observation interval on the movement path (t_{obs}). The latter, which is determined by the observer, may represent GPS locations in real data and may be applied to study the effect from varying strength of homing tendency from the ratio

$$\rho=t_{ret}/t_{obs} \tag{20}$$

For a given t_{obs} Eq. 20 gives the relative strength of site fidelity and thus an indirect estimate of t_{ret} under a premise of MRW-compliant space use (Gautestad and Mysterud 2013). Eq. 20 also defines the transition from autocorrelated to non-autocorrelated step series: $\rho<1$ leads to the latter. Intermediate return events at observational scale t_{obs} tend to remove spatial predictability of the next sampled location.

While constant t_{ret} over space and/or time implies environmental homogeneity, a spatially or temporally varying t_{ret} implies heterogeneity (Gautestad *et al.* 2013, Gautestad and Mysterud 2013). A varying movement speed (expressed by the parameter a in Eq. 13) may also reflect heterogeneity (Gautestad and Mysterud 2010b). If t_{obs} is kept constant and t_{ret} is smaller in one type of habitat than in another (more frequent returns), ρ will also decrease. Thus, transition to non-autocorrelated series and system ergodicity will happen at finer temporal resolutions of path sampling in environments where site fidelity is stronger. In ref. (Gautestad and Mysterud 2013) we interpreted the property of smaller ρ in Eq. 20 as an expression for stronger *space-use intensity* and thus stronger *selection* for this habitat. This aspect will be in focus later in the book, where aspects of MRW in the context of movement ecology are illustrated.

7.3.2 Constraining memory

In an alternative scenario, involving increased temporal heterogeneity and uncertainty of resource availability, a trailing memory window may be defined for simulations. In this case, the model animal may be considered subject to strong unpredictability of local resources outside its present perceptual field. For example, in the context of foraging ungulates, a memory horizon could represent a more or less gradual shift of space use by disregarding affinity to locations that were last visited further back than some time ago. If intrinsic fitness value of long-term returns is diminishing gradually with age of information, it does not pay to return strategically to locations after some time has passed. The basic effects of this kind of memory horizon on MRW output were first described in (Gautestad 2011). However, in generic MRW simulations I use infinite memory, which in practical terms means that the memory horizon is assumed to be larger than the time span for data collection and system analysis.

7.4 The Lagrangian aspect of MRW

7.4.1 Step distribution functions

A "step" in a path-generating algorithm equals the distance from one location (fix) to the next. In a statistical-mechanical context, the distribution function for a large set of such steps provides important information about the underlying process. At this level the actual spatial location for each step is not interesting and thus ignored in analyses. It is the set of steps or fixes that counts. For example, single-step information is lost upon collecting all steps of a given size interval ("bin") when creating the step length distribution. I will now summarize how this aspect of a set of fixes—the distribution function and the estimates of relevant parameters—may tell us quite a lot about animal movement from a statistical-mechanical perspective.

Studying the step length distribution implies a so-called Lagrangian system analysis (Turchin 1998). In the Lagrangian approach one may study the scaling properties of the long-tail part of a binned step length distribution, F(L), of successively observed displacements. The long-tail part is a jargon in theory of movement where the frequencies of steps larger than the median length are of particular interest.

It should be intuitively clear that the functional form of the distribution of step lengths from a MRW compliant set of fixes will depend on ρ in Eq. 20. If fixes are collected at large intervals, leading to ρ approaching or even becoming smaller than one, a power law distribution of step lengths (from Eq. 13) will be destroyed owing to intermediate return events. Thus, if the Lagrangian aspect of complex movement is analysed, ρ should be sufficiently large to allow for depression of site fidelity effects on the step length distribution (this influence—and how to measure and interpret it—will be described in more detail in a later section). Fixes from an individual at observation intervals t_{obs} in the domain of spatial autocorrelation may, for convenience, be called *meso-scales* if the temporal scale for fix sampling is sufficiently large to ensure the statistical-mechanical level. At these meso-scales resolutions, which are represented by the unit time increment t with respect to Eq. 13, the distribution of steps satisfies the properties of a Lévy walk:

$$F(L)=\alpha L^{-\beta} \qquad | L < \text{ʎ}, 1< \beta \qquad <= 3 \qquad (21)$$

Eq. 21 implies a scale-free, i.e., a power law, relationship between one step length bin (length interval) and the next (Viswanathan *et al.* 2000), as measured over the tail part (L is larger than the median step length, implying that smaller steps fall in the first bin). The "truncation scale" ʎ represents the approximate maximum displacement during an interval t_{obs} and typically reflects a maximum movement speed at this observational scale. Direct environmental constraint on step length (e.g., a "valley effect" or territorial borders) can also lead to truncation. If β is stable over a large range of L (for example, ten to a hundred times larger than the smallest class), the distribution satisfies a power law with fractal properties over the given scale range. Assuming a Lévy walk-like path at these temporal resolutions, the fractal dimension of the MRW path, D_{step}, can then be approximated by $D_{step}=\beta-1$ where $1<=\beta<=3$ (Viswanathan *et al.* 2000), as already specified in Eq. 15.

Scale-free distributions like Lévy walk and MRW (under condition of autocorrelated fixes) are typically characterized by $\beta\approx2$. This exponent means that $D_{step}\approx1$; i.e., a distribution that shows high frequency of short steps and also a presence of very long steps—in compliance with Eq. 13. On the other hand, scale-specific processes like Brownian motion and MemRW are characterized by β closer to 3 (D_{step} closer to 2) or typically much larger.

Brownian motion compliance is not scale-free with respect to the step distribution $F(L)$ (Figure 49). This means that the distribution is not satisfied by Eq. 21, since β is an increasing function of L. In this case the algorithm for scale-free space use in Eq. 13 needs to be replaced by an algorithm that generates *a negative exponential*—a Poisson-compliant pattern—rather than a power law distribution. The following algorithm generates a path that satisfies a Brownian motion at meso-scales:

$$L_{BM}=-\text{ʎ}[\ln(RND)] \qquad (22)$$

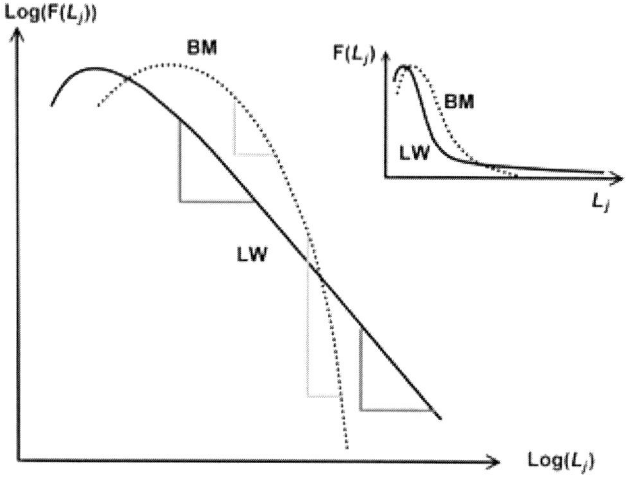

Figure 49. (a) Lévy walk (Lévy walk) and Brownian motion (Brownian motion) may be characterized by the tail part of the distribution F(L_j) of inter-fix steps lengths L, where L_j is a specific length range (the "bin size" of the distribution). In the case of scale-free movement like Lévy walk, a step of length aL (where a>1) is on average expected to happen $1/(aL)^{\beta}$ as often, with $1<\beta<3$. β is expressing how "steeply" the frequency of larger steps occur, relative to any chosen reference length L, over the range of j where β is relatively constant (implying a scale-free distribution over this range of the tail). The slope representing a functional fit of the distribution becomes linear on log-transformed axes as shown in the main graph. Since this ratio relationship between two magnitudes of L is independent of which absolute size of L we choose for comparison (e.g., "meters" in both numerator and denominator cancel each other, leading to a dimensionless number), the movement is scale-free over the range of L_j where β is constant. This is illustrated by the two ratio lengths on the LW line in the log-transformed graph. For a scale-specific kind of movement like Brownian motion (dotted line), the distribution of step lengths is characterized by a slope β that is increasing with increasing j (shown by the two ratio examples) in a manner which in practice leads to absence of very large steps relative to the average length.

Eq. 22 is thus the scale-specific counterpart to scale-free exploratory steps in Eq. 13. RND is again a random number between zero and one. Successive steps are calculated independently, reflecting a statistical-mechanical level of fix sampling. The parameter κ' in Eq. 22 may in statistical-mechanical terms be interpreted as $1/\kappa$ in Eq. 21; in other words reflecting the inverse of the average step length truncation scale for the moving animal.

In *2.2 From mechanistic to statistical-mechanical modelling,* this truncation scale was defined as the system's mean free path property, so a larger λ means a smaller λ' and thus a larger median step length in the actual Brownian motion path at implicit temporal resolution *t*. In the case of Brownian motion, extremely long steps are practically absent; there is an expectancy of a negative exponential compliant probability distribution for step lengths—a Poisson-compliant distribution—owing to the infinitesimal chance to avoid step truncation much larger than λ:

$$F(L_j)= \lambda' e^{-\lambda' Lj} \qquad (23a)$$

where λ' is inversely proportional to the average step length and the index *j* indicates bin sequence (larger *j* for larger step classes). Under log transformation this becomes

$$\log[F(L_j)]=-\lambda' L_j+\log(\lambda') \qquad (23b)$$

This kind of exponential tail is less "fat"—far less elongated towards large step sizes—than a power law tail with $\beta<3$. While a power law function as in Eq. 21 leads to a linear form of the tail part of the step length distribution in a double-log plot (both axes log-transformed), the negative exponential in Eq. 23a is linear in a semi-log plot (log-transforming the y axis only), as shown by Eq. 23b.

Figure 49 shows the distinction between scale-free and scale-specific movement. However, while MRW is also scale-free and thus similar to Lévy walk in the autocorrelated domain of t_{obs}, the parallel-processing postulate implies a qualitatively different space-use process.

7.4.2 The distribution hump from site fidelity under MRW

MRW is similar to Lévy walk with respect to β and D_{step}, except for a tendency for "hockey stick" pattern: at very large *L* the influence from return steps tend to inflate F(*L*) relative to expectation from a generic Lévy walk distribution. In short, one or more return events during an interval t_{obs} will tend to show a larger displacement than intervals without such returns. This effect may under some conditions create an extra "hump" at the fitted slope of the log-log transformed distribution of step lengths over the largest bins of step lengths (Gautestad and Mysterud 2006, Gautestad 2012).

The details on appearance of such a hump depend on the ratio between return step frequency and observational lag between successive fixes, ρ in

Eq. 20. The hockey stick form of the distribution tail strengthens as the ratio ρ is increasing beyond one. This is due to a weaker influence from return events when the series is moving into the domain of spatial autocorrelation. A supplementary file to ref. (Gautestad *et al.* 2013) explains the hump effect in more detail, but a visual illustration is provided in Figure 50 [reproduced from ref. (Gautestad and Mysterud 2013)].

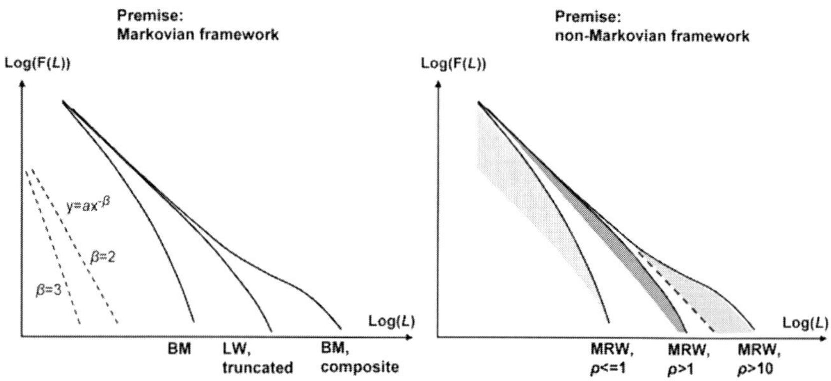

Figure 50. Depending on the process premise for interpretation of step length distributions, patterns showing $\beta>3$ may either verify Brownian motion (BM) under the Markov premise or MRW with $\rho\ll1$ under the non-Markovian (parallel processing) model premise. This principle is conceptually illustrated here for three typical variants of the F(L) function under log-transformation of both axes. The shaded areas indicate how the influence from return steps in a MRW may typically inflate F(L_j) under three strengths of the site fidelity ratio $\rho=t_{ret}/t_{obs}$. The dotted line for $\rho>1$ indicates the expectation either from very large ρ (leading to "pseudo-Lévy walk"), or from MRW under limited memory horizon (approaching the Markov condition, and thus expressing Lévy walk-compliant movement).

Figure 50 also summarizes the functional form of the distribution expectancies under the two system assumptions (a) Markovian space use and (b) parallel processing. The respective models that may explain the respective lines are mentioned below the x-axes. The reference to composite random walk is described as an alternative interpretation of a hump under MRW conditions, and this model will be described in the following chapter. The main additional information from Figure 50 is derived from simulations of MRW under variable site fidelity strength as expressed by ρ in Eq. 20. Large return intervals from weak site fidelity or high observation intensity relative to a given strength of site fidelity (large t_{ret} relative to t_{obs}) both lead to a power law distribution of step lengths with $\beta\approx2$ under generic conditions. An example is shown in Figure 51. However, as ρ is reduced from a very large ratio towards a smaller ratio but still larger than one—approximately in the interval $10 < \rho < 50$—a hump is typically seen in the distribution (coloured blue in Figure 50). As ρ is reduced further towards $\rho=1$, the hump becomes more broadly distributed over the range of step lengths, leading both to an inflated estimate of the power exponent β and less compliance with a power law (the slope also becomes less linear). For $\rho < 1$ the distribution may even resemble a scale-specific process like Brownian motion. Thus, the heated debate lately whether movement paths are verifying Lévy walk or not (Edwards *et al.* 2007, Humphries *et al.* 2010, Reynolds 2010, Humphries *et al.* 2012, Sims and Humphries 2012) may be based on an erroneous premise of a spatial memory-less process. Under this premise, Brownian motion and Lévy walk are the competing hypotheses, while movement under influence of spatial memory has generally not been considered. As shown in Figure 50 (right pane) variations in ρ between respective empirical results may easily lead to confusion that cannot be resolved by even the most sophisticated statistical protocols. Figure 52, reproduced from ref. (Gautestad and Mysterud 2013), illustrates the importance of accounting for the ratio between average return events and the actual observation interval when estimating compliance with scale-free versus scale-specific movement—given that a MRW process is simulating space use in a realistic manner.

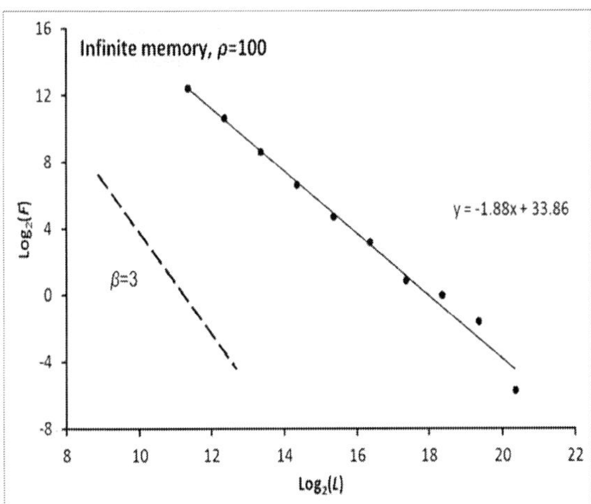

Figure 51. Path sampling of MRW under condition of very large intervals relative to average return interval brings the step length distribution in close compliance with standard Lévy walk.

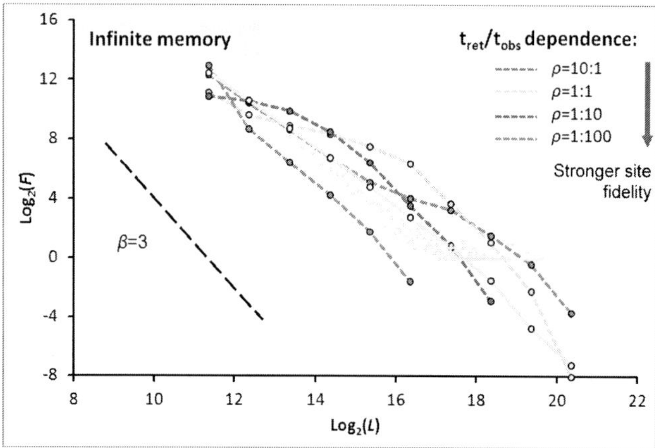

Figure 52. Depending on the process premise for interpretation of step length distributions, patterns showing $\beta > 3$ may either verify BM under the Markov premise or MRW with $\rho \ll 1$ under the non-Markovian model premise. Simulations with $\rho = 10$ and narrow memory horizon: estimated β is close to the true $\beta = 2.0$, owing to Lévy walk look-alike conditions. With $\rho = 10$ and memory horizon increased to 10,000 increments (line with "hump") a LW-like pattern with $\beta \approx 2.0$ is still apparent. Increased truncation is seen under reduced t_{ret} ($\rho <=1$). For $\rho = 0.01$ the slope looks like strongly truncated Lévy walk or Brownian motion (β close to 3).

To conclude, owing to the intrinsic system issues described in Figures 50-52, estimating the power exponent or testing for compliance with a power law or not is futile if (a) spatial memory effects are not accounted for or (b) memory is shown not to be a factor for the given individuals under the given conditions. How to test for these conditions will be outlined in the following sections.

7.4.3 Lagrangian and Eulerian system aspects

There is a close relationship between the Lagrangian parameter D_{step} of Lévy walk and the complementary Eulerian parameter D, which describes the scaling properties of the spatial scatter of fixes over the landscape (2001, Gautestad and Mysterud 2005, 2006). Thus

$$D_{step} \approx D \qquad (24)$$

Lévy walk is a movement process without long-term spatial memory, so space-use constraint is not expected in an open environment owing to lack of directed return steps. Further, step truncation at length scale λ (Eq. 21) can be expected to take place at fine scales—wit h a disrupted power law as a result. Hence, in this case the truncation level λ may, for example, reflect the average distance to a resource patch. Under strong step length truncation (relatively small λ), Eq. 21 will show a log-log linear relationship only over a narrow range of L, with steeper slope beyond this range. Further, Lévy walk does not show homing towards previous locations, and this property can be applied to distinguish Lévy walk from MRW in real data when estimates of D_{step} from Eq. 21 are found to be non-conclusive.

7.4.4 Truncation and mean free path

Truncation effects, which are scale-specific by nature, may interfere with the ideal power law property of a scale-free process also in more subtle ways, as for Lévy walk and MRW. For example, both movement speed (displacement pr. GPS fix interval) and the environment (e.g., a valley with steep slopes hindering free movement) may constrain the frequency of extreme step lengths, leading to cut-offs and thus a steeper slope (larger β) towards the larger classes of step lengths. The larger β may then lead to the impression of a larger D (larger β) than the underlying intrinsic process relative to how it would be expressed in the absence of physically or environmentally imposed truncation. In ref. (Gautestad and Mysterud 2013) we proposed the following supplementary and novel test to cast light on this issue.

Since power law distributions of step lengths are characterized by a sample-size (N) dependent second moment (non-stationary distribution) (Bouchaud and Georges 1990), increasing N may from this property be expected to show an increased influence from truncation with increasing N. In other words, an intrinsically *scale-free* pattern becomes subject to an increasing influence from *scale-specific* physical or environmental factors when the tail part of the distribution increases as a function of N. Conversely, a truncation effect on β should be expected to diminish when reducing N. Thus, by splitting a large-N series into sections containing smaller N and averaging over the sub-samples, a better interpretation of non-linearity in double-log plots of step lengths may be achieved if inflation of the exponent β with increasing L in Eq. 21 (a truncation effect) is linked to environmental constraint. This supplementary analysis was performed on GPS fixes from red deer data in ref. (Gautestad and Mysterud 2013).

7.5 Distinguishing MRW from composite Brownian motion

7.5.1 What is composite Brownian motion?

In Figure 50 (left pane) it was a reference to composite Brownian motion, which—by fine-tuning of parameters—may produce a step length distribution that may look like a power law process (Benhamou 2007, 2008). Before explaining this model construction in detail, let us first step through the following résumé of its counterpart, MRW.

Consider the movement path of a foraging mammal. After bouncing back and forth on a relatively local scale for a while, the animal suddenly takes a more directional and long-lasting displacement, bringing it to a relatively distant location within a few time increments. What has happened? From the perspective of small time increments, an observer will typically look for local causes: was the food patch depleted to a critical level, *sensu* the marginal value theorem (Charnov 1976)? Or had a predator suddenly emerged in the neighbourhood? From a coarser temporal perspective, was the individual just performing one of its more *strategic* moves (from the finer-scaled point of view), which just "waited to happen"—with no clear local causation whatsoever? However, the causation might have been easier to reveal from a coarser spatio-temporal level of analysis, where the finer-grained myriad of interactions could be averaged out as background noise. On the other hand, fine-grained mechanics also generate seeds for coarser-grained behaviour. Hence, the real challenge for a coherent theory regards this kind of multiple-scale complexity where processes at various scales interact in a dual-directional manner. Here I have termed this kind of complexity *parallel processing*. A power law pattern is indicative of an intrinsically driven scale-free process (Mandelbrot 1983), generating a distribution of step lengths as described by Eq. 21 whether memory-influence or not (but with ρ-complications under memory effects). To what extent Eq. 21 represents pattern of real animal movement has thus been surrounded by controversy, as reviewed by Viswanatan et al. (2011) and already referred to in connection with Figures 50-52.

The scaling controversy related to Brownian motion versus Lévy walk has been further enhanced by simulation results showing that *superposition of scale-specific processes*—for example, movement consisting of a mixture of two Brownian motion components with different scale constant and frequency of occurrence—may generate a Lévy walk-like pattern under specific boundary conditions, and hence appear statistically as a scale-free process (Benhamou 2007, 2008, Plank and James 2008, Reynolds 2008). The left pane in Figure 50 showed an example. Different scale constants imply different diffusion parameters for the respective components of the walk.

For a two-level composite, one of the components may be tuned towards frequent turns (fine-scaled habitat utilization) while the other produces longer step lengths that are executed less frequently (inter-patch motion). Fine-tuning the respective Brownian motion parameters has been shown to generate a Lévy walk-like step length distribution. In other words, while each of the processes in isolation produces a step length distribution characterized by a negative exponential function like Eq. 23, the superposition series shows a power law-like distribution (Eq. 21).

7.5.2 Applying the parallel shift method to verify scale-free movement

At fine temporal scales it is then difficult to distinguish a true Lévy walk or MRW pattern from a composite scale-specific process in disguise, but when the series are sampled at coarser lags than the model execution scale, the distribution swiftly loses its Lévy walk appearance and turns into a distribution with a negative exponential tail (Gautestad 2013a). However, the composite Brownian motion may consist of a mixture of many elements, covering landscape utilization at many resolutions. Thus, the power law look-alike issue from *tuning* of scale constants and parameters of a composite Brownian motion creates serious challenges for statistical verification of a true scale-free process for real data. In an analogy, a polygon may be made increasingly circle-like by adding a number of line segments and keeping inter-segment angles constant. However, in qualitative terms a polygon and a circle are still different.

The challenge adds to the fact that a power law compliance from a purely statistical perspective is notoriously difficult to verify due to influence of step length truncation (Edwards 2011), as expressed by $\varkappa$ in Eq. 21 (Lévy walk and MRW) and $\varkappa' = 1/\varkappa$ in Eq. 23 (Brownian motion). This truncation is a function of the mean path constraint under the Markov premise, and influenced by the maximum movement speed during the chosen time intervals for observation, t_{obs}. However, simulation results which I presented in ref. (Gautestad 2012) illustrated how this "power law in disguise" paradox from a composite Brownian motion may be solved by shifting the focus from a purely statistical analysis of step lengths to a more process-oriented kind of hypothesis testing, derived within the framework of statistical mechanics. The test method was termed *parallel shift*.

First some background for the parallel shift method. A scale-free distribution of step lengths from a scale-free process (Eq. 21) can—as already described—be constructed by fine-tuning a superposition of two Brownian motion processes with different scaling constants (a proxy for mean step length) and relative frequencies of occurrence. This composite Brownian motion approach from randomly mixing bouts of scale-specific movement at different scales is feasibly described as intra-patch movement mixed with less frequent inter-patch movement (Johnson *et al.* 2002, Benhamou 2007). Plank and Codling (2009) extended Benhamou's (2007) exemplification of Lévy look-alike processes by exploring a broader range of scale constants and conditions for the scale-specific components under a range of sampling lags on the generated series. They found that a strong directional persistence (correlated random walk) for the coarser-scale process in combination with low frequency of appearance of this component relative to the finer-scaled process led to a Lévy walk-like pattern. However, their methodology was criticized by Auger-Méthé (2011) [but see Plank and Codling (2011)]. In short, these approaches are stretching the Brownian motion paradigm under classical statistical mechanics to embed scale-free patterns by adding more Markov-compliant model components under the premise of *a superposition of scale-specific space use*.

On the other hand, a power law kind of pattern may be explored in the form of an explicitly postulated scale-free process, MRW, which extends the Lévy walk model by adding spatial memory. Scale-free space use—mimicked by a Lévy walk component in simulations (Eq. 13)—is supplemented by homing in the form of occasional returns to previously visited locations. The result is a self-organized home range (an emergent property of site fidelity) with fractal properties (Gautestad and Mysterud 2010b). The scaling and memory aspects are implemented under a postulate of parallel processing over a continuum of process rates rather than assuming Markov compliance at a specific temporal unit scale t for the simulations (Gautestad and Mysterud 2010a, Gautestad 2011). Thus, power law compliance is explained as *an emergent property from a complex cognitive process*, which may be mimicked in statistical mechanical terms by the MRW model design.

In short, three qualitatively different space-use processes—Markov-compliant and scale-specific composite Brownian motion, scale-free Lévy walk (time domain only), and memory-enhanced and scale-free MRW (over time and space scales)—may all explain the statistical power law pattern in Eq. 21. Hence, this field of research is currently in a state of confusion and controversy. Not only has the traditional approach towards testing for log-log linearity based on Eq. 21 (regression analysis) been questioned and more sophisticated statistical methods advocated (in particular, the Akaike model approach and the maximum likelihood method), but even these advanced approaches have been cast in doubt for some parameter range of β (Plank and Codling 2009). Thus, some alternative protocols for distinguishing between these processes from a statistical mechanical perspective, by studying fractal properties of space use, have been proposed (Gautestad and Mysterud 2010a, 2012).

In ref. (Gautestad 2012) an additional statistical-mechanically inspired protocol—as opposed to a more pattern-focusing statistical approach—was presented for the first time and illustrated by simulation results. This parallel shift method was based on analysis of the traditional double-log transformed distribution of step lengths and testing for power law compliance using linear regression, by some considered "obsolete" in the light of the maximum likelihood approach (Edwards 2008, Plank and Codling 2011). However, a supplementary aspect of this linear regression approach was invoked. Rather than struggling statistically to estimate the degree of (non-)linearity and the power law parameter β, the regression line's y-intercept and its dependence on observation lag t_{obs} were applied to reveal compliance with scale-specific or scale-free space use. While both correlated random walk and Lévy walk can show "super-diffusion," i.e., by satisfying that the root mean square deviation from a path's starting point expanding more rapidly than proportionally with square root of time, only a Lévy-like or MRW process (i.e., truly scale-free) will maintain this property over a substantial range of time scales. To visualize and analyse this distinction, a double-log plot where the step length distribution from several observational lags were superimposed was applied to estimate the behaviour of the scatter plot and the regression line (linearity and intercept). This statistical distinction not only provided a novel approach to test for scale-free space use, but pointed towards the deeper qualitative difference between Markovian- and non-Markovian-based statistical mechanics.

7.5.3 Simulating multi-level composite Brownian motion

To illustrate application of the parallel shift test, a Lévy walk "look-alike" was simulated as a superposition of four movement components (composite

Brownian motion with four scale levels), representing inter-independent (!) spatial scale levels of space use; j=1,2,3 and 4, with increasing characteristic step length λ_j^{-1} and decreasing relative frequency of occurrence (relative weights $0<w_j<1$). Successive steps of length L_{BM} were generated from $L_{BM(j)}=-\lambda'_j\ln(RND)$ (see Eq. 22, where j=1). For each of the respective levels j, this algorithm leads to a statistical step length distribution $F(L)=\partial e^{-\partial L}$ where $1/\partial$ is the mean movement length (Reynolds 2008). In other words, in a statistical-mechanical context, $1/\partial_j=\lambda'_j$ represents the mean free path of the moving object at spatial scale j and temporal scale t. A higher frequency of short-range steps at the expense of long ones is reflected in a larger ∂. In contrast, the power law function which defines Lévy walk does not contain the scale-specificity defined by $1/\partial$ and λ'.

Successive steps L_{BM} were randomly performed in compliance with scale level 1-4, with relative weights w_j (probability of occurrence), with proportionally smaller w_j for larger λ'_j (the product λ'_j*w_j=constant). Hence, the algorithm followed the superposition principle (see *2.5 Markov from another angle: the principle of superposition*). Under this premise the individual during a unit step interval t is postulated to obey the rule for one specific level only (as expressed by λ'_j), in order to obey the Markovian property. For example, when the individual is traversing in an inter-patch mode—by moving more persistently and unidirectionally in comparison to finer-level intra-patch foraging—it follows this coarser level's characteristic perception of its environment. Thus, *potential direction-influencing inputs at finer resolutions are filtered away*. In other words, to mechanistically mimic a difference in mean free path between the levels within the Brownian motion paradigm, the animal is assumed to turn a blind eye to environmental input at finer spatial resolutions than the level j characterizing the current step! If a foraging individual opportunistically interrupts its current high-j level step to take advantage of a smaller-grained food patch, the composite Brownian motion structure would not be maintained. Hence, a superposition of independent Brownian motion components puts strong constraints on the biological-ecological interpretation of the processing algorithm.

A Lévy walk (or MRW) process implies that the scale-free part of the distribution of step lengths originates from intrinsic (i.e., cognitive) causes alone, and not from a combination of the cognitive process (flipping between movement "moods" and environmental forcing, expressed by the product λ'_j*w_j and the respective resulting mean free paths) under the composite Brownian motion design.

In ref. (Gautestad 2012) a series of 26,563 steps, on average 20,000, were set to take place at level j=1, 5,000 took place at level j=2, 1,250 at level j=3, and 313 at level j=4 (in random order during step iteration, with κ'_1=100, κ'_2=400, κ'_3=1600, and κ'_4=6400; i.e., forced power law scaling from boundary conditions ($\kappa'_j * w_j$=constant). One series consisting of N=T=26563 steps was simulated, and all N steps were collected for analysis; t_{obs}=t=1 (unit lag). Three additional series were generated and collected at lag t_{obs}=10, 100, and 1,000, respectively, with a similarly increasing series length factor to compensate for sub-sampling. These series represented T increased by a factor of 10, 100, and 1,000, but with a constant N. This approach, by analysing space-use pattern over a range of observational scale, $1 <= t_{obs} <= 1000$, will influence specific parameter estimates. This variation may allow for a simple protocol to distinguish a look-alike scale-free process (composite Brownian motion) from a true scale-free process.

Four separate Brownian motion series of length N_{ind1}=20000, N_{ind2}=5000, N_{ind3}=1250, and N_{ind4}=313 steps were also run. The respective scaling constants were similar to the four-level composite Brownian motions, κ'_{ind1}=100, κ'_{ind2}=400, κ'_{ind3}=1600, and κ'_{ind4}=6400. These separate single-level series were produced to compare space use by a population of independently moving individuals where the Lévy walk-like pattern is apparent at the population level only (by pooling the set of movement lengths from several individuals, each showing classical Brownian motion in isolation but responding to direction-influencing interrupts at different scales). The total time period T was defined to be equal for the four individuals despite a difference in N, implying a difference in relative unit time due to difference in relative step frequency [see Appendix S1 in (Gautestad 2012)]. All individuals were postulated to move with the same speed, but due to a postulated longer interval without directional interrupt for the coarser-graining individuals, they will on average show a larger net displacement per unit time than their finer-scale responding conspecifics.

7.5.4 Simulating scale-free processes

Lévy walk was simulated as a set of successively independent steps with length $L_{\text{Lévy walk}}=a(\text{RND})^{-1/(\beta-1)}$ with α=1 and β=2 (see Eq. 13). It was implicitly assumed that during a given step at unit lag t, a sufficiently number of (virtual) micro-steps at finer spatio-temporal resolutions had taken place to ensure (a) an absence of inter-step directional persistence even for high-frequency fix sampling at t_{obs}=t, and (b) a Lévy-stable distribution of step lengths (Mandelbrot 1983). These two assumptions imply that a simulated path was conjectured to represent coarse-graining to the statistical-mechanical level of system abstraction even at the simulation's temporal execution scale. Four series were run at $t=t_{obs}=1$, and additionally sampled at lag $t_{obs}=10$, 100, and 1,000, respectively, for a final sample size of 20,000 steps from each series (thus, the fourth series was run for 20 million steps to compensate for larger t_{obs}). A maximum step length was set to 40,000 length units at t=1, to mimic a simplified step truncation effect.

MRW was simulated and sampled under the same conditions as for Lévy walk, but extended with its algorithm expanded with a rule for spatial memory-dependent strategic return steps: on average at every 100th time increment ($t_{ret}=100t$), the individual jumped back to a randomly chosen previous location in the series.

When generating successive steps from a movement process at coarse lags (the assumption behind the present simulations) a Lévy walk pattern is expected to be similar to a Lévy flight pattern due to self-similarity, and thus allowing for Eq. 21. However, influence of step length truncation is expected and should be accounted for (Reynolds and Rhodes 2009). By assuming that the simulations' execution scale reflects coarse-grained lags, one also circumvents the chronically difficult issue of defining "move length" (between successive reorientations) at the mechanistic level t of the true movement process. What is a reorientation event? Reynolds & Rhodes (2009) and others have proposed various rules of thumb when analysing real data, but these remain subjective. By defining the simulations' execution scale to represent a coarser statistical-mechanical level, a given step distance is normally the result from a sequence of what in mechanistic terms would have been defined as independent, successive moves. Thus, at this coarser scale successive step directions become randomized and uniformly distributed over 0-2pi. This principle was applied to the simulations to allow for absence of directional persistence between successive steps.

7.5.5 Simulation results of scale-specific versus scale-free processes

With reference to (Gautestad 2012), Figure 53 verifies that a four-component superposition of Brownian motion under the chosen conditions generates a Lévy walk look-alike distribution of step lengths. After double-log transformation, the slope is approximately linear with a slope close to -2.2, which satisfies $\beta \approx 2.2$ in a Lévy walk parameterization (Eq. 21). As shown, the power law pattern is very resilient to coarse-graining over an observational range $1 <= t_{obs} <= 1000$. The linear regression slopes for the various levels of t_{obs} right-shifted by a factor of ca. $\log_2(\sqrt{10}) \approx \log_2(3.2) = 1.7$ when t_{obs} is increased by a factor $n=10$. In other words, when the inter-observation interval is increased n-fold, the animal on average performs $\sqrt{n}$ times as large net displacement for all bins in the set $\boldsymbol{L}$. Crucially, this "diffusion rate" regards all four levels in the composite BM, leading to maintenance of the log-log linear relationship in the superposition series. This diffusion rate (as observed at a given scale t_{obs}) is compliant with a classical Brownian motion-compliant process, a "Fickian diffusion." In other words, the mean square net step length for all bin classes of L, $\boldsymbol{L}^2$, expands proportionally with time, despite Lévy walk-like (rather than a negative exponential) distribution of step lengths at any given observational scale t_{obs}. Figure 53 extends the illustration of this generic and well-known statistical mechanical law of diffusion, by showing that an individual moving in accordance to a superposition of independent Brownian motion components follows the same law, under a premise of maintenance of the relative magnitudes Λ_j and relative frequency of occurrence w_j during T. Thus, the overall impression of a power law step length distribution is also maintained.

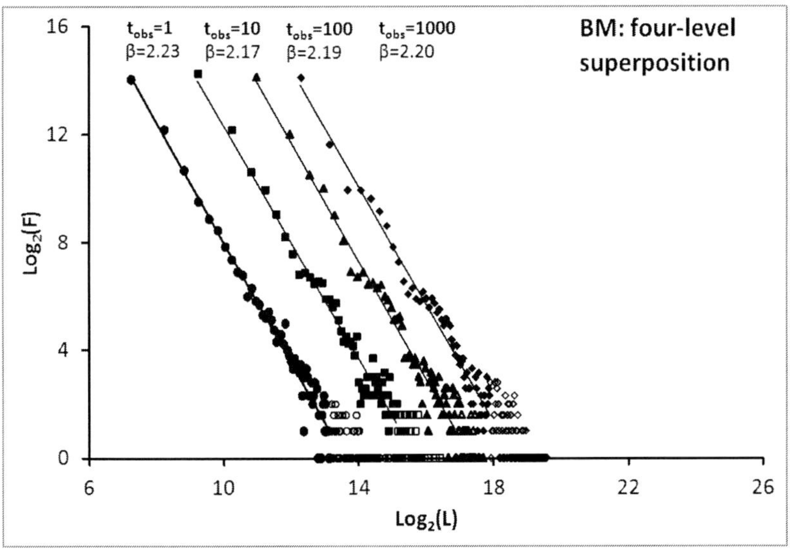

Figure 53. The parallel shift test. Step length distribution of composite Brownian motion, fine-tuned to satisfy a Lévy walk look-alike. However, sub-sampling the path at different time resolutions reveals the process' intrinsic property of scale-specific movement, albeit flipping between respective characteristic movement scales over the total sampling time. See main text for details.

Figure 54 shows how a Lévy walk process [truncated at length scale $\log_2(40{,}000)=15.3$ length units at execution scale t=1] also right-shifts its step length distribution under a coarser observational lag. However, contrary to the composite Brownian motion in Figure 53, the diffusion rate for Lévy walk reflects the expectation from a true scale-free movement: net displacement changing proportionally with observational scale t_{obs}, i.e., the set of step lengths $\boldsymbol{L}$ is changing (showing parallel shift in the histogram) proportionally with observational lag. For example, a tenfold increase in t_{obs} leads to a tenfold increase in net displacement for a given step length bin class. With reference to the $\log_2(L)$-axis in Figure 54, $\log_2(10)=3.3$. This diffusion rate, mean square deviation ($\boldsymbol{L}^2$ in the present context) increasing proportionally with time squared, characterizes Lévy walk in general (i.e., super-diffusion). Figure 53 also illustrates that the step length truncation is not visible for $t_{obs}<1000$, under the given conditions ($\beta=2$ and N=20,000). At the level $t_{obs}=1000$, the truncation becomes apparent as a reduced frequency in bin classes for steps close to and larger than $\log_2(40{,}000)=15.3$, relative to the expectation from a non-constrained LW. The frequency of step lengths larger than 40,000 is not zero at $t_{obs}=1000$, since there is a slight probability for two or more displacements of this magnitude among any given intermediate sequence of 999 steps (executed at scale t) between each sampled location.

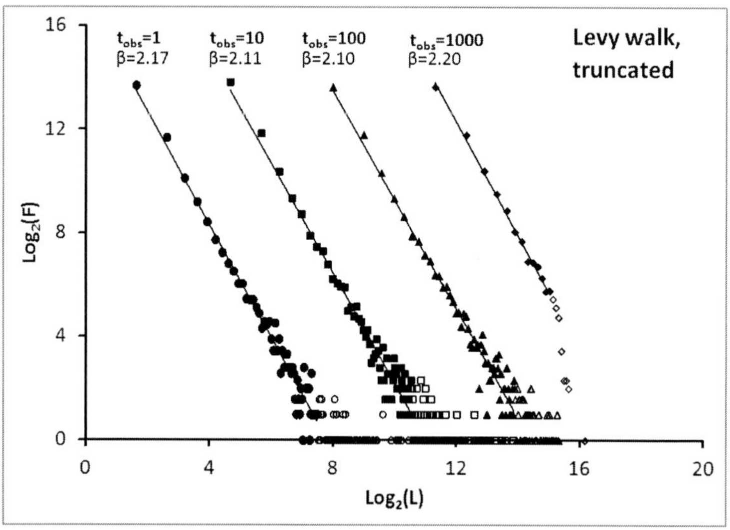

Figure 54. The parallel shift test performed on true scale-free movement (parallel processing from truncated Lévy walk). See main text.

Figure 55 illustrates a scale-free process under the influence of long-term memory (MRW); a truncated Lévy walk-like space-use process with a superposition of return events to a previous location at frequency 1:100 relative to execution scale t=1. For observational levels t_{obs}=t=1, i.e., t_{obs}<<100 (to avoid step truncation due to return steps), this MRW condition reflects a step length distribution similar to a generic Lévy walk. However, a tendency for a "hockey stick" pattern (Figure 45)—an elevated frequency of large steps relative to the expectation from a power law—is increasing in magnitude with t_{obs} approaching 100t, the defined scale for return steps. As t_{obs} is increased beyond 100t, the step length distribution from the return step process—in combination with the step length truncation—dominates the overall process. The power law aspect of the movement—when expressed through Eq. 21—then becomes "hidden" at finer temporal scales, as previously shown by in ref. (2005) and explained in *7.4.2 The distribution hump from site fidelity under MRW*. Hence, if an animal space-use process is both scale-free and memory-influenced, it is crucial to estimate the level of spatial autocorrelation prior to testing for power law compliance. Unless t_{obs} is chosen substantially smaller than the transition level between spatial autocorrelation and non-autocorrelation (t_{obs}≈100t in the present simulations, with spatially non-autocorrelated fixes at t_{obs}>>100t), the step length distribution may (a) either appear hockey stick-like due to the increasing dominance of the return step component for the large-step classes at these levels of observation, or (b) appear Brownian motion-like (closer fit to a negative exponential, as seen at levels t_{obs}=100 and t_{obs}=1000 in Figure 55), due to influence from step length truncation of both intrinsic ("mood" to return) and extrinsic origin.

Hence, Figure 55 shows a double-interference with the power law pattern—a combination of the influence from return events (a superposition of a return step distribution, which leads to a hockey stick pattern when t_{obs} → 100t) and an influence from a truncation of steps beyond a given size (physical limit to movement speed at scale t, and/or environmental constraint like territorial borders). In combination, these independent factors may lead to either a steeper or shallower slope of the step length distribution for the largest bin classes relative to unconstrained Lévy walk, or they may even more or less cancel each other out at the chosen t_{obs}.

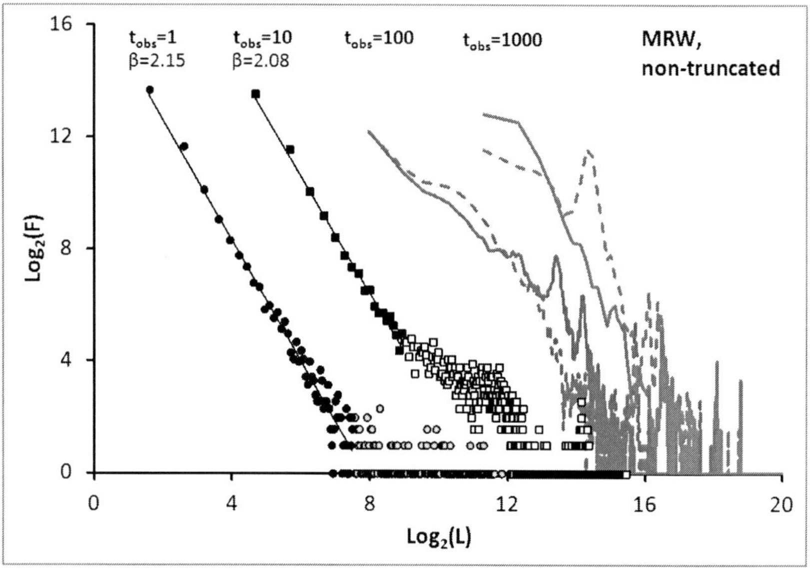

Figure 55. The parallel shift test performed on MRW data, non-truncated step lengths.

Figure 56 illustrates how MRW without influence from step truncation may make the extreme part of the step length distribution more "noisy" in the extreme part of the tail. This jaggedness of F(L) for large L is mainly due to chance effects from the early phase of a given simulation series. Return steps lead to a self-reinforcing use of some locations at the expense of others (i.e., a positive feedback), and the emergence of a multi-modal distribution function of density of fixes (Gautestad and Mysterud 2010b). Initial chance effects with respect to relative distance between the "winning" clusters from the early-phase self-organization of space use will influence the distribution of return step lengths: of the dominant clusters that happen to appear relatively far apart, return steps of length similar to this inter-mode distance will tend to be over-represented (Gautestad, unpublished results). As shown in Figure 56, these inter-mode distances will vary stochastically from one simulation sequence to the next due to early "pattern locking" from the positive feedback process, and the peaks in the step length distribution will vary accordingly.

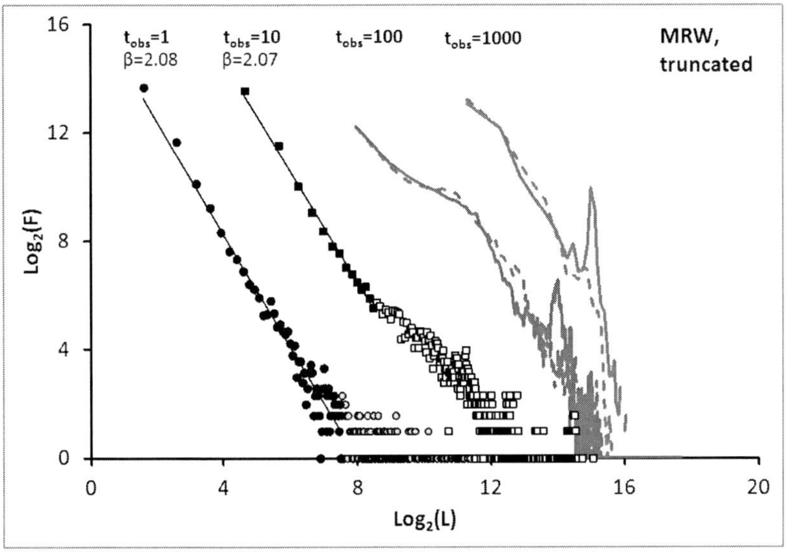

Figure 56. The parallel shift test performed on MRW data, truncated step lengths.

7.6 Testing for parallel processing: red deer GPS fixes

The parallel shift method has recently been tested on real data; a set of eighteen summer season GPS fix series from eighteen female red deer, *Cervus elaphus*, moving in a northern forest ecosystem (Gautestad and Mysterud 2013). Figure 57 is reproduced from this paper, and shows compliance both with a scale-free kind of movement ($\beta \approx 2.2$) and a parallel shift of magnitude in compliance with expectation from the theory above. The parallel shift of $F(L_j)$ of the pooled set of step lengths by changing t_{obs} by factor of 10 was 5.7 (blue arrow in Figure 57), which deviated only 7.5% from the theoretical expectation 6.16 [rms(X) $\approx X^{1/(2.26-1)} = X^{0.79} =$ 6.16], based on the estimated $\beta = 2.26$ from the slopes. This diffusive rate is 80% larger than the expectation for composite Brownian motion; with expectancy $\beta >= 3$ [rms(X) $<\approx X^{1/(3-1)} = X^{0.5} <= 3.16$].

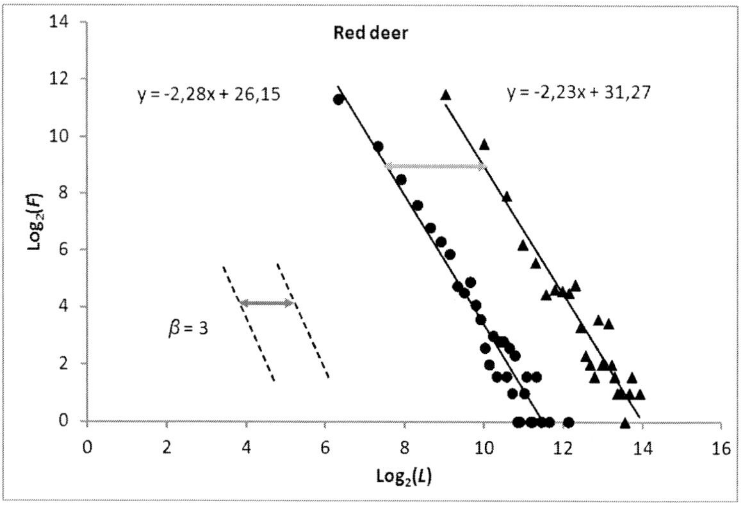

Figure 57. Space use by red deer in Norway recorded by aid of GPS, averaged over 18 individuals. Lagrangian view; measuring parallel shift from increasing t_{obs} by a factor of 10, from 1 hr to 10 hrs. The similar magnitude of β and the magnitude of the parallel shift (upper right arrow) both confirm scale-free movement, which may be LW-like (Markov-compliant) or MRW-like (memory-influenced).

This result confirms super-diffusion, as expected from scale-free movement, which may be caused both by memory-influenced MRW observed from $\rho \gg 1$ (Eq. 20) and by spatially memory-less Lévy walk, as already outlined in previous sections. To distinguish between these two conditions, analysis of I(N) (Figure 58) revealed compliance with MRW, not Lévy walk, owing to slope z close to 0.5 rather than z close to one (according to simulation results in Figure 44). An extended set of fifty-two red deer series gave a similar slope, z=0.43 (Gautestad *et al.* 2013). The observed slope $\beta \approx 2.26$ from step length distributions (Figure 57) should, according to Eq. 15-17 and supplemented with Eq. 24, give $z = 0.37$ from $D \approx \beta - 1 = 2(1-z) \rightarrow z = 0.37$ if D =1.26 as derived from the step length distribution.

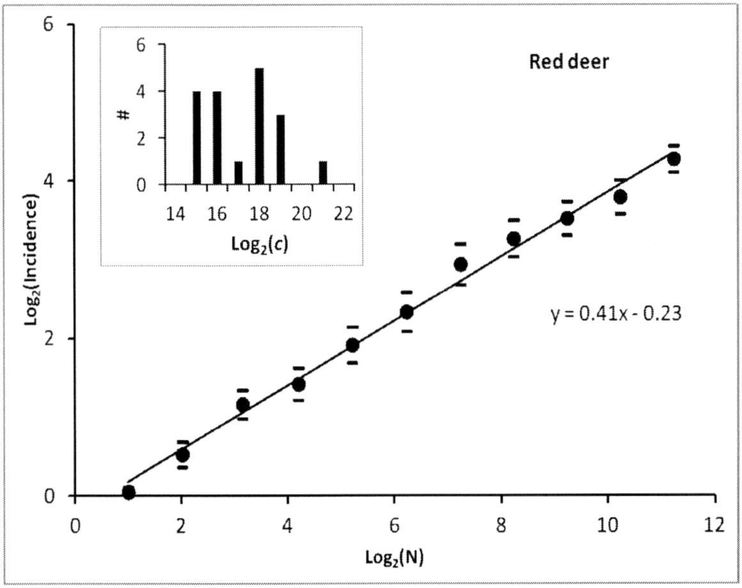

Figure 58. Analysis of space use by red deer confirms scale-free movement in compliance with the MRW model. The inset shows individual variation of the y-intercept, log(c), which is interpreted as the individual's characteristic scale of space use during the given period of sampling and in the given environment.

Figure 59 shows a direct estimate of D from the box counting method. Also this method showed a somewhat larger D than expected D≈1 from the generic parallel-processing model.

The compliance between D derived from the Lagrangian approach (step lengths) and the Eulerian approach (spatial scatter of fixes) is thus very good in this empirical test of parallel processing. Hence, both a somewhat larger β relative to the theoretical optimum β =2 for parallel processing (equal weight of kinetic energy and space use over a scale range), a somewhat smaller parallel shift owing to this larger β, a somewhat larger D from a direct estimate from box counting over grid resolutions, and the somewhat smaller z =0.41 relative to the optimal z=0.5 all comply well theoretically. In short, by analysing data from four angles—step distribution, parallel shift, box counting, and N-dependent space use I(N)—the degree of scale-free movement and direction of deviance from optimum for equal weight over scales are all consistent with theoretical expectations.

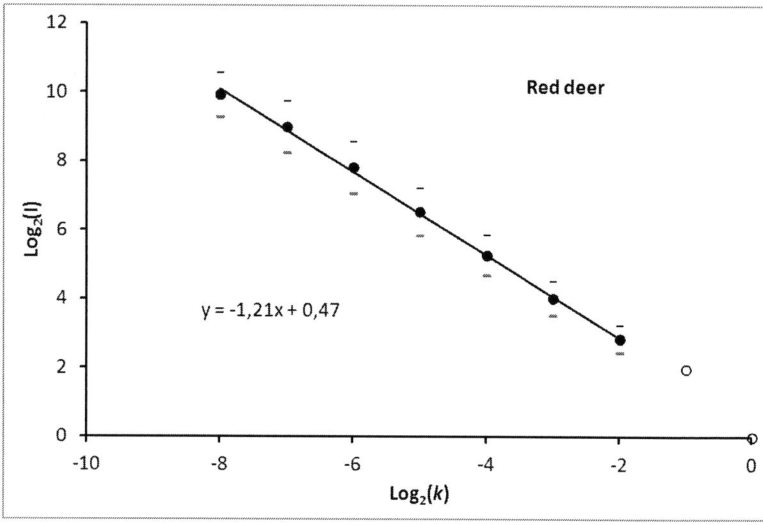

Figure 59. Fractal analysis of the scatter of fixes of red deer movement confirms that space use on average is fractal-compliant (D≈1.2) rather than area-compliant (D≈2).

Figure 58 also shows an inset pane with distribution of magnitudes of the characteristic spatial scale, c , as outlined in Eq. 18 where intensity of scale use was expressed by $1/c$. The red deer individuals showed interesting variability with respect to c, but owing to lack of sufficient GIS representations of local habitat quality at the scale of the respective home range demarcations, we could not estimate at that time to what extent the range of c reflected the respective inverse of habitat quality under the assumption that larger c implies less strongly selected habitat on average within the home ranges.

7.7 Parallel shift and the statistical-mechanical approach

The results in Figures 53-58 lead to four main take-home messages. First, one should pay closer attention to the process behind the statistics. A purely statistical approach, for example, by defining a scale-free process as "*...a Lévy flight is a random walk for which each movement step is independently drawn from a probability distribution that has a heavy power law tail*" and then using this property as the sole criterion for testing (Edwards 2011) may contribute to an elevated level of confusion and controversy in this field. This definition is an excellent description of the expected pattern from the ideal model, but misses the dynamic aspect of the real-life process.

With respect to Figures 53-56, the physical aspect describes the degree of deviation from an ideal power law pattern in the form of truncation of the tail of the step distribution, and underscores the importance of distinguishing a Lévy flight (instant displacements) from a Lévy walk. Only the latter has relevance for animal movement (Reynolds and Rhodes 2009). A Lévy flight differs from a Lévy walk by showing immediate displacement regardless of distance between successive fixes, while a Lévy walk describes a movement process with physical realism in the context of observing animal paths at fixed intervals (Auger-Méthé *et al.* 2011, Viswanathan *et al.* 2011). For example, if an individual has a maximum movement rate of 5 km/h, a scale-free kind of movement *sensu* Eq. 1 will have a step distance of 5 km as the upper limit for successive fixes. This time-dependent aspect of step length truncation, L_{max}, will thus influence the double-log linearity for bin sizes approaching this limit, and supplement the truncation effects from spatial memory (if the scale-free distribution arises from MRW) and environmental border effects (e.g., a fish in a lake). Thus, if Eq. 21 and double-log linearity are used as criteria for testing for power law compliance, the test will produce a false negative if the magnitude of t_{obs} is large enough to impose truncation effect on the observed pattern. Further, as illustrated by Benhamou (2007) for a two-component composite Brownian motion model and in Figure 53 for a four-component model, a superposition of scale-specific processes with inter-independence and toggling between "modes" may produce a Lévy walk in disguise pattern under specific combinations of scale constants. In this case, a false positive may be found from a purely statistical approach. Thus, for a true Lévy walk process, Eq. 21 should always be expected to show strong dependence on t_{obs} with respect to the power law compliant range from L_{min} to L_{max} (Plank and James 2008). A purely statistical approach to the power law aspect does not explain the influence of observational scale on the observed pattern (e.g., the shape-shifting of a MRW step length pattern from Lévy walk-like, to hockey stick-like and finally towards Brownian motion-like by increasing t_{obs}; see Figures 50-52).

Second, confusion from studying the step length distribution from a statistical rather than a statistical-mechanical approach also was shown in Figures 50-52, where the intensity of site fidelity relative to fix sampling interval in a memory-influenced process will produce conflicting results if this aspect is not accounted for. The notoriously difficult statistical verification (or falsification) of a true scale-free space-use process may be simplified by estimating the constancy or variability of the diffusion rate over the set of step length classes L, as a function of observational scale, t_{obs}. The difference expressed between Figure 53 and Figures 54-56 underscores the potential from this approach. The statistical signal from the parallel shift as a function of t_{obs} is supplementing the compliance with true double-log linearity, and may even survive temporal coarse-graining better than linearity compliance. Thus, analysing space-use data (for example, in the form of GPS fixes) at several interval sizes t_{obs} to observe the process over a temporal scale range may provide a clearer distinction between composite Brownian motion (including correlated random walk variants) and Lévy walk through multi-scale estimates of diffusion rates. This approach also provides a test to distinguish between Lévy walk and MRW by looking for hockey stick and truncation in the transition zone between spatially autocorrelated and non-autocorrelated steps.

Third, by analysing the data with different t_{obs} relative to system-specific conditions, two observers may reach very different conclusions with respect to step length compliance with a negative exponential or a power law. Both may in fact be right! In particular, if the animal in question has utilized its habitat under influence of long-term memory, the observed pattern at temporal level t_{obs} may shape-shift from power law, through a hockey stick pattern, to a truncated power law pattern, and ultimately to a negative exponential (Brownian motion compliance) if t_{obs} is chosen large enough. This aspect was illustrated in Figures 50-52. Hence, this paradox—from the perspective of classical Markov conditions—may to some extent be rooted in a relative difference in observational scale between the respective studies. The parallel shift method may contribute to test this assumption.

Fourth, the advice to study both the spatial scaling (distribution of step lengths) and the temporal scaling (a parallel shift of the double-log regression line as a function of t_{obs}) is rooted in an explicit consideration of statistical-mechanical properties of movement. Statistical mechanics—which regards the dual nature of process and statistics from a meso- and macro-scale perspective—is key to a deeper understanding of (and solution to) the paradox of two observers finding qualitatively different patterns in the example in the foregoing paragraph. The statistical-mechanical approach implies paying closer attention to classes of meso- and macro-scale laws that emerge from micro-scale interactions but where these interactions are analysed from the perspective of a coarser observational scale $t_{obs} >> t$. Contrast this with a mechanistic approach, which dominates individual-based modelling (Grimm and Railsback 2005), puts the focus on behavioural mechanisms (algorithmic rule-execution close to level t), and looks for statistical patterns that can be compared with similar analyses of data from real individuals. As an example of the coarser-perspective statistical-mechanical perspective and what it offers as a supplement to the mechanistic analysis, consider a drunken walker—the random walk archetype. In the long run (or walk), walking with a variable step length and randomly chosen direction for successive steps will be found to bring the person a distance from the starting point which (from statistical-mechanical expectation) increases proportionally with the square root of time T from start to the point of relocation. When averaging over many series of walks, a 100 times larger T will bring the walker $\sqrt{T}$ times further away. Then contrast this with a very sober walker, who carefully determines direction and goal for successive steps, based on available information at the given point in time. This information is then churned through an extensive set of "if-then" kind of movement rules, and out comes the step decision. The next step is then taken independently of previous step decisions, in compliance with a Markovian system description. The various rules under the algorithm may be purely deterministic, purely stochastic (like the drunken colleague scenario), or a combination. The well-established fact from statistical mechanics is that—again in the long run—the sober and the drunken walker produce *the same* expectation of root mean square distance after time T. Hence, a walk with a simple rule of random steps follows the same statistical mechanical law as a walk with very detailed and more or less deterministic movement algorithm (Gautestad 2013b)! What they have in common is the Markovian, memory-less processing of successively independent movement-influencing events. Emerging from the hidden world of zillions of micro-scale step (in)decisions at mechanistic level t come the meso- and macro-scale observables at level $t_{obs}>>t$, like the exponential or power law function of step length distribution (telling us about main classes of movement), the overall movement rate (mean square deviation as a function of T, offering important biological and ecological information) and a method to distinguish between a scale-free process like LW or MRW and a look-alike scale free process (composite scale-specific walk).

7.8 Effect from area constraint on model parameters

So far I have presented individual-level simulations under condition of an open environment. How does strong area constraint influence the estimates of fractal dimension (D) and compliance with a power law expansion of area as a function of sample size N, with exponent z? Figure 60 illustrates dispersion of 10,000 spatial fixes from nine independent numeric simulations of an individual's space use under area restriction equal to the square borderlines shown in respective panes. The patterns are generated from Brownian motion [random walk-like, row 1; from choosing β=3 in Eq. 13 (D_{step}=2 according to Eq. 15) and no return events], Lévy walk (row 2; β=2 and no return events), and MRW with return step frequency 1:100t (row 3).

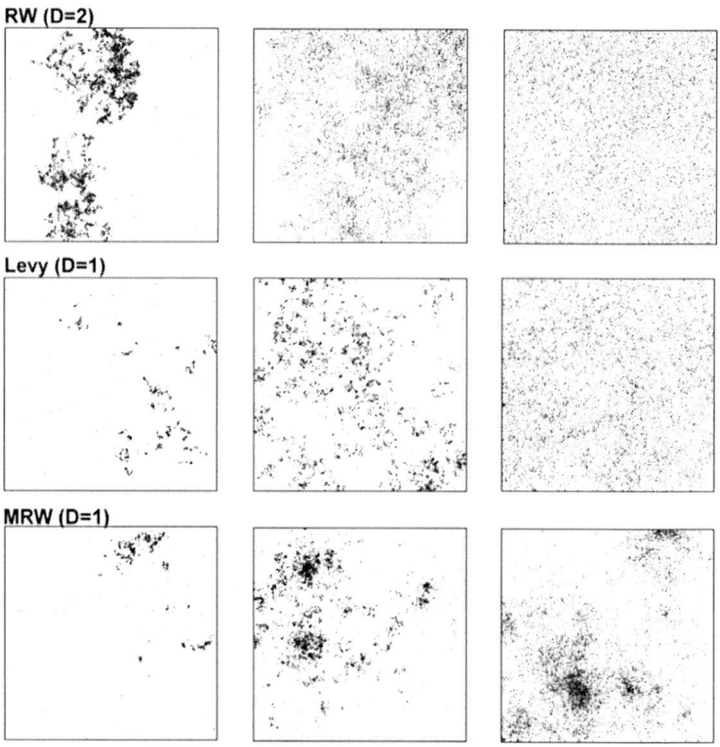

Figure 60. Simulation outputs in limited space from varying movement models. See main text.

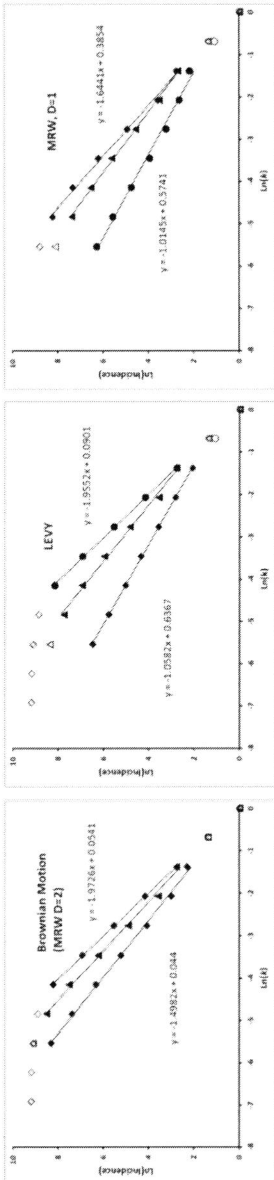

Figure 61. Fractal analysis (box counting) of the patterns in Figure 60.

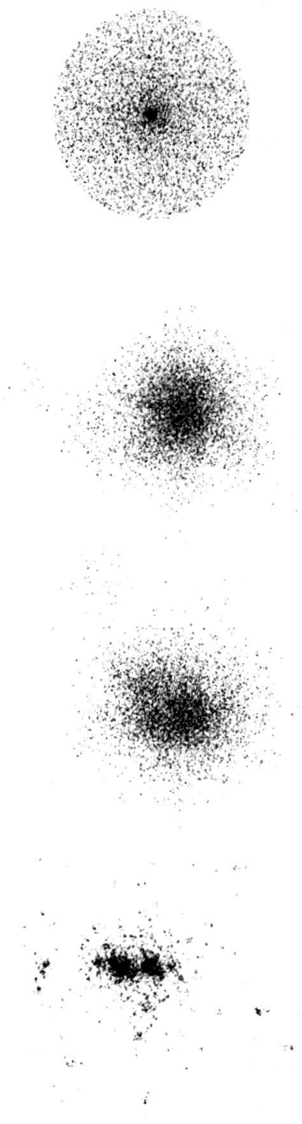

Figure 62. Additional model conditions for movement. See main text.

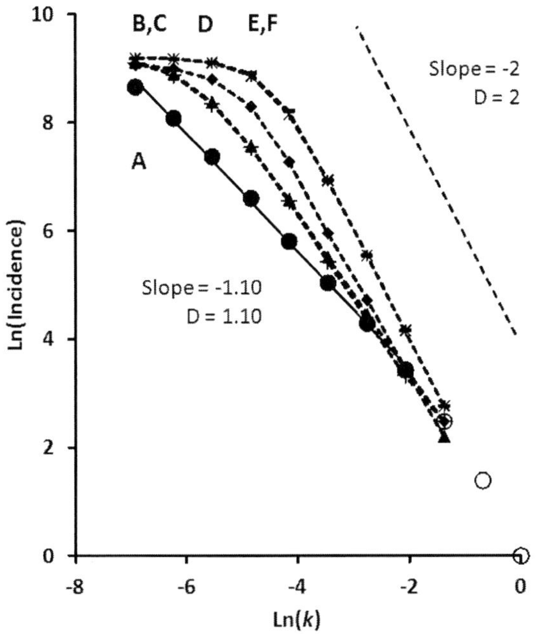

Figure 63. Fractal analysis of the respective space use patterns in Figure 62.

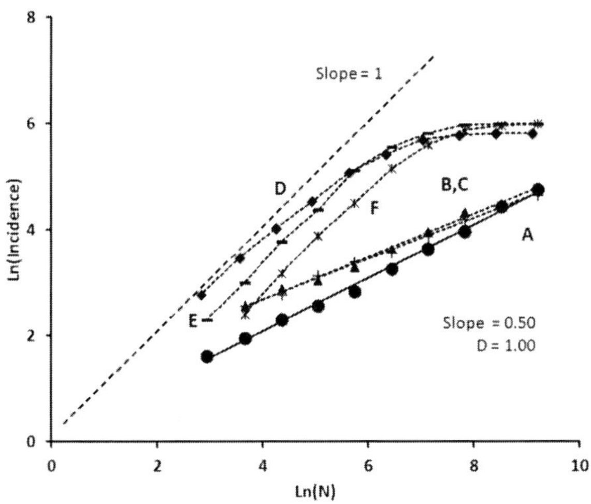

Figure 64. A(N) analysis of the data in Figure 60 and Figure 62.

The generated series were sampled at three time scales: unit time interval for the respective simulations (column 1), ten times larger interval (column 2), and 100 times larger interval (column 3). Proportionally larger series were generated for collecting at larger intervals, to maintain 10,000 fixes in all nine fix samples.

The respective series' fractal dimension, D, can be estimated from the linear least square regression slopes when plots are log-transformed. Figure 61 shows how the fractal dimension is affected by the respective conditions in Figure 60. As expected, the Brownian motion series were not influenced at unit-scale (t_{obs}=1) sampling, since the area constraint was not limiting space use substantially at this time scale. Thus, as shown by the lower regression line in the left pane (log-transformed axes), D=1.5 as expected for Brownian motion in open environment. However, at t_{obs}=10 and t_{obs}=100, the slopes have become successively steeper. t_{obs}=100 shows slope -1.97, which implies D≈2. Thus, as verified visually in Figure 60, space use is filling the area. However, at the finest resolutions the sample size is not sufficiently large to avoid the dilution effect (zero slope), which was defined in a previous section. Lévy walk shows a similar tendency towards larger D as space use becomes more area-constrained; D=1.05 is increasing towards D=1.96. MRW is also influenced by borders at largest fix sampling interval, but to a lesser extent than Lévy walk (D=1.01 increasing to 1.64). The latter result is clearly a result of the slower area expansion of MRW relative to Brownian motion and Lévy walk; with reference to Eq. 14 and Figure 44, z≈0.5 for MRW while z≈1 for Brownian motion and Lévy walk.

A supplementary set of simulations which illustrate four types of area-restricted space use is shown in Figure 62. Also these results are based on 10,000 spatial fixes in a homogeneous environment. From left to right: MRW (standard condition, β=2), two instances of MemRW with exploratory steps from (a) Brownian motion and (b) MRW with β=3, and a random dispersion of locations in compliance with a circular utilization distribution with central ove rweight [the classical home-range model, see ref. (White and Garrot 1990)]. The set of locations were collected from the domain of spatially non-autocorrelated successive fixes (ρ<1; see Eq. 20).

Figure 63 shows the results of estimating D from box counting on these data sets; MRW (**A**: estimated D=1.1), MemRW (**B,C** overlapping: D_{steps}=2, estimated D≈1.5 in the linear range), classical home range (**D**: estimated D=1.8 in the linear scale range), Brownian motion from Figure 60's row 1, column 3 (**E**: estimated spatial D=2.0 in the linear scale range), area-restricted Lévy walk from Figure 60's row 2, column 3 (**F**: estimated spatial D=2.0 in the linear scale range). All series except free-ranging MRW show asymptotic approach towards zero slope at very fine spatial scales (small *k*), due to the dilution effect. However, the general result is an increased D from physical area constraint. D is also increased relative to generic MRW under MemRW conditions when a classical Brownian motion is invoked for simulations.

What about area expansion as a function of sample size? Using the same data as in Figure 60 and Figure 62, Figure 64 shows z=0.5 for MRW decreasing to 0.39 under the two MemRW conditions. According to Eq. 17 this means D=1.0 for MRW (**A**: z=0.5) and D≈1.2 for MemRW (**B, C**: estimated z=0.39 and 0.41). The other series—**D**, **E,** and **F** for classical home-range, Brownian motion under area constraint and Lévy walk under similar constraint—are all dominated by the dilution effect over a large range of N (z≈1, D≈0), and an asymptotic approach towards z=0 (D=2).

To summarize, Figures 60-64 show one main result relative to previous simulations: area constraint on space use tends to decrease the depth of the scale-free dispersion of fixes. N fixes are filling available space—owing to the path being *extrinsically* (environmentally) forced to return rather than self-organizing themselves into a fractal pattern from *intrinsic* forcing (memory-based returns). Under classical conditions, area constraint is based on an assumption about territorial borders or some other kind of unspecified influence on otherwise unconstrained movement, owing to lack of memory under Brownian motion conditions. Without area constraint from extrinsic forcing, space use will become larger. This means that a given grid cell size *k*' will appear smaller relative to the final space-use extent from N fixes, as N is progressing. Thus, the scale ratio (extent/*k*') in an open environment is then becoming successively larger with increasing N, and will show compliance with undistorted emergence of the fractal pattern of fixes under respective conditions. Under area limitation, a movement path is repeatedly folding back onto previously utilized space, meaning a denser space utilization and thus a larger D.

7.9 Density distribution of fixes

The distribution F of steps of length L for a Lévy walk with $D_{step} = \beta - 1 = 1$ is characterized by $F(L) \propto L^{-2}$ (Viswanathan *et al.* 1999) (Eq. 21). Thus

$$F(L^2) \propto L^{-1} \qquad (25)$$

In other words, the relative number of displacements to a k times larger neighbour cell (size changing proportionally with step length squared) at the next higher (logarithmically scaling) level k+1 is 1/k as frequent (Gautestad and Mysterud 2005, 2006). As already described, Eq. 21 and thus Eq. 25 are expected to apply also for MRW. If return steps are substantially larger than the fix sampling interval (ρ>>1, see Eq. 20 and Figure 51), the power law distribution is expected to be maintained with relatively little influence from return events. Eq. 25 was illustrated in the lower right histogram of Fig. 23, where all bars have the same area, but the next bar (magnitude class for L^2) has k times larger width and 1/k times as large height.

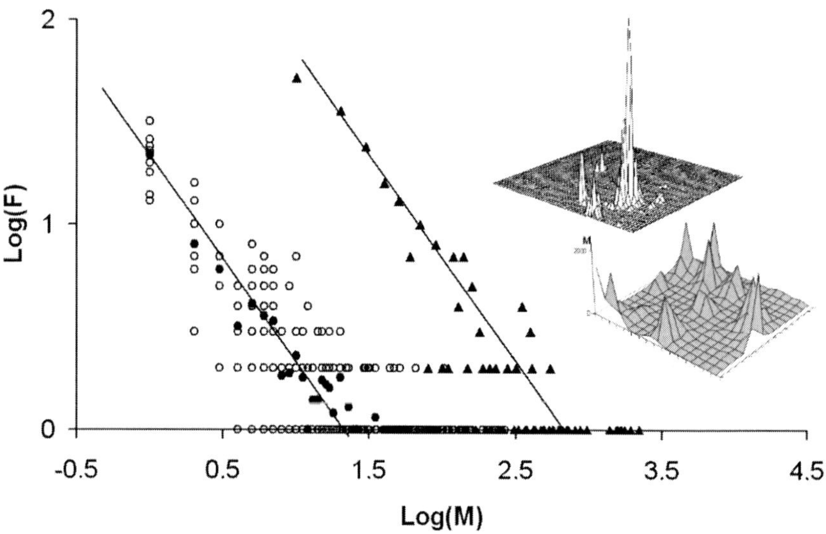

Figure 65. A congregated population dispersion emerges from a superposition of utilization distributions from 10 inter-independent individual MRW simulations in a homogeneous habitat (lower inset to the right). A single-series distribution of fixes is also shown for comparison (upper inset). Log-log-transformed frequency (F) histogram of local variations in densities of fixes (M) satisfies the theoretical slope —1 (superimposed, best fit), both for individual sets (circles, bin size 1) and for a superposition of the 10 independent sets (triangles, bin size 10). Filled circles give average F for a given class M, for cells with one or more locations in at least 5 of the 10 individual sets at the given resolution of analysis (16x16 grid cells), while the open circles show values for all sets.

A Brownian motion step length distribution is characterized by an exponentially declining frequency of steps (Eq. 23). However, MRW with β=3 rather than the ideal β=3 in the step-generating algorithm in Eq. 13 will produce a spatial pattern of fixes that resembles a Brownian motion despite the still scale-free process (see Figure 52). MRW produced from β=3 satisfies $F(L) \propto L^{-3}$. Consequently, $F(L^2) \propto L^{-2}$. In other words, the distribution becomes concentrated over a narrower range of L, and larger steps are almost absent, as in Brownian motion.

Defining $k \propto L$, there is a direct relationship between the expected distribution of local grid cell density, M_k, of a sample of fixes and the distribution of L^2 of a Lévy walk (Mandelbrot 1983, Hagen *et al.* 2001). This means that

$$F(M) = qF(L^2) \qquad (26)$$

where q is a scaling parameter. Substituting $F(L^2)$ owing to Eq. 25 gives

$$F(M) \propto L^{-D} \qquad | \, D = \beta - 1 \qquad (27)$$

With respect to a sample of fixes from Lévy walk and from Brownian motion, the spatial density of locations in a two-dimensional arena is expected to show fractal dimensions D = 1 and D = 2, respectively (Mandelbrot 1983, Viswanathan *et al.* 2000, Hagen *et al.* 2001). Simulations have confirmed that Eq. 25-26 also apply to MRW (Gautestad and Mysterud 2005, 2006). An example is shown in Figure 65 [reproduced from ref. (Gautestad and Mysterud 2006)]. The average slope for log(M,F) plots for fixes from ten independent MRW paths in spatial superposition satisfies D=1 in Eq. 27. However, while the ten simulated individuals in Figure 65 had only weakly overlapping core areas with largest local density of fixes, a similar scenario with stronger spatial overlap showed that the power law structure for the collection of fixes broke down, both for MRW with β = 2 and β = 3 (Figure 66). Interestingly, if the ten individuals shared a small part of spatial locations rather than returning to individual, completely independent sets of historic locations, the power law in Eq. 27 was restored (Figure 67). Thus, Eq. 27 may provide a method to test whether respective individuals with overlapping home ranges are utilizing their habitat in a manner that is independent of conspecifics, or if they show a tendency for conspecific attraction. So far this aspect has not been tested on real animals.

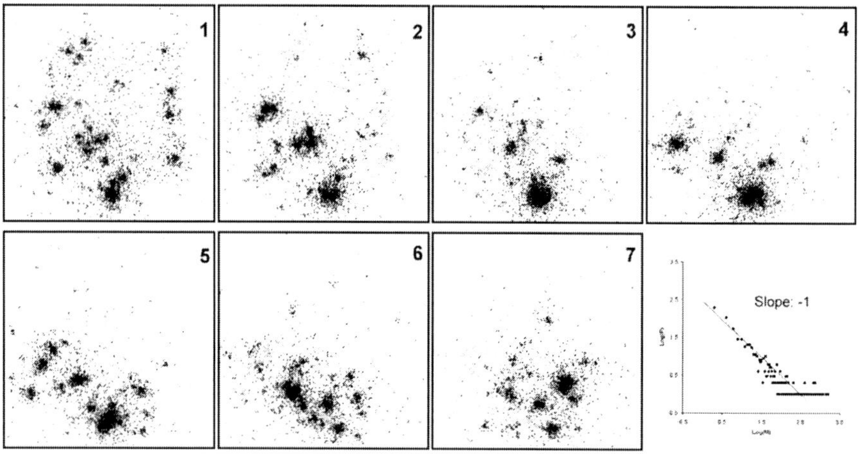

Figure 66. Distribution of local density of fixes under different conditions. See main text.

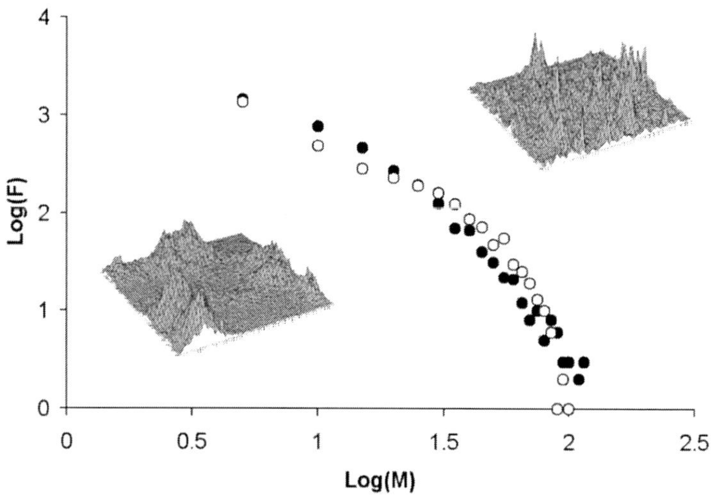

Figure 67. Seven successive MRW series, where the next series is dependent on previous series space use to a small degree, owing to sharing a small number of visited locations. Local fix densities in the first and last series have become locally independent. Lower right corner: A log-log transformed histogram of log(M,F) from the merged set of fixes confirms a power law distribution with slope close to the expected -1 from Eq. 27.

7.10 The next step

The reader has now been introduced to a wide range of theoretical concepts, both from the classic paradigm and from the extended biophysical framework. In the following chapter it is time to wrap up the development what regards the biophysics of animal space use, in particular with respect to ordering the various models into eight major universality classes of individual movement.

8. Movement Classes at the Statistical-Mechanical Level

8.1 Testing protocols for movement class compliance

Figure 68 summarizes statistical expectation from three methods ("protocols") to test for respective classes of movement—based on series of fixes—under standard, open environment conditions (Gautestad *et al.* 2013). Together with the parallel shift method (Figures 53-56) and a study of how fix or population variance behaves under various degrees of spatial coarse-graining (Figures 35-40), this book provides a range of approaches to test for (a) scale-free versus scale-specific kinetics, (b) memory-free versus memory-influenced kinetics, and (c) Markovian (mechanistic) versus non-Markovian (parallel processing) kinetics. In *7.6 Testing for parallel processing: red deer GPS fixes,* I presented a short summary of the most extensive empirical test of MRW and the parallel-processing conjecture we have performed to date (Gautestad *et al.* 2013, Gautestad and Mysterud 2013). This was a crucial test, since red deer move in an environment where they frequently meet various kinds of path-interrupting elements. Thus, according to current consensus on Lévy walk versus Brownian motion, long displacements during a given t_{obs} should be expected not only to be rare but in fact absent (for the red deer data, t_{obs} was one hour). In other words, the series of obstacles during a given interval t_{obs} should lead to a fine-grained mesh for the mean free path, which consequently should lead to strong truncation effect on successive displacements. This again should mean an expectancy of Brownian motion when sampling at the scale of t_{obs}=1 hour, whether the path complied with Lévy walk or Brownian motion at finer temporal resolutions; t_{obs}<1 hour. Thus, owing to the strong support for MRW when applying the three tests in Figure 68, red deer apparently were able to execute several goals at different temporal scales in parallel in a scale-free (log-linear) manner. This capacity was conceptually presented in Figure 24. Red deer also related to their habitat in a way that verified spatial memory execution at different spatial resolutions, owing to $\beta\approx2.2$, $D\approx1.2$, and $z\approx0.4$, in compliance with MRW expectations and with only ca 20% deviance from the respective ideal parameter scores for equal weight of kinetic energy allocation over the actual scale ranges: $\beta\approx2$, $D\approx1$, and $z\approx0.5$. Thus, classical model expectations were falsified for these individuals' space use.

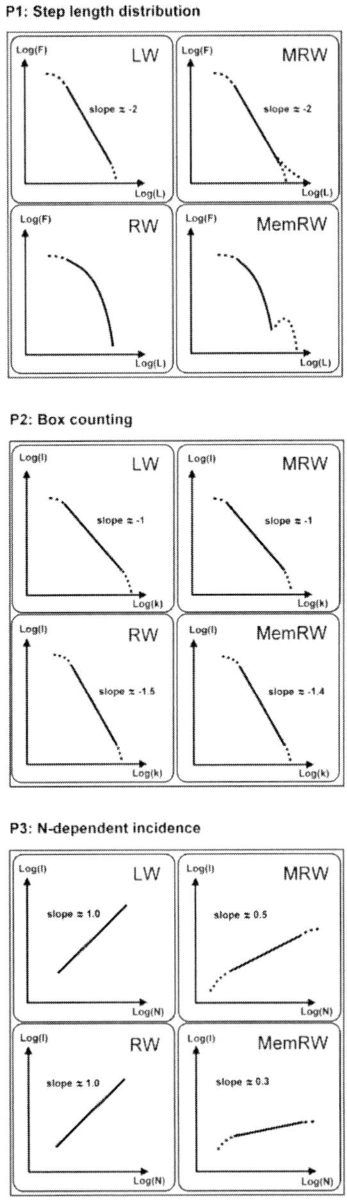

Figure 68. Expected statistical results under respective protocols and space-use models.

While Figure 68 regards the statistical "fingerprints" of respective movement classes, their statistical-mechanical framework in conceptual terms has previously been presented as a two-dimensional continuum consisting of spatial scaling and degree of long-term memory capacity (y- and x-axis, respectively; Figure 69). In this geometry parallel processing and mechanistic processing are presented at the same x-y plane, albeit with dotted lines as demarcation.

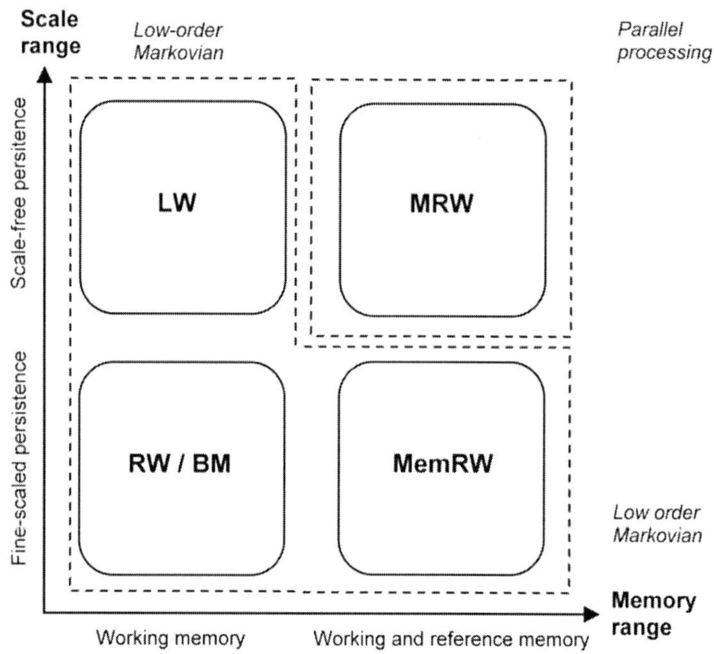

Figure 69. Statistical-mechanical variants of space use may be organized along two independent dimensions. The four movement classes represent arbitrarily defined discrete intervals over a continuum from approximately scale-specific to (theoretically infinite) scale-free movement, and over a continuum from no memory or short-term memory (working memory-influenced movement; remembering over a short or narrow time horizon) to a mixture of working and reference memory (remembering over a broad, and potentially infinite, time horizon). The combination of scale-free space use and long-term memory (MRW) is postulated to require parallel processing—a spatio-temporal expansion of the low order Markovian framework.

However, two additional movement classes are "hidden from view" within this two-dimensional geometry. First, there is a movement class missing in Figure 69: composite Brownian motion. Referring back to upper left pane in Figure 50, from the perspective of the step length distribution composite, Brownian motion (or composite correlated random walk) may—by parameter fine-tuning—appear Lévy walk-like and thus confusingly overlap with genuine Lévy walk in the upper left corner. Second, MemRW's position in the lower right corner presents it as a spatially scale-specific process with respect to exploratory steps. There is thus a need to make room also for MemRW with an exploratory step composition like composite Brownian motion or composite correlated random walk (at statistical-mechanical resolutions these variants become indistinguishable). By fine-tuning of parameters, this could lead to pseudo-scale-free exploratory steps while still maintaining the mechanistic and Markov-compliant property. In Figure 69 this would confusingly place this variant in the upper right corner, together with MRW.

In addition, Lévy walk is defined to belong to the non-Markovian domain owing to the long-term temporal memory needed to define a path's progress under condition of a scale-free distribution of step lengths. On the other hand Lévy walk may also be considered Markovian, owing to missing spatial memory and statistical independence of successive steps. This duality is connected to a similar issue with respect to continuous time random walk [see *4.3 Site fidelity under continuous time random walk (CTRW)*]. Lévy walk is thus spatially memory-less but is temporally scale-free with respect to long-term memory. The latter thus implies that Lévy walk may be considered compliant with parallel processing in the time domain, but not in the space domain. While Lévy walk implies a cognitive process leading to a multi-scaled distribution of step lengths, it lacks directed returns to memorized locations. Lévy walk has no site fidelity. Thus, I hereby extend the number of movement classes with parallel-processing compliance by including Lévy walk and simultaneously making a distinction between temporal and spatial memory, not explicitly stated in Figure 69. In short, Figure 69 is not exhaustive with respect to movement classes within a statistical-mechanical system abstraction, and some additional system diversification is needed to allow for Lévy walk in the parallel-processing domain under the modified classification as partly non-Markovian.

8.2 The scaling cube

I propose that the issues regarding overlapping location of specific movement classes in the two-dimensional geometry of Figure 69 may be resolved by extending the framework with hierarchical scaling as an independent z-axis. This allows for an explicit distinction between degree of scale-free temporal memory (y-axis) and degree of scale-free spatial memory (the x-axis), and simultaneously diversifies parallel-processing compliance from the Markovian counterparts. In this context "scale-free" has two qualitatively different meanings in a statistical-mechanical sense: system kinetics may be truly scale-free, as defined by the parallel-processing conjecture (marked as the **PP** plane in Figure 70), or it may be scale-free in a Markovian sense, like in composite Brownian motion (CompRW). The former regards an intrinsically driven movement and space-use *process* that generates the scale-free pattern. The latter regards a superposition of scale-specific processes, which are processed in separate and non-overlapping time intervals, and which in sum lead to the scale-free pattern. In other words, this "pseudo-scale-free" statistical pattern is achieved by specific parameter tuning for the respective terms in the composition (Benhamou 2007). It is still a Markov-compliant process and thus on the plane marked as **M**: it is modelled as a sum of independent terms for scale-specific movement at respective characteristic scales. With many terms and proper tuning, CompRW may *statistically* resemble a Lévy walk with β<=2<=3 and thereby defend its position as shown in the cube, but the process is very different from a PP-compliant kind of Lévy walk. The Markov variant (LW_M) regards a process where movement at any point in time is executed in accordance to either of the scale-specific terms (representing movement modes), but not several in tandem as is allowed for—by postulate—in LW_{PP} (Gautestad 2011). In other words, since the respective terms in a composite random walk model are executed in respective time intervals, the terms that represent coarser scale movement do not influence the movement at finer resolutions. In short, the respective components, which in superposition generate a Lévy walk look-alike with respect to the step length distribution, are executed successively in a statistically toggling manner. By tuning the parameter for median step length and relative contributions of each term in the superposition, the combined step length distribution may appear scale-free in the long term (after many toggling events). In contrast, LW_{PP} implies that various goals may be executed simultaneously but at different temporal scales, owing to difference in process rate for the respective goals (this will be elaborated on later). When Lévy walk in previous sections is referred to as Markov-compliant, it regards the specific variant CompRW—LW_M—where the respective terms in the "sandwich" of independent RW modes are parameter-tuned towards a scale-free distribution of step lengths. In *7.5.5 Simulation results of scale-specific versus scale-free processes,* I showed how LW_{PP} may be distinguished from LW_M in real data by applying either the coarse-graining test (Gautestad 2013a) or the parallel shift method (Gautestad 2012).

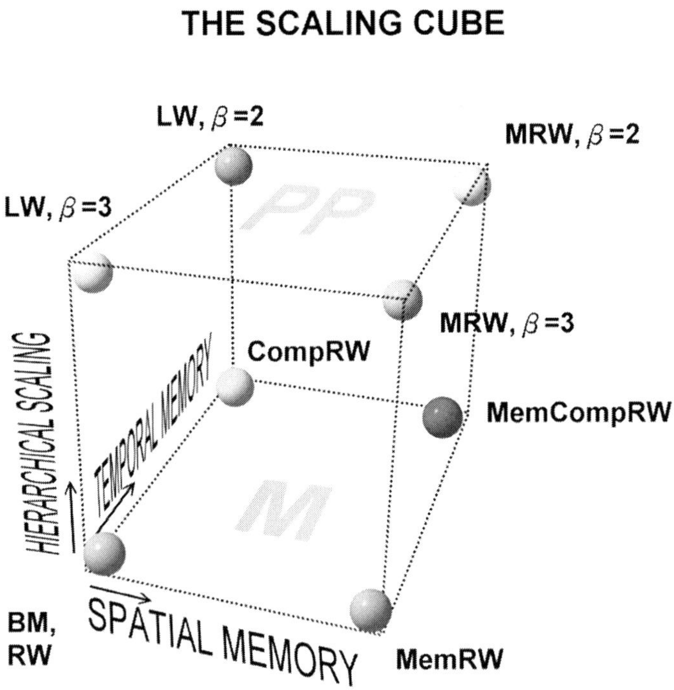

Figure 70. A conceptualization of respective movement classes along three independent dimensions: scaling compliance for spatial memory, scaling compliance for temporal memory, and hierarchical scaling. Small and large scaling compliance (distance from the lower left corner) regard the extent of equal weight of kinetic energy distribution over the actual scale range over space, time, and hierarchical depth (temporal scale range); as explained in *6.4.6 Dispersion of kinetic energy over a scale range*. A statistical-mechanical system representation is assumed for all eight movement classes in respective corners, and intermediate classes will emerge from parameter tuning.

Such a modified statistical-mechanical structure—hereby named *the scaling cube*—is presented in Figure 70. It contains 2x2x2=8 distinct movement classes, but transitions are implicitly acknowledged. In a physical context, these corners represent canonical (ideal) universality classes of movement. The eight classes regard discrete ideal models of respective continua. The z-axis reflects *the hierarchical range*—the "depth" of the actual parallel processing of space use—given by the degree of scale-free compliance with homogeneous dispersion of kinetic energy over temporal resolutions (Eq. 11). However, according to the parallel-processing conjecture, there is also an unspecified magnitude of potential energy involved, representing the lurking influence from temporally coarser scale goals on finer-scale goals. Instances where this potential energy transforms to kinetic energy (and thus is percolating from a coarse scale to a finer scale) will appear as a "surprise event" from the perspective of the finer scale behaviour (more on this below). The kinetic energy aspect was formulated in Eq. 11. If this scale range for parallel processing is small or absent (position on z-axis approaching zero), scale-free cognitive processing of a memory map is Markov compliant. In Figure 70 the z-axis thus introduces the parallel-processing concept more explicitly than in previous descriptions (Gautestad *et al.* 1998, Gautestad and Mysterud 2005, Gautestad 2011, 2012).

To summarize, parallel processing implies balancing frequent short-term goals with less frequent longer-term goals in a hierarchically structured cognitive processing (Gautestad and Mysterud 2010a). For example, while optimizing resource search at a minute-to-minute scale, there may be more coarse-scaled optimization going on—in parallel at longer intervals—like a goal for the next hour or more. In contrast, a Markovian counterpart version of parallel processing will execute fine- and coarse-scale goals sequentially; i.e., in separate intervals of time. Hence, Markov compliance implies that only the aggregative pattern (embedding many toggling events with behavioural shifts between fine- and coarse-scale goal executions) may express a statistical pattern that resembles a true scale-free process.

As will be described in more detail later, parallel processing is in this manner embedding memory with both backwards- and forwards-including dynamics: the behaviour is on one hand reflecting past experiences and on the other hand setting goals (which involve the future) at different temporal resolutions. This generalized definition of parallel processing, referring to the temporal memory structure with no reference to spatial memory, deliberately does not diversify spatially memory-less LW_{PP} from spatio-temporally memory-influenced MRW. Lévy walk of variant LW_{PP} has temporal memory (scale-free distribution of relaxation times in the time domain), meaning that directional persistence of the path may occasionally last for a long time. MRW has the additional memory capacity involving spatially explicit goals (targets for returns). One such goal could be reaching a spatially explicit target within an approximate time frame. If $\beta \approx 2$ ($D_{step}=D \approx 1$), both classes belong to respective upper corners of the parallel-processing plane in the cube. As β is increased towards $\beta=3$, kinetics under these two classes are sliding from respective corners towards zero temporal memory scaling, as temporal goals are becoming increasingly compressed towards a specific time scale but still scale-free. From a statistical perspective, the slope in the log-log transformed step length distribution becomes steeper but is still linear. With respect to Eq. 11a: $E_{k+1}/u_{k+1} < E_k/u_k$. In this manner small y-axis value implies less degree of "equal weight" of goal processing at different time scales. These instances may thus in some respect resemble statistical mechanics of Brownian motion and correlated random walk (the condition of Lévy walk with $\beta >= 3$), and MemRW (MRW with $\beta >= 3$).

While both LW and MRW (when the latter is sampled under condition of large ρ, i.e., $t_{ret}/t_{obs} >> 1$) produce scale-free step length distributions even under simulation condition $\beta=3$, BM, RW, and MemRW produce scale-specific Poisson (negative exponential) distribution of step lengths. Hence, LW and MRW with $\beta=3$ produce BM- and MemRW-*like* processes, respectively. Similarly, the corners for composite BM and memory-enhanced composite BM (MemCompRW) under specific fine-tuning of respective parameters are producing LW_{PP}-*like* and MRW-*like* processes, respectively; i.e., they represent mechanistic "look-alike" counterparts of true scale-free processes. While LW_{PP} implies parallel processing (limited to the time domain), composite Brownian motion under modulation towards its statistically scale-free class (LW_M) is executing movement mechanistically.

LW_{PP} and MRW with $\beta \approx 2$ maximize the equal weight principle in scaling terms, and thus belong on the corner for maximized value on the y-axis in the scaling cube.

LW_M means that a path is self-crossing only accidentally, as for non-composite (classic) Brownian motion. It leads to scale-free distribution of step lengths because a coarser-scale *temporal* goal may lead—on average—to maintaining directional persistence over a longer time span than a temporally finer-scale goal (a LW is assumed to maintain a constant movement speed, as discussed in *7.2.2 The step-generating algorithm: scale-free movement*). However, LW also implies that the actual path is susceptible to extrinsic (environmental) interrupts of the directional persistence, owing to lack of an explicit spatial target with respect to the actual goal under execution. Consequently, LW—whether of the LW_M or the LW_{PP} class—is difficult to maintain over physical space, owing to influence from the mean free path property under the given conditions. Thus, long steps will tend to be prematurely terminated, which leads to truncated LW. Presence of scale-free distribution of step lengths in a heterogeneous environment is therefore a strong indication of MRW, owing to this process' ability to produce a movement path that is "back on track" towards a target-specific goal after a temporary directional disturbance. This property was described in detail in Figure 24.

MemRW is represented by two classes in Figure 70; both classes are characterized by spatial memory, but they are distinguished by degree of superposition of movement modes and thus more or less LW_M-like with respect to the temporal scaling aspect of movement. The variant of MemRW with temporal memory scaling regards composite Brownian motion with a capacity to return to previous locations. A composite MemRW in this case is flipping between periods of different diffusion speed and is simultaneously utilizing spatial memory in a mechanistic manner (see *3.3 Coarse-graining the mechanistic memory model*).

The three-dimensional cube in Figure 70 is assumed to contain all kinds of Markovian (mechanistic) and non-Markovian processes, classified into eight main movement classes. These universality classes represent the statistical-mechanical conceptualization of objects and their motion through time and space, and over scales. The traditional (so-called Boltzmann-Gibbs) statistical mechanics belong to the lower left corner in the scaling cube, under the special (additional) condition of a closed system that allows for statistical-mechanical equilibrium. In this corner a well-developed theory exists, elegantly linking probability theoretical models for the system's particles (e.g., gas molecules) to the macro-level system properties (thermodynamics). Some of the other parts of the scaling cube (x=0, y>0, z=0) are only partly resolved theoretically, and the remaining parts are not biophysically explored by others at all, to my knowledge. Thus, the present book describes the first stepping stones towards a statistical-mechanical theory for the remaining corners.

Any output (generated movement path in a simulation) from any deterministic, semi-deterministic, or purely stochastic movement model will be represented by a specific (x,y,z) point or three-dimensional volume somewhere within this cube if the path is sufficiently coarse-grained (sub-sampled) to pass the transition from a dynamic to a statistical-mechanical representation of the process. Beyond this transition, a further sub-sampled BM, LW, or MRW is still a BM, LW, or MRW. For MRW it is the additional transition from autocorrelated to non-autocorrelated fixes to account for, as explained in *7.4.2 The distribution hump from site fidelity under MRW*.

For MRW Figure 70 implies both (a) scale-free temporal memory (parallel processing) and (b) scale-free spatial memory. Such a combination of scaling over both time and space means a potential for returns both to specific locations (return steps; memory utilization involving future goals) and to more spatially extended localities through cognitive coarse-graining (Figure 71).

Thus, MRW deviates qualitatively from MemRW, whether the latter is fine-tuned towards a scale-free look-alike pattern (CompMemRW) or not. Both MemRW and MemCompRW are mechanistic processes, which means that spatial memory utilization regards reference to specific locations and their respective attractiveness. A future goal to reach a given spatial location is not implemented; attraction towards such a target follows for a series of mechanistic steps where the direction towards this target is reaffirmed in a moment-to-moment manner. In simulation terms this means a mechanistic moment-to-moment weighing of respective known locations' attractiveness when deciding—from the resultant vector for the weighing representing directional bias—in what direction to move next. MRW, on the other hand, is postulated to have cognitive capacity both to relate to specific locations (providing potential targets for return steps) and to spatially coarse-grained representations of locations (overall conception of environment).

The latter aspect is conceptualized by difference in spatial scales of the neighbourhood in Figure 72 and how these are cognitively referred to over a range of temporal scales (the parallel-processing conjecture). As explained from various angles in *5. Parallel processing: the concept*, parallel processing allows for different goals being executed simultaneously owing to difference in process rate, and with increased autonomy the larger the time scale difference between respective goals. For example, a short-term decision to explore a local food patch may not conflict with a coarser goal to rest before sunset. Thus, intermediate goals may target localities that may involve temporary moves with independent step directions relative to the direction towards coarser-scale targets (for example, direction for local patch exploration relative to direction to the resting site). However, coarser goals impose constraint on the former: too

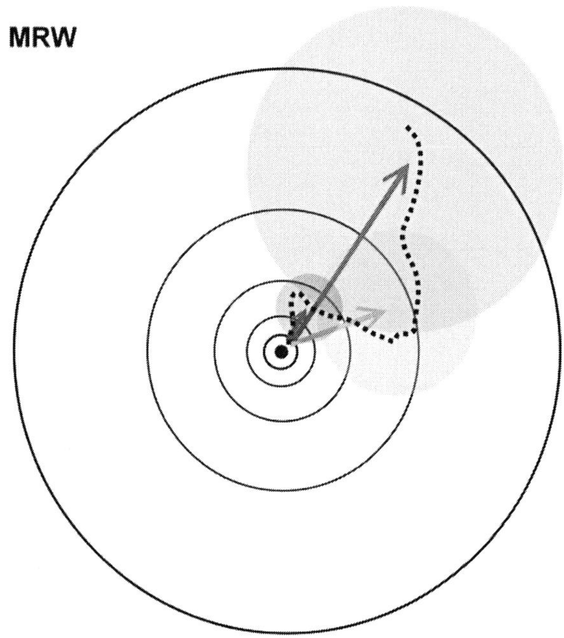

Figure 71. MRW simulates a scale-free kind of memory-enhanced movement process, where scale-specific process constraints (small-scale circle radius for spatial range and short-term memory horizon for temporal range) are expanded over both space and time by ability to coarse grain the memory map information in a hierarchical, power-law like manner (in this example, by power of 2: next larger circle is twice the size in linear terms). The respective circles in this cognitive scaling hierarchy are all representing the closest "neighbourhood-distance" to the individual's current location at different levels of coarse-graining. In other words, each circle represents a neighbourhood as cognitively perceived from the perspective of the respective scales. As the individual moves (dotted line), the realized path is the result from a complex inter-scale interference, here illustrated by different spatial targets (arrows) at three different spatial scales. Since a coarser scale target is typically accompanied by a lower frequency execution scale in the time domain, there is only a small chance that this coarse goal is actually executed during any given small time interval that characterizes target executions at finer-scales (higher frequency).

much short-term rumbling and tumbling (from the perspective of a coarser scale goal) may be limited as soon as the longer-term goal is threatened. Contrast this with mechanistically structured memory utilization like MemRW, which by necessity invokes only a very narrow time scale for processing (small value on the y-axis). Thus, the resultant vector determines successive step directions, which in case of targeting a distant goal by necessity implies a drift-like movement. As described in *3.2 Spatial memory from a mechanistic premise*, in mechanistically executed simulations of memory-influenced space use, it has been customary to apply biased correlated random walk for semi-stochastic variants of movement.

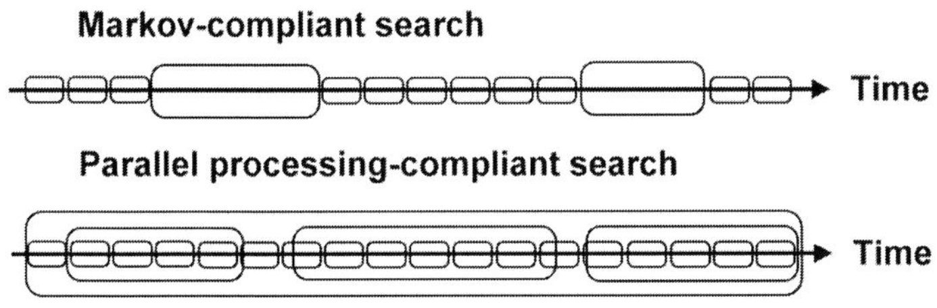

Figure 72. Conceptual difference between classical (mechanistic) movement and parallel-processing-compliant movement. See main text.

The MemCompRW variant of Markov-compliant movement is in some sense mimicking MRW's concept of moving towards targets at potentially different directions. MemCompRW involves a superposition of "modes," each leading to a mode-specific mean free path under respective time intervals of execution. With respect to memory-less and scale-specific (classical) random walk models, mode 1 could be area-restricted search, which is a common concept in movement ecology. An individual then tends to change direction frequently to effectively explore and utilize a resource-rich patch upon encountering it. Mode 2 could be a coarser-filtered behaviour with respect to what environmental inputs to relate to. This mode is, in composite Brownian motion models, referred to as inter-patch movement. Typically, the model animal switches to this mode after energy gain in a resource patch has deteriorated to a critical level, invoking new search at a coarser environmental scale. Thus, modes 1 and 2 may compute different directions for the next step at a given point in time. However, even if a superposition (sum) of modal influences is influencing movement in a scale-specific process, at each time increment there is only *one* mode active, owing to dependency of the mechanistic (Markovian) premise. This principle is illustrated in the upper row in Figure 72, where rounded rectangles represent goals at respective time scales that are reflected by the rectangle's width.

Finally, I refer to *6.4.7 Redefining statistical mechanics* with respect to accepting the term "statistical mechanics" in the context of non-mechanistic systems. In precise terms, the more general "statistical physics" should be used, while "statistical mechanics" should be referring to the Markovian/mechanistic class of dynamics. However, for simplicity I continue using the well-established term "statistical mechanics" for both classes of dynamics.

8.3 Parallel processing: planning for the future

The mechanistic principle for scale-specific processes of one mode active at a time also applies when the field of perception is increased to allow for spatial memory effects under the premise of mechanistic modelling (see *3.2 Spatial memory from a mechanistic premise*), referred to in Figure 70 as MemRW and MemCompRW. Thus, there may be a Markovian-compliant rule deciding when to flip from one mode to the other [for example, based on summed foraging gain over the last intervals, see a good mechanistic model example provided by Spencer (Spencer 2012)]. In contrast to MRW's parallel-processing postulate for memory implementation, there is no cross-scale constraint from goals at a small process rate (coarse time scales) on goals at a large process rate (fine time scales), nor the ability to execute goals at different process rates simultaneously. These MRW principles are illustrated by the lower row in Figure 72, which thus embeds both types of non-Markovian movement; Lévy walk and MRW. The principle of scale-free (log-linear) temporal coarse-graining is illustrated in Figure 73. While a Markov-compliant process may be considered relating to a limited time range relative to present time both with respect to past and future, parallel processing implies a kind of temporal coarse-graining that increases log-linearly with temporal distance from present. Parallel processing under Lévy walk (LW_{PP}) or MRW regards dynamics involving scale-free time axis towards both the past and the future.

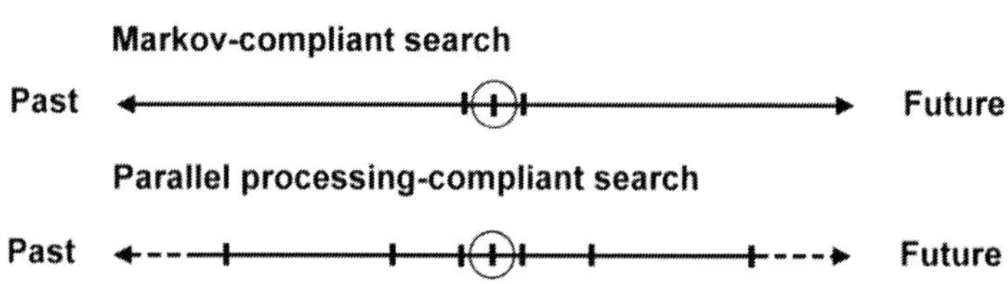

Figure 73. The time dimension under mechanistic (Markov-compliant) and non-mechanistic (parallel processing) process conditions. See main text.

The latter aspect—you may call it memory utilization that includes a potential to plan the future (at coarser temporal resolutions)—is particularly important when comparing parallel processing with other non-Markovian or pseudo-Markovian memory-influenced model designs, like continuous time random walk (CTRW) and MemRW. For example, CTRW implements memory as the past influencing the present but not with potential to influence the future. Likewise, MemRW regards spatially explicit memory where the "memory horizon" goes backwards to the past but not including goal-planning towards the future: next step is reconsidered at successive moments, with influence from stored information. A MRW system means that parallel goals (which regard the future, based in utilization of information in the past) are postulated to be executed at different temporal resolutions, as illustrated in Figure 73. The time axis is scaling: in a parallel-processing system, β=2 is expressing the "equal cross-scale utilization weight" conjecture (Eq. 11).

8.4 Summary of individual-level test protocols

In previous chapters I described a general, statistical-mechanical approach to model memory-influenced animal movement and population kinetics, MRW, where the animal, according to a specific postulate—parallel processing—relates to its environment strategically by utilizing collected information of its environment beyond its instant perceptual field in a scale-free manner over both space (Figure 71) and time (Figures 72-73), with cross-scale memory influence involving both the past and the future. Contrary to alternative movement classes embedded in the scaling cube (Figure 70), simulations based on MRW were able to reproduce the empirically observed space-use statistics of free-ranging sheep (Figure 3), black bears (Figure 4), red deer (Figures 57-59), and other data sets. Contrary to alternative movement classes, the MRW approach was thus able to resolve what appeared paradoxical from the perspective of alternative approaches, including the paradigm representatives Brownian motion, correlated random walk, and Lévy walk.

To verify MRW compliance in relocation data sets, I have outlined a simple approach whether the data are synthetic (simulation-based movement paths) or real GPS series from animal space use: first, coarse-grain the series to a statistical-mechanical level if it is not there already (in home-range situations, to level $\rho = t_{ret}/t_{obs} < 1$ to approach ergodicity; see Eq. 20). Then test for "home-range ghost" compliance with respect to accumulated space use as a function of N (Eq. 1-2, Eq. 14, or Eq. 17). The null models are in this case either asymptotic approach (z approaching zero at large N) or $0.5<<z<=1$ at grid scale close to the characteristic scale of space use [$\log(c)\approx 0$; see Eq. 18 and Figure 45]. In other words, compare the result with Lévy walk or Brownian motion, whether under condition of area boundaries or in open space. Avoid the KDE method to estimate area, due to the intrinsic statistical artifacts described in *1.2 Pinpointing a feasible reason for ignorance*. Additionally, test for fractal dimension of the spatial scatter of fixes, and confirm if D is significantly smaller than expectation from scale-specific processes like Brownian motion and MemRW. Third, for $\rho>>1$ (strongly autocorrelated series), apply the parallel shift method (Figures 53-56) to test for compliance with a scale-free process like Lévy walk or MRW. Contrary to Lévy walk, generic MRW should *simultaneously* satisfy both $z\approx 0.5$, $D\approx 1$, $\beta\approx 2$ and also pass the parallel shift test, given that the effect from site fidelity is accounted for (ρ larger or smaller than 1), as illustrated in Figure 50 and Figures 53-56.

8.5 Path analysis and cross-scale complexity

The shortcomings of the KDE method—called a statistical subtlety or "nuance" in Chapter 1—have been recognized by many ecologists. Hence, another approach has rapidly gained popularity: along-the-path estimates of space use (Hjermann 2000, Fortin *et al.* 2005, Coulon *et al.* 2008). One popular approach in this respect is the "Brownian bridge" method, which means a process obtained by conditioning a Markov process X to start in some state x at time t=0 (representing a given spatial location of the animal) and arrive at some state z at a later time (the next location). The successive calculations along the observed path from GPS fixes produce another kind of (more or less narrow) space-use demarcation of expected local whereabouts for the individual between observed locations. Habitat selection may then be estimated from this trailing area demarcation (Byrne *et al.* 2014, Kranstauber *et al.* 2014).

However, a similar issue as described for the KDE approach may surface also in path-based analyses if movement involves long-term memory. Both KDE and all path analysis protocols depend on Brownian motion and classical diffusion principles and thus do not account for memory effects. An old challenge in analyses of habitat selection regards the fact that some path segments may simply represent commuting between preferred sites. The animal may then frequently traverse over generally non-preferred habitat, and environmental conditions along these path segments may thus confound the analyses and inflate the level of statistical noise. It is therefore a problem to distinguish intensity of space use where commuting and non-commuting path segments are lumped together, whether path analysis or a more classical use vs. availability method is applied. In the latter approach, density of relocations pr. area unit is compared between habitat types to infer strength of selection for specific environmental attributes.

For example, in one study that applies the use/availability approach (Mysterud *et al.* 2014), it was found that coarse-scale representations of local productivity of forage in a generally high-quality environment correlated positively with intensity of space use (density of GPS relocations) by free-ranging sheep *Ovis aries*. On the other hand, in an adjacent area with generally low productivity, a negative correlation was found between forage availability and local space-use intensity. The sheep apparently and paradoxically selected against local-scale areas with higher quality where coarser-scale environmental conditions were poor. The discrepancy between the two areas was hypothesized to emerge from a higher degree of finer-scale patch utilization of a resource-rich but strongly fragmented habitat type in the poorer environment. One may then reasonably assume that a larger part of the sheep relocations in the generally poor environment regarded inter-patch commuting between small high-quality patches. This patch type went unnoticed in the medium-scale estimates of forage dispersion, owing to lack of sufficient spatial resolution of reliable resource mapping. It was concluded that without higher-resolution mapping of available resources, it is difficult to interpret habitat selection results.

Would a finer-grained habitat mapping produce a better coherence with expectation in both high- and low-productivity areas? The specific study by Mysterud et al. raises interesting general questions with reference to both theoretical and applied ecology. From a standard Markovian frame of reference, quantification of difference in movement speed or stronger directional persistence while an animal is traversing sub-optimal habitat may to some extent reveal selection differences when using density of GPS fixes as proxy variable for local space-use intensity. Observe that "commuting" under this premise is replaced by "traversing", which—contrary to commuting—does not depend on site fidelity and memory. A traditional Markovian-based analysis could thus bring better coherence with expectation also at finer scales by explicitly quantifying traversing versus intra-patch movement, and potentially resolve the paradox observed in the generally poor environment. Many Markov-based model structures exist to describe this kind of "intensive" (fine-scale) versus "extensive" (coarse-scale) movement modes, like area-restricted search [ARS models (Bell 1990, Benhamou 2007)] and the so-called Lévy flight foraging hypothesis [switching from scale-free to scale-specific movement mode when entering an environment with a high resource level (Viswanathan *et al.* 2011)]. On the other hand, it would have been very interesting to apply a memory-based, non-Markovian method as an alternative approach in studies on habitat selection and then compare the respective levels of statistical noise in the results. If commuting fits observed multi-scaled movement behaviour better than traversing, a memory-implementing model should be expected to be more applicable.

According to the MRW model, which combines scale-free and memory-influenced movement, directional persistence may be both high (during long-step executions) and low in both high- and low-quality environment. As was illustrated in previous chapters, under the MRW premise for space use, other proxy variables than locally varying path tortuosity, movement speed, and density of relocations are thus invoked to quantify intensity of space use. Tortuosity, speed, and relocation density are all linked to Markovian and thus memory-less quantification of habitat selection strength. Studies like Mysterud et al. (2014) do illustrate the scale challenges with respect to quantifying habitat selection, and these kinds of data should be ideal candidates for comparing selection estimates from two qualitatively different theoretical platforms.

8.6 The next step

So far the chapters have been biased towards space use by vertebrates, in particular mammals. The progression is now ripe to switch towards invertebrates to gain further biophysical insight into animal space use. After all, invertebrates are not only more abundant in ecosystems relative to vertebrates. They are also offering larger datasets involving large populations, which in particular invites to further analysis of the complexity of population abundance.

9. IS PARALLEL PROCESSING APPLICABLE TO INVERTEBRATES?

9.1 Classifying path integration in insect movement

So far the MRW model has only been tested on space-use data of vertebrate species, which for a long time have been acknowledged to have cognitive capacity to utilize a memory map (however, MRW extends the accompanying theoretical framework by proposing a parallel processing kind of space use). What about invertebrates? For example, limpets were a long time ago shown to have home ranges and thus a cognitive capacity to express site fidelity (Frank 1964). Flying insects, like the emperor dragonfly *Anax imperator*, have been tracked and confirmed to show a home range (Levett and Walls 2011). Male neotropical orchid bees (several genera) have also been shown to perform site fidelity, probably based on memory map utilization (Pokorny *et al.* 2014). Therefore, in the following I switch the focus from mammals to insects.

Many insects seem to follow a specific memory mechanism based on path integration and local landmark guidance. Sophisticated simulations based on neural network processing has shown that navigational skills of insects like ants and bees can indeed be mimicked without invoking a memory map in the vertebrate sense (Cruse and Wehner 2011). With respect to the scaling cube, under which movement class does path integration belong? Insect navigation seems to contain both a strategy (decision to reach a goal, like a previously visited food source) and a finer-scaled level of tactical moves (incrementally fine-tuning the path integration process based on local cues and context-specific information recalled from memory). This kind of memory-enhanced space use thus seems to be compliant with a low-order Markovian kind of movement, since the displacement is defined by vector addition and a resultant vector for shortcut returns. Regarding the latter, this successively updated vector pointing towards a distant goal (whether "home" or some other target based on spatial memory) represents next-step's magnitude and direction. It is defined by adding together strength and direction of the various movement-influencing factors at the given point in time. In other words, it may resemble the mechanistic kind of dynamics termed MemRW. In contrast, MRW involves parallel processing of goals at different spatio-temporal scales, executed at different process rates (Gautestad and Mysterud 2010a) (Figure 24). In other words, MRW might be expected to represent the space use of dragonflies and neotropical orchid bees (Levett and Walls 2011, Pokorny *et al.* 2014), but not insects performing vector-based navigation where successive steps can be understood as a vector sum of movement-influencing factors. In the latter case the strategic goal (executed by path integration) is weighted and executed at a similar time scale as the "fine-tuning" of tactical factors (Cruse and Wehner 2011). The principle for such a characteristic time scale in Markov dynamics was described in *2.4 Markov compliance and the characteristic scale.*

9.2 Exploring complex population kinetics in sycamore aphids

Can a statistical-mechanical analysis of insect movement and population kinetics indirectly verify that there is more to insects' cognitive path integration process under memory influence than has been revealed so far? With reference to the scaling cube—are there other kinds of movement classification that fit some groups of insects better than path integration, as indicted by the results from studying dragonflies and orchid bees (Levett and Walls 2011, Pokorny *et al.* 2014)? The null model in this respect is that movement is Markov-compliant and mechanistic at micro-scales and leading to MemRW or MemCompRW at statistical-mechanical scales (Figure 70).

In the following data analysis, I show that the local dispersion of a population of sycamore aphids *Drepanosiphum platanoides* (Figure 74) seems to support scaling properties of MRW and Lévy walk. As is indicated already by visual inspection of the leaf population in Figure 74, aphids show congregation. As will soon be illustrated, a similar aggregative pattern was present also among leaves and among group of leaves: some leaves and leaf groups having substantially larger number of aphids than other leaves and leaf groups.

The following tests are distinguishing scale-free from scale-specific space use, with results indicating that the process is embedding site fidelity (return events) where local abundance of aphids seems to be the critical "resource" leading to attractive patches at respective spatial resolutions. I include the analysis as a pilot test, since it is missing a detailed documentation of ecological conditions. Despite this shortcoming, the results hint at additional

Figure 74. Sycamore aphids under the leaves.

support to the generality of scale-free space use from intrinsic forces rather than environmental heterogeneity. This is particularly interesting since the system regards spatio-temporal dispersion of an insect species rather than a vertebrate species.

As already mentioned, it is difficult to imagine a path integration algorithm that leads to a scale-free distribution of fixes for a given individual. The argument against path integration in the context of the sycamore aphids is even stronger if a scale-free *population* distribution can be verified. For example, Figure 66 showed how the scale-free nature of fix dispersion broke down when independent sets of fixes were spatially superimposed. Thus, if aphids were moving inter-independently whether in a scale-free manner or not, the population dispersion should—in an approximately homogeneous environment—approximate a Poisson distribution in the limit of large M: $V = aM^b$, with $b=1$. The present results will be shown to verify a Z-paradox-compliant pattern at the population level, $V = aM^b$, with $b=2$, and may thereby be considered giving support both to conspecific attraction and utilization of spatial memory with equal weight over a scale range (see *6.4.6 Dispersion of kinetic energy over a scale range*). Conspecific attraction, expressing itself as scale-free population dispersion in the aphid data, lends support to a parallel-processing kind of movement (Figure 72) where the individual aphid consequently should be able to relate to a spatially explicit mapping of conspecifics, coarse-grained to different spatial scales (conceptually described in Figure 71). According to the scaling cube, the movement class that comes closest to this kind of space use in statistical-mechanical terms is MRW.

The sycamore aphid results provide food for thought and provocative model proposals with respect to what kind of population kinetics are linking individual movement to population density fluctuations over space and time in this species. In *6. Modelling parallel processing,* I described a kind of self-organized population dispersion based on the multi-scaled, memory-dependent positive feedback process referred to as the parallel-processing postulate. Below I compare the predictions from this model with the spatial dispersion of individuals in the local population of sycamore aphids. The results are then compared to expectation from kinetic processes under classical population dynamics. In a theoretical follow-up chapter, I will then return to the aphid results and perform a deeper analysis from the perspective of scale-free population structure.

Animal population density is often found to be more variable than one should expect from equilibrium theory, where individuals disperse themselves in accordance to some variant of ideal free distribution. An "ideal free distribution" is a way in which animals distribute themselves among several patches of resources. The theory states that the number of individuals that will aggregate in various patches is proportional to the amount of resources available in each patch (Fretwell and Lucas 1970). However, in the following I will show patterns that are problematic to explain from classical approaches, whether it is ideal free distribution theory or some sort of trivial constraint on intra-population mixing (for example, metapopulation structure or limited dispersal power (Salt and Hollick 1946, Nielsen 1954, Taylor 1986, Stamps 1988, Smith and Peacock 1990, Stamps 1991, Brown *et al.* 1995, Lane *et al.* 2001, Brewer and Gaston 2002). In particular, many populations are found to be dispersed under influence of site fidelity and conspecific attraction (Yasukawa and Searcy 1985, Stamps 1988, Beletsky and Orians 1989, Schieck and Hannon 1989, Smith and Peacock 1990, Stamps 1991, Dobson and Poole 1998, Pöysä *et al.* 1998, Courchamp *et al.* 1999, Stephens and Sutherland 1999, Välimäki and Itämies 2003), rather than just showing a pre-emptive distribution (Pulliam and Danielson 1991) generated mainly by negative feedback effects close to local carrying capacity. As has been shown repeatedly in this book, space use under MRW is driven by positive feedback, in the sense that visited patches will tend to become targets for even more returns in a self-reinforcing manner (in an upcoming chapter I will show how the positive feedback may turn negative under specific ecological conditions).

The sycamore aphids' general ecology, and experiments on the displacement behaviour and dispersion pattern at finer scales than one leaf, has been described previously by Kennedy and Crawley in 1967 (Kennedy and Crawley 1967). What I find particularly interesting in their study is the spacing behaviour, which they called "spaced-out gregariousness." When some individuals were placed in standard Petri dishes (homogeneous environment), complex aggregations began to form but in combination with individuals respecting a minimum inter-individual distance approximately equal to the length of the aphid's antenna. The spaced-out gregariousness phenomenon is clearly visible in Figure 74.

Petri dishes are very limited in spatial extent, which inspired me to test for spaced-out gregariousness in the aphids' natural and open environment. By studying the aggregation pattern at the scale of sycamore leaves and beyond—by grouping aphid counts over several leaves—I extended the scope to test for gregariousness by analysing conspecific attraction over a scale range. The null models are the classical premises based on a scale-specific kind of movement or scale-free movement under a premise of lack of spatial memory.

9.3 Sycamore aphid data

A spatial transect of a local population of sycamore aphids consisted of leaves at the lower part of two sycamore trees at the campus of the University of Oslo, Norway. Sycamore aphids are relatively mobile throughout the summer, and they occasionally fly from one leaf, branch, or tree to another. I did not perform dispersal range estimates, but flying aphids were frequently seen arriving and leaving trees and subsections of trees in a more or less continuous manner. In simple terms, the population should be subject to strong "mixing" and thus satisfy the important ergodicity premise when testing for complex kinetics. Therefore, I assume that the sycamore aphid populations on the various trees on the campus belonged to the same local population, and that a given aphid could potentially relocate itself to any leaf within the system during the actual time increment of one day or longer.

As will be shown below, the half-leaf and leaf scale densities of aphids at the first and last day of the observation period were not strongly correlated. Owing to the high individual mobility of the aphids, this could be assumed to be primarily caused by aphid movement, not local birth and death, owing to the high mobility rate relative to birth and death rates.

Aphids passively feed on sap of phloem vessels in the leaves, but foraging success rates at respective leaves were not studied, so local density variation of aphids could not be tested against this important variable. However, my impression was that density variation at particular leaves through the observation period was primarily caused by conspecific attraction. Leaves with high density normally tended to uphold high density over the weeks, showing that even a high rate of aggregated resource consumption at these leaves was not depleting individual foraging success to the extent that these leaves were abandoned. Observations showed that other leaves that had none or very few individuals could suddenly have gained higher aphid density over a day or a few days (documented in the respective leaves' time series). Perhaps the strongest argument for assuming resource availability being secondary to conspecific density when aphids determine where to target their moves is given by the fact that left and right parts of the same leaf could contain very different aphid density. According to reasoning based on, for example, ideal free distribution, one would expect a homogenization of individuals between the respective half leaves at leaves with high density. Further, half-leaf densities were fluctuating relatively independently over time. In summary, the density fluctuations at the scale of half leaves and leaves were strong over both space and time, and the process seemed to be attributed to individual displacements and driven by positive feedback like conspecific attraction.

Forty leaves from each of two sycamore trees were chosen randomly, surrounding the trees' periphery at a height of about two m. The distance between the trees was about 200 m. The stalks of the leaves were carefully numbered from 1-40 and 41-80, respectively, without disturbing the aphids. Twenty days during the period from July 9 to August 13, 1990, aphids on the left and the right half of each leaf were counted without disturbing the individuals (Figure 75). The counting of one leaf took about thirty seconds on average, and all 160 half-leaf populations were counted within an hour before noon. Local populations on neighbour half leaves were only weakly positively correlated (linear regression: $y = 0.36x + 0.90$; $R^2 = 0.13$, $y = 0.47x + 0.93$; $R^2 = 0.23$ for the two trees, called sampling station 1 and 2, respectively). Thus, half leaves were treated as separate and independent sample units in the analysis. Sets where one or both half-leaf counts contained zero were disregarded. Counting aphids from each half leaf over the five-week observation period gave a series of 160 sample units. In addition to spatial coarse-graining, the dispersion pattern by grouping leaves into larger sample units, temporal coarse-graining, was also performed by summing or averaging aphids at a particular sample unit over a number of sampling days. The temporal variability at the scale of each tree is shown in Figure 76. The fluctuations were substantially stronger at finer spatial scales, like leaves and half leaves, in comparison to the scale of trees (stations). An example of fluctuations of two leaf groups of five leaves each is shown in Figure 77. Notice that even if Figure 76 and Figure 77 show a generally declining population over time, this may be a statistical artifact from non-random choice of leaves to be monitored. Most of the eighty leaves had zero abundance of aphids at day 1, which may reflect a subjective choice.

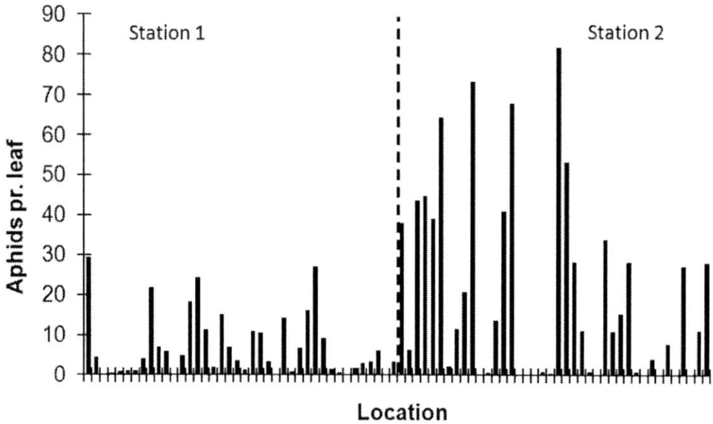

Figure 75. Time series of sycamore aphids at scale of single leaves. Station 1 and 2 represent leaves at different trees.

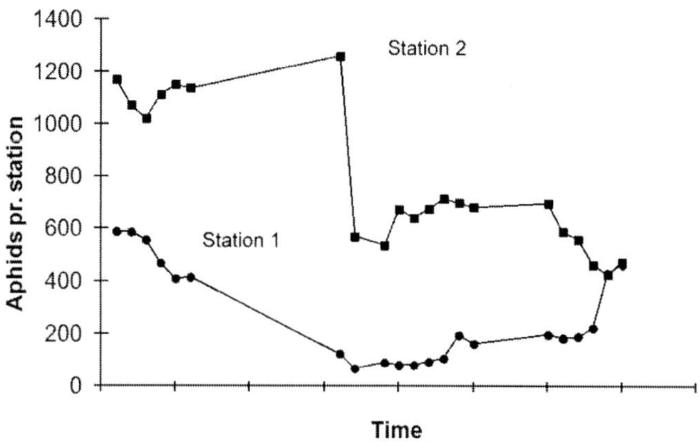

Figure 76. Time series of local aphid populations at the scale of respective trees (stations) during the study period.

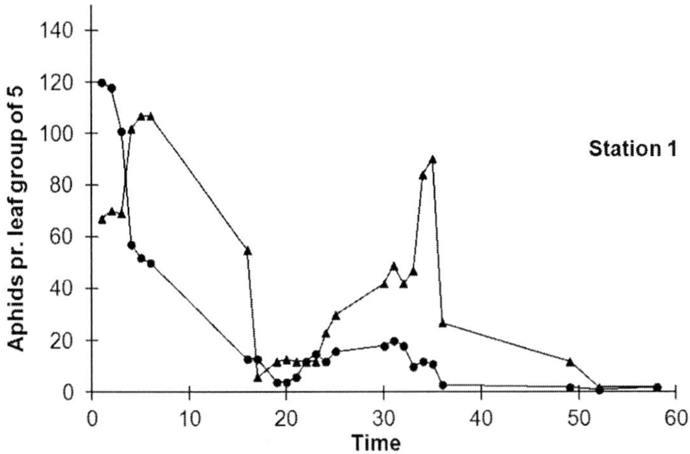

Figure 77. Example of time series of aphid population density at spatial scale of five leaves.

9.4 Hypotheses for local dispersion of aphids

9.4.1 Outline for the study

A central limit theorem assumption—Poisson distribution of individuals over spatial scale (V changing proportionally with M, where M is the average number of individuals at a given grid scale)—can be used as a general null model for spatio-temporal variability, based on a premise of Markov-compliant aphid movement under the classical paradigm Brownian motion and diffusion. Based on theory provided in previous chapters, the Markov premise may be tested at the population level from several angles. Here I use three approaches.

First I study the distribution of local population density as such; the frequency of high-density versus low-density patches, and how this distribution is sensitive to spatial scale (sample units consisting of leaves or groups of leaves). Second I study how population variance varies as a function of average number of individuals pr. sample unit over space and time. Again I study sensitivity to spatial scale, and I relate the result to the Z-paradox. Third, I perform a simulation where I test whether local geometric growth/decline of population density may explain the variance in the sycamore aphid population. Geometric change, where number of individuals is a stochastically calculated small percentage larger or smaller than the number in the foregoing point in time, implies the null hypothesis that randomly fluctuating birth and death rates—rather than individual movement—dominate the production and maintenance of strong variance.

9.4.2 Method 1: the density distribution

With respect to the sycamore trees, the leaves that were monitored seemed to provide a similar level of foraging success for the aphids (owing to the aphids' behaviour, see above), and grouping of leaves should tend to further smooth inter-leaf variability in food availability. In *7.9 Density distribution of fixes,* I described the expected scale-free relationship between local number of individuals, M, in grid cells at a given spatial grid resolution k (cell size k^2) where k was set to be proportional to the *relative* number of individuals M. Eq. 27 described the statistical distribution $F(M) \propto L^{-D}$ with the fractal dimension D=1 from the premise of Lévy walk or MRW (strong population congregation over a range of scales), and 1.5<=D<=2 from the premise of a Brownian motion (less congregation). For example, Figure 63 verified these expectations for D under respective movement classes. According to Eq. 27, if the individuals distributed themselves towards a scale-free pattern at scale k, one should expect sample units with k^2 times larger M to appear 1/D as frequent, where D is the fractal dimension. D≈2 is expected from classical premises (Poisson distribution and Markov compliance), while D≈1 is expected from parallel processing.

D=2 implies a very low appearance of high-M sample units relative to D=1. Under the premise of classical population kinetics (illustrated in Figure 15 and Figure 30), close to uniform distribution of individuals is expected at spatial resolutions where the habitat is uniform when coarse-grained to this level. On the other hand, complex population kinetics lead to scale-free aggregation, under condition of conspecific attraction (Figure 33).

To understand the spatio-temporal dispersion of M, one needs to consider the population's kinetics: how the processes of birth, death, immigration, and emigration are forming the respective magnitudes of M_j, where j=1,2,3,... sample units at observation scale k. Immigration and emigration are in this respect determined by movement out of and into respective sample units' area segments. Both Lévy walk and MRW are expected to produce D≈1 (β=2) if kinetic energy and thus "space-use effort" is distributed evenly over scales (Eq. 11), but Lévy walk regards movement without spatial memory. *Thus, if D≈1 and conspecific attraction can be verified, MRW is supported at the expense of Lévy walk.*

9.4.3 Method 2: the variance-mean relationship

A study of variance relative to the mean provides the second approach to test for model compliance in the sycamore aphid population. In *6.4 Population kinetics under the parallel-processing premise,* scale-free congregation was quantified by variance as a function of M. Compliance with $V = aM^b$ with b=2 implied a strongly aggregated pattern, satisfying (in a homogeneous environment) Lévy walk and MRW with step distribution exponent β=2. The null model is again Markov-compliant space use at the individual level, leading to b≈1 if (M,V) plots regard average M at different spatial scales. If b≈2 is maintained at different spatial resolutions, and population kinetics is primarily driven by movement rather than locally varying birth and death rates (to be tested in Method 3), the pattern is inexplicable from the classical Markov premise for population kinetics (the Z-paradox).

9.4.4 Method 3: testing for dominance of movement rate

To test the assumption that movement rate dominated over birth and death rates in the aphid population, I performed a simple simulation based on the following line of arguments. Consider a specific spatial dispersion of individuals, whether the dispersion is homogeneous or aggregated at the start of a time series. Under influence of conspecific attraction and compliance with parallel processing, individual movement should then (from zooming-out and zooming-in events over a range of scale levels k; see Eq. 8) lead asymptotically to a strongly congregated pattern. From Eq. 27 (the distribution F of local magnitude of M) and starting with transect data that is satisfying D=1 (strong aggregation), the null hypothesis from classical premises is a drift towards 1.5<D<≈2 (weak aggregation, see Figure 63), even when starting from a distribution showing D<<2. However, the preceding arguments regard influence from the movement aspect of population kinetics. How to differentiate the effect from local movement rates from effect from birth and death rates?

To illustrate the null hypothesis, I performed a simple simulation based on the local aphid abundance at half-leaf scale, starting with the first day's abundance at station 2, excluding zero counts, and letting each half leaf's successive abundance change according to a random positive and negative growth rate that had sufficient magnitude to give independency between first and last abundance at respective simulated time series over twenty iterations, representing days. Another condition was that average spatial abundance at the end of the simulation should be approximately similar to the average abundance in the start condition, which was given by real data. By defining zero migration between sample units, spatial and temporal variability of M is determined by birth and death rate alone, which is simulated stochastically by the net growth rate.

By comparing F(M) from the start (real data) to the end of the iterations (simulated result), and averaging over ten independent, a change in D over time may then be studied. Locally varying net growth rate owing to locally varying ecological conditions will tend to lead to high variance of M over a given transect, owing to what is called environmental stochasticity.

9.5 Results

9.5.1 Local correlation of abundance

Figure 78, upper pane, verifies that local abundance of aphids at respective half leaves of a given leaf was weakly dependent. Half-leave aphid abundance at neighbour leaves was completely independent (Figure 78, lower pane).

A comparison of aphid abundance M at scale of half leaf with the average abundance at their respective full leaves revealed that for leaves with high M the aphids tended to be concentrated at one of the two half leaves (Figure 79). The dashed line shows the expectation from random dispersion of aphids between the respective two half leaves for each leaf. In this case M would be expected to be twice as large [$\log_{10}(2)=0.3$] at the scale of a leaf relative to its half leaves. Since leaves with $M>\approx 10$ showed a tendency for smaller M at leaf-scale than the dotted line (observe log-transformation), this reflects a concentration of individuals at one of the half leaves. In the extreme case with all individuals on one half leaf, the plot should be located at the continuous line where $M_{leaf}=M_{half\ leaf}$. Interestingly, many leaves did in fact show this pattern. The aggregative tendency can also be sees at leaves with $M_{half\text{-}leaf}<10$ [$\log(M_{half\text{-}leaf})<1$]. Since M_{leaf} over this range was much larger, it means that most individuals were located at the other half of these leaves.

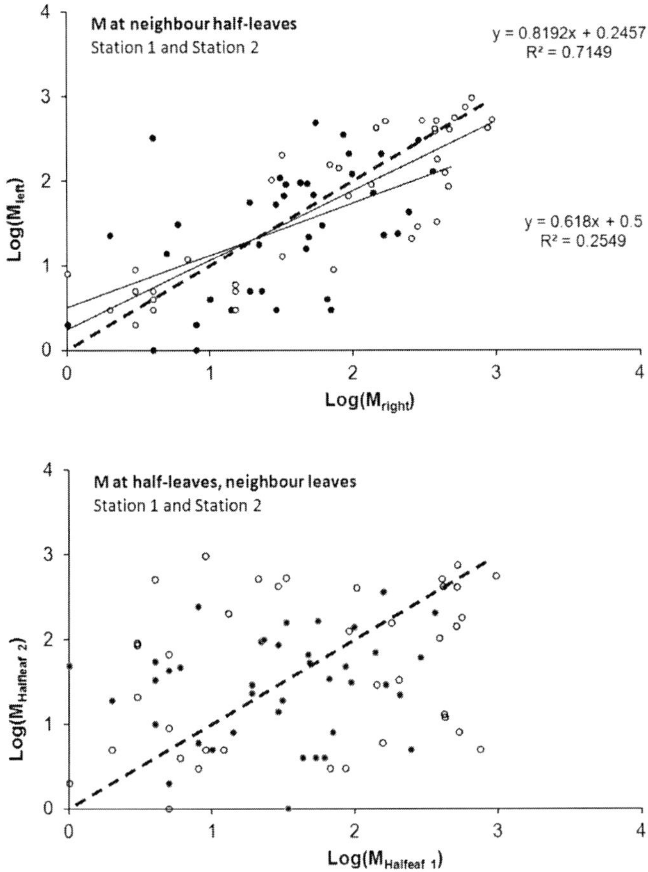

Figure 78. Comparison of local density of aphids at spatial scale half leaf and leaf. See main text.

Consequently, Figure 79 verifies that the aphids showed conspecific aggregation by preferring the local half leaf with highest abundance of conspecifics. By visual inspection it was apparent that the aphids confirmed a minimum inter-individual distance about equal to the length of their antennae, as could be seen also on the snapshot in Figure 74.

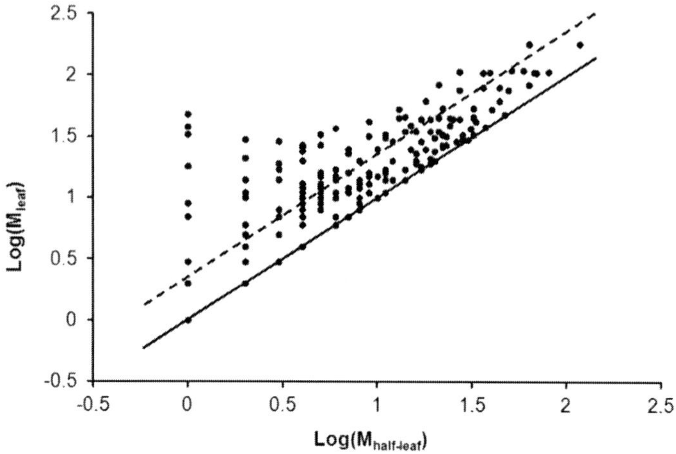

Figure 79. Comparison of aphid density at scale of half-leaves and respective leaf densities.

9.5.2 The density distribution

The aphid data from the two sycamore trees showed a power law distribution of local densities at scale of half leaves, in compliance with Eq. 27, with slope close to -1 (Figure 80). Abundance over the twenty days was summed, to reduce temporal noise and simultaneously demonstrating that temporal coarse-graining does not reduce a strongly aggregated distribution of M over space. D=1 implies that double magnitude of M appeared in half as many half leaves, over a range of M of magnitude $2^{7.5}$=181. This means that the aphid population contained individuals that were distributed in a self-similar (scale-free) manner over more than two orders of magnitude, with fractal dimension D=1.06, from the resolution of half leaves to the perimeter of a tree (the extent of the respective transects at the two stations).

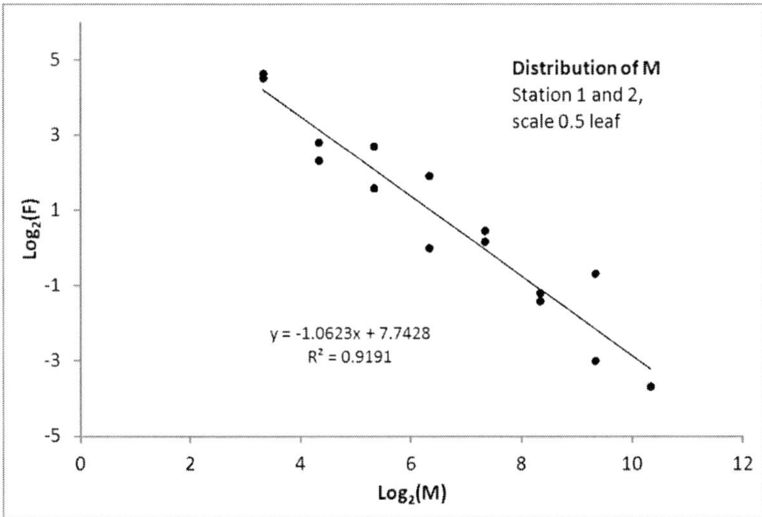

Figure 80. Distribution of aphid density at scale of half leaves.

9.5.3 The variance-mean relationship

Figure 81 shows a close agreement with Taylor's power law, with variance squared changing approximately proportionally with mean M over (a) respective transect counts of aphids over 160 half leaves, repeated over twenty-one days; and (b) averaging M over twenty-one days rather than over half leaves for the respective 160 half leaves. Plots with log(V)<0 were discarded (open symbols), owing to statistical artifacts for such small (M,V) scores. Figure 82 shows a similar result with the coarser scale of one leaf (eighty sample units).

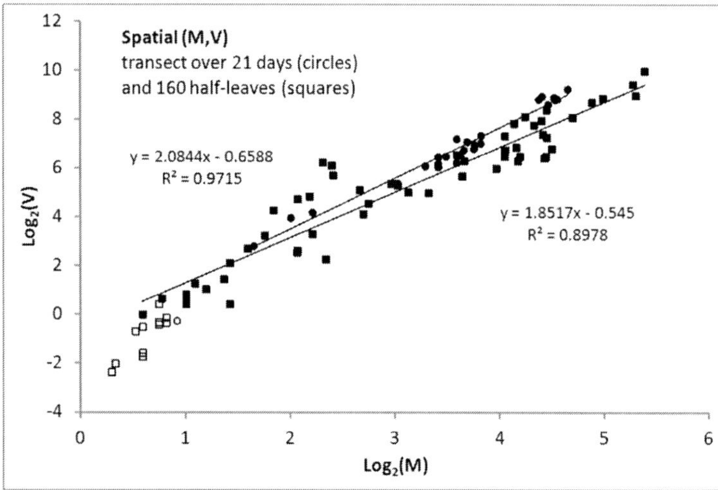

Figure 81. Variance-mean analysis of aphid dispersion. See main text.

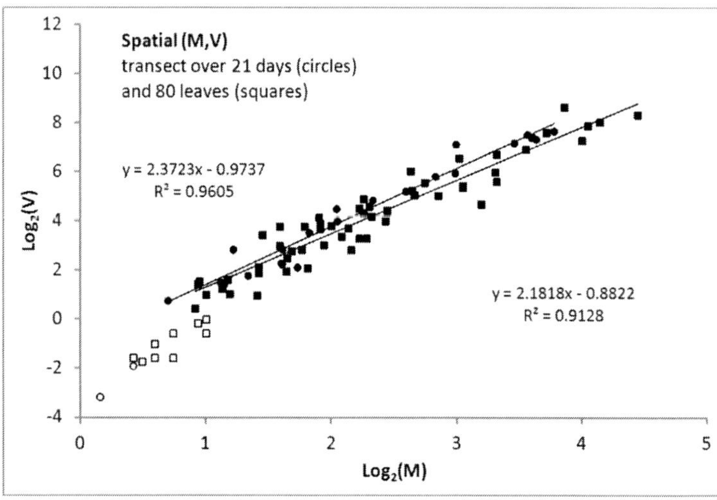

Figure 82. Variance-mean analysis of aphid dispersion. See main text.

Figure 83 shows how variance depends on M over a range of spatial scales. M gives the average abundance at respective scales (zero counts omitted), averaged over all sampling days. Transects from Station 1 and 2 were treated as a single sample (filled circle). Averaging over all sampling days gives a better statistical-mechanical representation of the process than treating days separately. In an analogue, the point count of specimens in an insect trap gives a better reflection of local abundance if the trap is open for (for example) ten hours rather than one hour. Alternatively, ten hourly counts could be summed or averaged to give the same result as from ten-hour continuous sampling. Similar argument holds for the pooling of sample units from the two stations. Larger sample reduces small-sample artifacts.

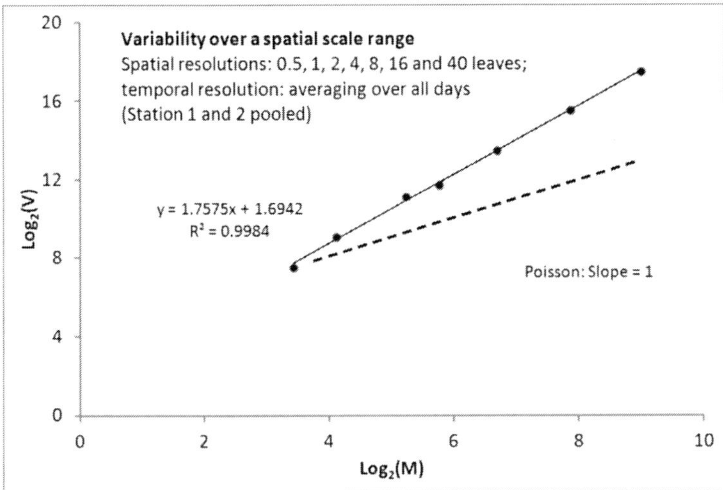

Figure 83. Variance-mean analysis of aphid dispersion. See main text.

The result shows an astonishing close compliance with a power law ($R^2 \approx 1$) and thus a scale-free spatial population structure on average over the sampling period of five weeks. The slope b=1.76 is somewhat smaller than the generic scaling model for equal kinetic energy over the scale range (equal weight expectation: $b \approx 2$). According to the parallel-processing conjecture, the aphids seem to have somewhat larger space utilization at finer spatial resolutions relative to coarser scales, but the 12% smaller slope than the equal weight expectation shows that this tendency is weak. The population aggregation is also substantially (76%) stronger than expected from the null model with b=1 (scale-specific space use, Poisson-compliant kinetics).

Despite relative and absolute discrepancies from uniform dispersion of sample units over space, the power law where M represents abundance at different spatial scales did not seem to be affected. Half leaves at the same leaf were closer together than half leaves at different leaves. Leaves at the respective stations had different inter-leaf distance. Larger groups of leaves represented an arc along the respective trees' perimeter, reducing the distance (linear scale) from first to last leaf in a group relative to the distance along the perimeter. Additionally, the distance between leaves on Station 1 and Station 2 were ca 200 m. I will return to this interesting aspect—apparent insensitivity of the scaling pattern to distance between sample units—in Chapter 10: independence between scale (resolution) and Cartesian distances points towards an additional paradox from classical expectations.

9.5.4 Testing for dominance of movement rate

Owing to compliance with $V=aM^b$ with $b>>1$, both over a range of M at a given spatial scale (Figure 81-82) and M over a range of spatial scales (Figure 83), the sycamore aphid population showed a congregation pattern that satisfied the fractal MRW dispersion from the scale of half leaves to trees. Crucially, this scale-free kind of "shifting mosaic" of aphid abundance developed gradually over the five-week period without changing the fractal dimension D or the variance-mean relationship in the overall statistical pattern. The result from the random growth simulation in Figure 84 shows that the aphid abundance properties were not due to a classical dispersion process. Starting from an aphid density distribution satisfying $D \approx 1$ at day 1 (Station 2), the simulated distribution has after twenty iterations of geometric population change drifted to $D = 1.53$, which is within the theoretical expectation from $1.5<D<2$. The significant change from $D \approx 1$ to $D = 1.53$ by simulation environmental stochasticity in the absence of migration falsifies the null hypothesis that the scale-free dispersion of sycamore aphids was generated by varying birth and death rates over space and time. Consequently, neither classical population dynamics nor classical movement processes could explain the aphid data.

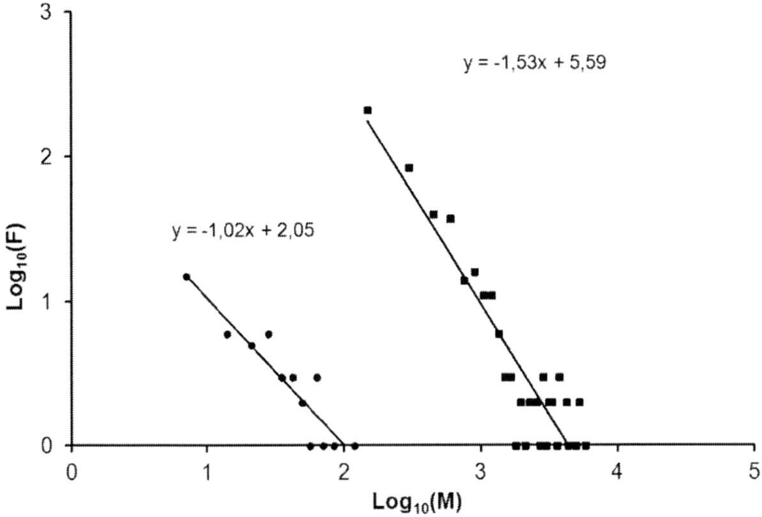

Figure 84. Density distribution of aphids (left) as a template for simulated geometric growth (right). See main text.

9.6 Summary and concluding remarks

Figures 79-84 show that the aggregative tendency is scale-free from resolution of half leaves and upwards towards coarser scales. While "*spaced-out*" gregariousness implies a negative feedback (individuals are aggregating until locally uniform dispersion is achieved at spatial resolution ca 1 cm^2), *scale-free* gregariousness is—according to the parallel-processing hypothesis—maintained by positive feedback (Eq. 8). If sycamore aphids moved in accordance to MRW in statistical-mechanical terms, two kinds of processes were thus dominating at different scales: spaced-out gregariousness at micro-scale (ca 1 cm^2) and scale-free aggregation at coarser scales from half leaves to (at least) tree size. While spaced-out gregariousness has a clear behavioural mechanism at the scale of the aphid's field of perception, scale-free gregariousness is a statistical-mechanical concept.

The results from this pilot study on a local sycamore aphid population lend strong support to MRW and parallel processing as the most feasible statistical-mechanical process for this species' population kinetics. Alternatively, the pattern might also be interpreted as aggregation from a Lévy walk-like space use at the individual level, since fractal dimension is close to D=1. However, while both Lévy walk and MRW share this magnitude of D, only MRW implies spatial memory utilization. A tendency for scale-free conspecific attraction is a feasible property from MRW but not for Lévy walk, which is memory-less in spatial terms. Figure 67 illustrated a feasible algorithm for conspecific attraction at the individual level, where respective individuals shared a few spatial attraction points where they had bumped into each other in the past.

Hypothetically, if the aphids were attracted to population aggregations based on olfactory cues (smell), which do not require a spatial cognitive representation of an individual's environment, it is difficult to imagine scale-free emergence of $V=aM^2$ where M represents abundance at various spatial scales. Advection (wind) and diffusion will rapidly tend to destroy otherwise scale-free concentration of attractive odours in the neighbourhood of aphid concentrations. A set of individuals following parallel processing under absence of spatial memory (Lévy walk) could thus be imagined to show aggregation in a more classical manner: a superposition of independently moving Lévy walkers may show D=1 in open environment, but introducing a repeated "fold-back" of their respective paths onto a common area of attraction should be expected to lose the scale-free dispersion of relocations and thus inflate the fractal dimension of local population dispersion. This was illustrated by simulation results in Figure 66.

In general there is growing empirical support for homing tendency among animals even during natal dispersal (Delestrade *et al.* 1996, Conradt *et al.* 2000, Travis and French 2000, Matthysen *et al.* 2001, Välimäki and Itämies 2003). The aphid results lend additional support to this behaviour.

9.7 The next step

As a natural follow-up to the foregoing analysis of sycamore aphids, new concepts are in the next chapter introduced and interpreted under the parallel processing conjecture. Prepare for 1/f noise, also called flicker nose and pink noise! To my knowledge, for the first time this very troublesome concept in both ecology and physics is hereby linked to the variance-mean theory of animal space use. May some long lasting controversies and paradoxes finally approach a solution under the parallel processing conjecture?

10 INTRINSIC COMPLEXITY: 1/F NOISE AND THE SCALING PARADOX

10.1 Population availability over space and time

One feature of natural population fluctuations is that the variance, V, in a time series of measurements on abundance—number of individuals in a sampled population over time, $M_t = \{M_1, M_2, M_3...\}$—tends to increase with the length of the series (Pimm and Redfearn 1988). Originally, this increase was seen as the consequence of extrinsic causes, whether the mechanism was density dependent [for example, a variable carrying capacity (Nicholson 1954) or not, for example, climatic perturbations (Andrewartha and Birch 1954)]. Longer series simply covered more temporal variations in the environment (Brown *et al.* 1995). A similar argument has been invoked to explain spatial variation. A larger transect contains larger embedded variation in the sample. Spatio-temporal covariance is called the Moran effect, or Moran's theorem (Moran 1953). It states that the time correlation of two separate populations of the same species is equal to the correlation between the environmental variability where they live, in particular with respect to climate. Strong effect implies strong spatial synchrony.

The question I will continue to address below is whether it is possible to separate this kind of extrinsically (environmentally) driven density fluctuations over time and space from the intrinsic causes of variability, with weight on the spatial aspect. Relative to the protocols in previous chapters, I will study the elephant from yet another angle, using power spectrograms. A spectrum has dimensions of power per frequency range, or energy per this range. The spectrum decomposes the content of a stochastic process into different frequencies present in that process, and helps identify periodicities and strength of respective cycles (power equals the squared magnitude of the amplitudes).

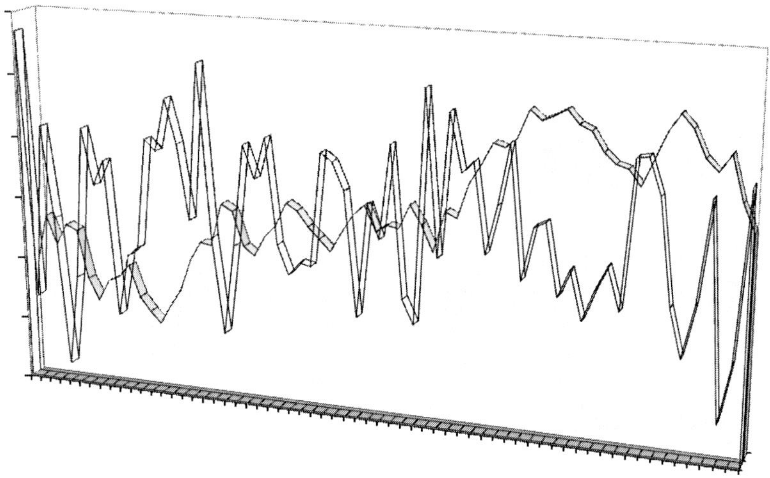

Figure 85. Two simulated series of abundance with the same overall variance, V, and mean number of individuals, M, may have very different fluctuation patterns. The shaded series is a Brownian motion (classical random walk); the next value equals the foregoing value, plus a normal-distributed increment with a zero mean expectation (i.e., 50% chance for a positive and negative change for successive increments, regardless of lag). For the series as a whole, this creates a strongly autocorrelated trace and a strongly negative slope, b = -2, for log(power) versus log(frequency) in a power spectrogram. In other words, the spectrum is reddened. The transparent series has stronger increments, which are normal-distributed around mean M for the series. Thus, the power spectrum is white, giving zero slope in a power spectrogram (b=0). Since overall V and M are equal for the two series, the summed (integrated) variability over the range of lags in the series (resolutions from lag 1 to lag N/2, where N is the series length) is equal. The difference lies in how this total magnitude of variability (power) is distributed over the range of lags. In the Brownian motion series, most of the power is concentrated at the largest lags, with little power (small amplitude of cycles) at finer lags. In the white noise series, power is uniformly distributed over the range of lags.

A basic but little-emphasized premise for stochastic variation of the *classical* type is that the process is stationary. The frequency variable should then be associated with a stationary stochastic process. A classical system is—as defined in previous chapters—based on the Markov premise. Consider a time series of population abundance at high-frequency sampling, meaning that the effect from birth and death rate may be *ignored* at statistical-mechanical scales and only individual-level movement remains to explain the kinetics at this resolution. Fluctuations under this classical premise then regard movement in and out of sample units, depending on the number in the previous point in time plus a random (positive or negative) perturbation. The increments satisfy a Gaussian distribution. Such a series is temporally autocorrelated, meaning that two successive densities are more similar than two distant ones (Figure 85). This is called a first-order autoregressive model (AR1), which is another name for a first-order Markov model. A density ceiling, appearing under equilibrium conditions [a situation that has proven less abundant than previously thought (Andrewartha and Birch 1954, Pimm and Redfearn 1988)], tends to reduce the temporal autocorrelation due to more dampened fluctuations. In a spatial series rather than a time series, variability regards change over space at a given moment of time (a snapshot). Alternatively, a spatial series may consist of units where the respective values at a given moment of time are replaced by the average magnitude of a given sample unit's values, from averaging the unit's abundance value over a period of time.

Under classical conditions—a stationary series—the successive increments are statistically independent (for example, the difference M_{t+1}—M_t is independent of M_{t+2}—M_{t+1}). In a spatial context, like the leaf-to-leaf abundance of aphids in Figure 78 with low or no spatial autocorrelation, one should from *classical* assumptions expect a similar compliance with the stationary variability premise. However, both the aphid results and other results in the foregoing chapters showed examples were simulated and real data showed non-stationary variability over time, space, and scale. The mean, the variance, and the variance as a function of spatial resolution all violated the Markov expectation from a non-autocorrelated series. Spatial fluctuations in local abundance, M, were scale-free, and variance increased non-linearly with physical size of sample units over a scale range from half leaves to trees.

The latter was particularly astonishing, since it was not explicable from local correlation of abundance: auto-covariance of M was weakly positive only between half leaves of the same leaf, and absent between neighbour leaves and upwards in scale. Consequently, the magnitude of V for a given magnitude of M was independent of physical distance between sample units at a given spatial resolution. Thus, the traditional first guess or assumption—increased variability as a consequence of transect length or scale as such (longer series embedding more space in Euclidian terms and thus more environmental variability)—did not hold. This was expected, since resource availability on leaves, groups of leaves, and different trees probably represented quite homogeneous environment. Still, V increased non-linearly as a function of increased spatial scale. Below I bring this set of "odd" results from a classical system perspective a step further by elaborating on the theoretical aspects, in particular with respect to the relationship between variance and power spectra of population variability.

Power spectra provide an alternative approach to study such complex (non-Markovian) fluctuations and thus a method to quantify both trivial (classical) and non-trivial population variability in statistical-mechanical terms. When and where the premise of a stationary process is not met, heated discussions and creative models have emerged to explain the discrepancy from classical systems theory. However, the controversy persists in most fields of science where non-compliance has been documented—and these fields are many and still growing.

10.2 Chaos and self-organized criticality

Two alternative system categories show inflated variability if intrinsic origin needs mentioning: low-dimensional chaos and self-organized criticality. Chaotic fluctuations in a time series were a hot topic in theoretical ecology twenty to thirty years ago. Emergence of so-called low-dimensional, deterministic chaos in difference equation models with sufficiently strong non-linear terms is the now-classic example of intrinsic complex variability. A chaotic system in this context is low-dimensional, primarily because spatial effects are absent. With reference to previous chapters, a population showing chaotic fluctuations is assumed to be both Markov-compliant and well mixed with respect to its kinetics (strong movement rate at the individual level, like in a stirred bioreactor), and thus driven mainly by birth and death rates. Not only do the fluctuations in a chaotic system prevail, but they have a tendency to increase V in a series the larger the series is made. It simply takes some time to experience the largest cycles (in terms of power spectrogram analysis, since a chaotic trajectory is not cyclic). The term "chaos" is typically linked to the property "critical dependence on initial conditions." Starting a system from almost the same initial state may show similar trajectory through phase space for some time increments but then depart into independent paths with an exponentially increasing divergence. This is the famous "butterfly effect" that characterizes chaotic systems.

Prevailing fluctuations from intrinsically driven complex dynamics are found also in another class of models. Numerous computer simulations have shown that for spatially extended systems (lattice models), complex dynamics may emerge and prevail even in a homogeneous environment

(Solé and Bascompte 2006). Environmental variability then comes in addition to the intrinsically driven fluctuations. In physical systems the time to stabilization of fluctuations can be extremely long under very broad model conditions (Bak *et al.* 1987). If these conditions are satisfied, variance also increases as we extend the density observations in a series over more space in a similar manner as when we include more time. Thus, the system is no longer low-dimensional with respect to axes showing the evolution of the variable's trajectory in phase space. With reference to animal abundance and individual space use, I will refer to such non-Markovian systems as expressing intrinsic complexity (IC).

If a chaotically fluctuating population is studied on a spatial scale (resolution of the sample units) that is larger than the scale for "full mixing," it will show proportional response to environmental changes, in contrast to finer scales where functional responses are non-proportional (non-linear) and erratic due to the local butterfly effect. In short, the chaotic influence on the variability is averaged out at coarser scales (technically, the system is no longer "low-dimensional"). On the other hand, IC regards a scale-free process over a larger range of scales. There seems to be an absence of a characteristic scale for full mixing, as expected for a Markovian system. Here I will focus on IC rather than Markov-compliant and deterministic chaos, since variance, V, in natural populations tends to increase over space, time, and scale in complex and little-understood ways (Taylor 1986) and in a manner that does not satisfy expectations from low-dimensional (deterministic) chaos.

While deterministic chaos may be understood within the framework of classical statistical mechanics, IC embeds a broader kind of complexity, sometimes referred to as a "mild chaos," connected to the concept of self-organized criticality (SOC). Like the parallel-processing concept, neither SOC fits into the framework of a mechanistic/Markovian system. Therefore, the term "mild chaos" is not indicating a transition between a chaotic and a non-chaotic system in the context of classical deterministic (and thus predictable) evolution of the system state. Rather the term "mild" refers to the magnitude of the so-called Lyapunov exponent, which is large in chaotic systems (strong butterfly effect) but closer to zero in mildly chaotic systems (SOC). The classical example of SOC is a pile of sand. Adding new grains will eventually lead to avalanches. If grains are continuously added one by one, the slope of the pile will gradually be "driven" towards a critical state where a new grain may just reach a relaxed position somewhere on the pile after some tumbling or—occasionally—ignite an avalanche of *a priori* unpredictable size from domino effects. Interestingly, in this slowly driven state, by adding grains, potential energy is gradually increasing (slowly increasing the angle of the slope). This accumulation of potential energy is released by many small or some occasionally large avalanches.

While the single-grain effect is unpredictable (hence underscoring that a system's state may change in an even more disrupting manner than in a low-dimensionally chaotic system), the statistical-mechanical system properties are highly predictable. For example, the distribution of avalanche sizes will tend towards a scale-free distribution, conceptually similar to what we have already seen in previous chapters: scale-free step length distribution (Lévy walk and MRW), scale-free distribution of local population density, and so on. Interestingly, both physicists and mathematicians studying SOC tend to invoke "long-term memory" as a feasible term to explain the complex "behaviour" in such systems. It is of course not real memory, since a cognitive object is not involved, but it is applied as a heuristic explanation to describe what is going on over time, space, and scale. In short, these systems are both far from (classical) equilibrium and still show properties that are predictable at the statistical-mechanical level but not at the micro-scale level. The concept of SOC has even been applied to interpret complex cognitive processing in mammalian brain (Bédard *et al.* 2006). Clearly, SOC is non-Markovian, and not explicable within the mechanistic framework.

The observant reader will now immediately see how the purely physical kind of complexity—SOC—may be closely linked to the concept of parallel processing! An individual's cognitive map for space use and site fidelity regards complex processing of information in the brain, which, from a completely different approach that invokes theory of SOC, has been verified to show bursts of activity (Bédard *et al.* 2006). I dare to speculate that a key to linking these two aspects at a deeper physical or biophysical level may one day be revealed.

As it stands now, SOC regards empirically derived relationships with still unclear system details with respect to the physics of these broad-ranging classes from piles of sand to brain activity—and thus perhaps memory map utilization. What regards the latter, conceptualized here as parallel processing; it is currently a postulate for scale-free spatio-temporal cognitive dynamics, invoked as a premise for MRW-compliant space use at the individual level. In *6.4 Population kinetics under the parallel-processing premise,* I showed coherence at the individual and population level by applying this postulate at both levels, with support from sycamore aphids in *9.2 Exploring complex population kinetics in sycamore aphids*. While SOC theory invokes the memory concept as a metaphor to describe the scale-free nature of the process, parallel processing invokes memory as a concrete concept. The latter is linked to (for example) how kinetic energy is dissipated over a range of scales as a consequence of cognitive processing of spatially explicit knowledge of the individual's environment (Eq. 11 in *6.4.6 Dispersion of kinetic energy over a scale range*). Both approaches, parallel processing and SOC, regard a statistical-mechanical level of system abstraction.

The fact that parallel processing is a postulate rather than being verified from first principles at this stage is not necessarily a critical weakness. Boltzmann's famous definition of entropy under condition of equilibrium in classical-statistical mechanics was also based on a postulate; in its simplest form $S = k_B*\ln(\Omega)$, where S is the entropy, k_B is Boltzmann's constant, and Ω is the number of microstates. Consequently, entropy of sub-sections of a system at equilibrium scales logarithmically rather than arithmetically. In fact, this description is still only a postulate, but it represents a cornerstone of this branch of physics. It has indirectly been proven right from so many angles that it is considered a law of nature. In short, "it works." To what extent parallel processing also works when tested under a broader range of individual data and theoretical development than presented at this stage remains to be seen.

10.3 Linking alternative descriptors of complexity

Even if population variability under IC may be less correlated with extrinsic factors when measured from traditional variables from classical population dynamical modelling (like correlation with local fluctuations of resources), population abundance is obviously still linked to changes in the environment. However, such correlation requires alternative statistical variables for proper quantification. In previous chapters I described several such "complexity" variables linked to IC both at the individual and the population level: fractal dimension D, the power exponent β in step length distributions, the power exponent b in variance-mean relationships, the exponent z and the intercept c in the home-range ghost context, and many others. For example, the red deer home ranges showed a wide range of c-estimates; observed space use from a given number of fixes, N, varied between individuals. One could thus imagine $1/c$ (larger $1/c$ implying smaller space use for a given N) being an alternative expression for space use relative to the traditional quantification "home-range size," which is a very fuzzy concept based on Eq. 1. Larger *1/c* implies stronger space use intensity (fixes are less "spread-out" in overall terms). Further, as documented in the population model with zoomers, I showed how a specific change in the environment at one site can in an unpredictable way lead to no apparent response locally and in the short term, or it can trigger a cascade of fluctuations over large areas and time periods. Thus, a specific change in density on a site may be linked to events at other places a long time ago by domino effects, and correlation with local environmental conditions becomes blurred.

The resemblance with SOC seems strong, but has so far not been properly explored theoretically. These cascade effects may be inexplicable from a local perspective, but explicable at coarser scales (the parallel-processing concept). Non-proportional functional responses and long-term and long-distance interactions are in fact the two major fingerprints of non-linear system dynamics in general (West and Shlesinger 1990) and SOC in particular. *A complex system seems to be associated with no preferred spatial or temporal scale when relaxing after a perturbation.* Energy is "dissipated" over a range of scales, as was explained for a pile of sand above. Similarly, the sycamore aphids showed many small and a few large fluctuations in a scale-free manner. Individuals show a complex, scale-free distribution of many small and a few large displacements. If the physics of potential energy (a system's power to perform work in the future) can be further linked to the context of spatially explicit memory—including the kinetic energy concept in Eq. 11—SOC and parallel processing may be interwoven.

10.4 The power spectrum

10.4.1 Fourier transforms and reddened spectra

An alternative way of demonstrating the increase of V over spatial or temporal scales in a series is to decompose population variability into cycles using techniques like Fast Fourier Transforms (FFT) and then study how the *power* (squared magnitude of amplitudes) of those cycles varies with their frequency. The frequency equals the inverse of cycle length, and the power is an expression for cycle-specific variance. A larger frequency means a smaller cycle length. Cycle length is proportional with the lag in the series, which is the time interval or space distance between successive measurements we use for calculating the power. In the spectra this analysis produces, the longer cycles typically have the greatest power (larger squared amplitude). Consequently, the power spectrogram is termed "red."

The spectrum of the colour red has more variance at larger wavelengths, or smaller frequencies, than at shorter ones. A reddened spectrum is often described as emerging from a superposition of several sine waves with different wavelength and amplitude. An increased series length may in this case lead to increased variance, because more sine waves may be included and the waves with the largest wavelength—in the red part of the spectrum—may have the largest amplitude.

To summarize the red noise phenomenon in classical terms, variance and power are larger when we measure them over larger lags. When the power is independent of cycle length—meaning that the spectrum is "flat"—we call it a white spectrum. In this case no colour or specific cycle length is dominating the fluctuations. Serial autocorrelation is then absent. White noise is also expected to appear if we look at the differentiated series where the original series shows a reddened spectrum (Figure 85). Differentiating means that each number equals the difference between successive original numbers: M_2-M_1, M_3-M_2, etc. M in the present context is the population abundance at time t or spatial sample unit u in our series. The spectrum from the array of such increments then describes the colour of the time series, which for a Markovian system is white if the original series is red.

If temporal or spatial variance is independent of the series length (white noise in the power spectrum), we have simple population kinetics from a *stationary* process, based on Brownian motion theory. In this case the "motion" is fluctuation of the dependent variable in a time series or a spatial transect, not the physical path of a single individual. As already explained, under classical conditions of a stationary process, M_{t+1} equals M_t plus a random positive or negative number with zero mean. Thus, the increments describe white noise. Brownian motion is non-complex, because the variance in a series of first differences is constant under increasing series length. The other way around, integrating white noise series produces a Brownian motion with reddened spectrum. Sometimes such an integrated spectrum is called "brown" rather than reddened (despite the fact that Brownian motion refers to the botanist Robert Brown, not the colour).

The power spectrum is the integral, P, of the variance of increments in a variance versus lag regression, using frequency (the inverse of lag) as the independent variable:

$$P = Q/f^{\eta} \qquad (28)$$

Q is a parameter and f equals frequency. "The integral, P, of the variance of increments" means that Eq. 28 is expressing a superposition of inter-independent cycles' contribution at a given frequency, and where the sum of each cycle's amplitude contribution at a given frequency is squared. Compliance with the principle of "superposition" or "sum" was explained conceptually in *2.5 Markov from another angle: the principle of superposition*, and described as a necessary property of a Markov-compliant process. Differentiating a reddened series—for example, an autocorrelated series with Markov-compliant fluctuations with exponent η=2 in Eq. 28—reduces η to η-2 and leads to white noise (η=0; successively independent power levels over frequencies; see Figure 86). Conversely, integrating white noise increases η by two and leads to strongly autocorrelated and Brownian motion-like fluctuations. The slope of the power spectrum then becomes close to -2 when both axes (integrated variance and frequency) are log-transformed. The dynamics behind such a pattern is well understood within the framework of linear statistics and classical-statistical mechanics. "The total variance equals the sum of the parts" in compliance with the central limit theorem and Markovian processes. Complexity arises when and where a non-autocorrelated series shows $0<\eta<2$; for example, $\eta\approx1$ as in 1/f scaling.

10.4.2 Flicker noise

On one hand, in a time series successive values may be made independent by differencing and analysing the series of increments. On the other hand, a spatial series may *a priori* consist of independent values beyond a given resolution. For example, in the population of sycamore aphids, the left and right half-leaf abundance was weakly positively correlated, while abundance on neighbour leaves and beyond was independent (Figure 78). The theoretical difficulties emerge when the variance increases stronger than proportionally with spatial or temporal scale, and successive values are *independent* and thus should behave statistically—under classical conditions—as a differenced time series as in Figure 85. With respect to Eq. 28, the classical expectation for a differenced series is η=0 (Figure 87; observe log-transformed axes), while non-classical pattern is verified if $0<\eta<2$. If this is the result of differentiating a time series, the original series has a "black" spectrum (Figure 86). Black spectra have, for example, been found in the variability of ice thickness in the arctic (Rothrock and Thorndike 1980):

> *When the spectral exponent is -3 or greater, the functions display properties which may seem pathological on a first encounter, most*

strikingly the ability to be continuous but not differentiable. Perhaps these functions seem strange because they do not appear in the studies of calculus and differential equations. They may be unfamiliar, but they are not unnatural. In fact, the pathological properties of these functions correspond with observable properties of many natural objects.
Rothrock, D.A. and Thorndike, A.S. (1980), p. 3,958.

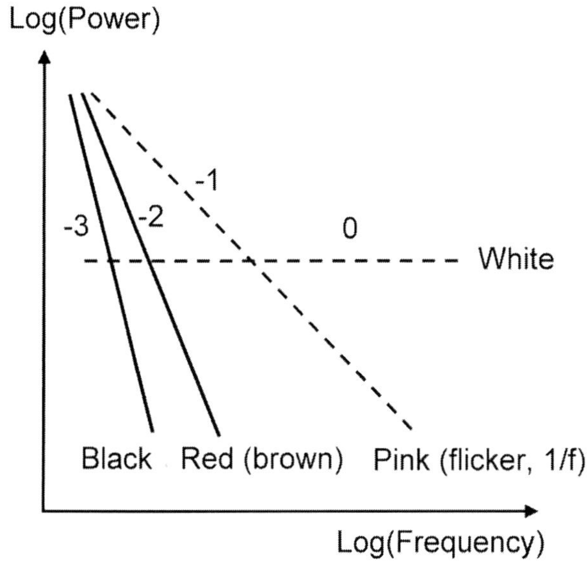

Figure 86. Typical regression results from log-log transformed power spectrograms. Scattered lines illustrate noise "colour" of differentiated or *a priori* non-autocorrelated series (increasing slope exponent by 2 relative to the integrated counterpart). Complexity is linked to black and pink noise.

Observe that the lag is defined as the distance in space or time between successive values (sample units), not the series length as such (sample extent). Thus, for situations where $0<\eta<2$, the red/white noise concept needs—in an apparently innocent manner—to be diversified. $0<\eta<2$ is referred to as pink noise, flicker noise, and 1/f noise.

... there are no simple, even linear stochastic differential equations generating signals with 1/f noise. The widespread occurrence of signals exhibiting such behaviour suggests that a generic mathematical explanation might exist. Except for some formal mathematical descriptions like fractional Brownian motion (half-integral of a white noise signal), however, no generally recognized physical explanation of 1/f noise has

been proposed. Consequently, the ubiquity of 1/f noise is one of the oldest puzzles of contemporary physics and science in general.
L. M. Ward and P. E. Greenwood (2007), Scholarpedia, 2(12):1537.

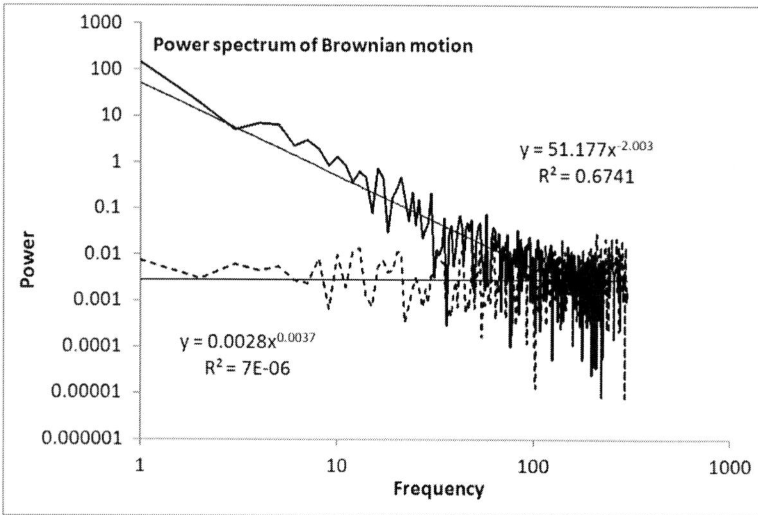

Figure 87. Power spectrum of Brownian motion and its increments; simulated data.

Flicker noise in a time series or in a spatial transect means that the variance of density differences of a population tends to increase (on average) with increasing lag in a power law manner close to a specific power exponent, $\eta\approx1$. The flicker noise concept is on one hand linked to $0<\eta<2$ for a time series and $-2<\eta<0$ for a time series of first differences ("blue" spectrum, with more power at large frequencies) and on the other hand a spatial transect where successive values (like population abundance, M) are uncorrelated but still show coloured spectrum with $0<\eta<2$ ($-2<\eta<0$ in the differentiated series, shown as a dotted line). With respect to the "Zoomer" model for complex population dynamics, Figure 88 shows the lack of significant auto-covariance at level k=1 already at lag 2 (comparing log(M) in two sample units two cells apart; R^2=0.16), and very weak auto-covariance in neighbouring cells (lag 1; R^2=0.38). Figure 78 showed a similar lack of autocorrelation for aphids. Figure 89 shows the power spectrogram for the aphid abundance at scale 0.5 leaf, revealing close compliance with 1/f noise ($\eta\approx1$) and the differentiated series ($\eta\approx-1$). Despite lack of substantial autocorrelation, the V-M statistics are non-linear, non-Markovian, and thus far from "innocent." As shown by the citation above, 1/f noise is still an unsolved mystery and a central theme in complexity research in general. The Zoomer model describes one path towards 1/f noise, and simultaneously describes the link between the population level and the individual level.

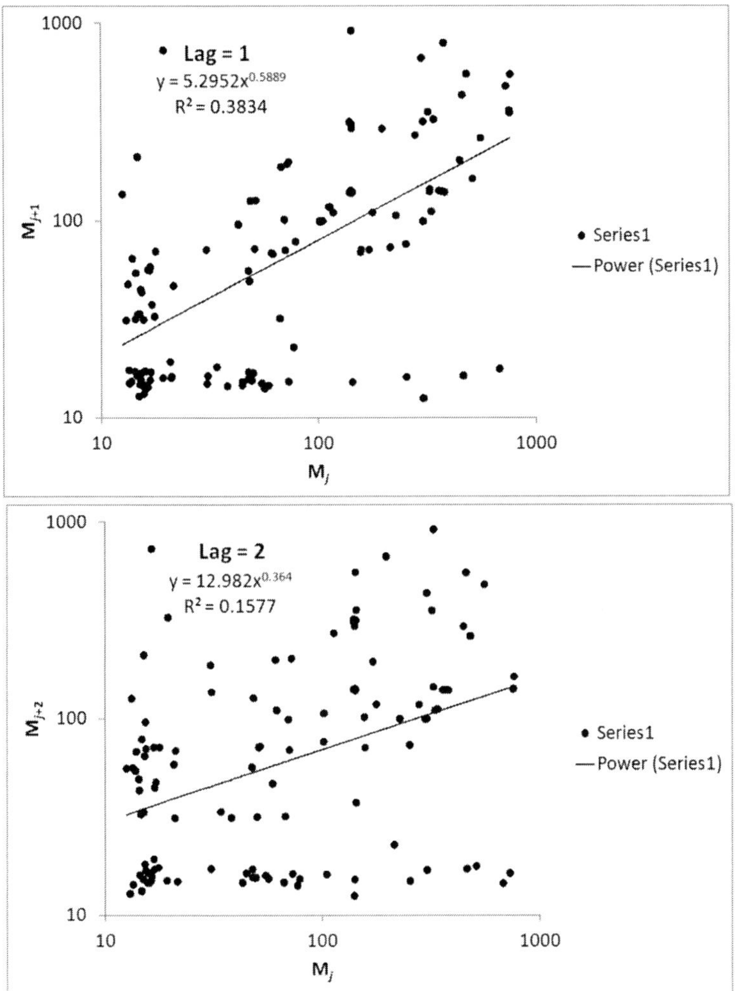

Figure 88. Inter-level comparison of population density, based on Zoomer model data. See main text.

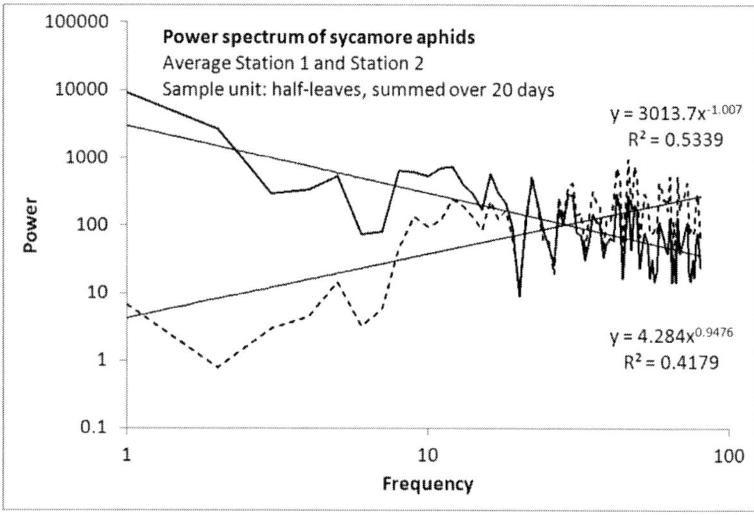

Figure 89. Power spectrum of sycamore population data, summed over 20 days (large hidden layer of system detail ignorance).

10.4.3 Invoking a grain/window conjecture to quantify complex spectra

Q represents the noise level within the space and time period covered by our series (similar to the intercept *a* in $V=aM^b$). *P* is a function of *f* for a given *Q* and series length. Non-autocorrelation implies that we are studying the colour of the noise in the series. If the series shows strong autocorrelation [like the series in Figure 85 in a spatial context, or the underside of ice as referred to above (Rothrock and Thorndike 1980)], it should first be differentiated to allow for a study of the statistical noise. However, in the following I will give *f* an alternative definition, complex f_c, representing the independent variable as a replacement of the inverse of *f* in Eq. 28 to quantify variability of population abundance over space:

$$P_c = Q_c/f_c \propto u/U \quad | \; u \ll U;\; u = \mathbf{Z},\; U = \mathbf{Z} \qquad (29)$$

P_c and Q_c represent respectively the power and scale level parameter in a system expressing flicker noise ($0<\eta<2$, like the aphid data in Figure 89), which in the current context is termed intrinsic complexity (IC). The numerator u equals observational grain (size of the sample unit), and the

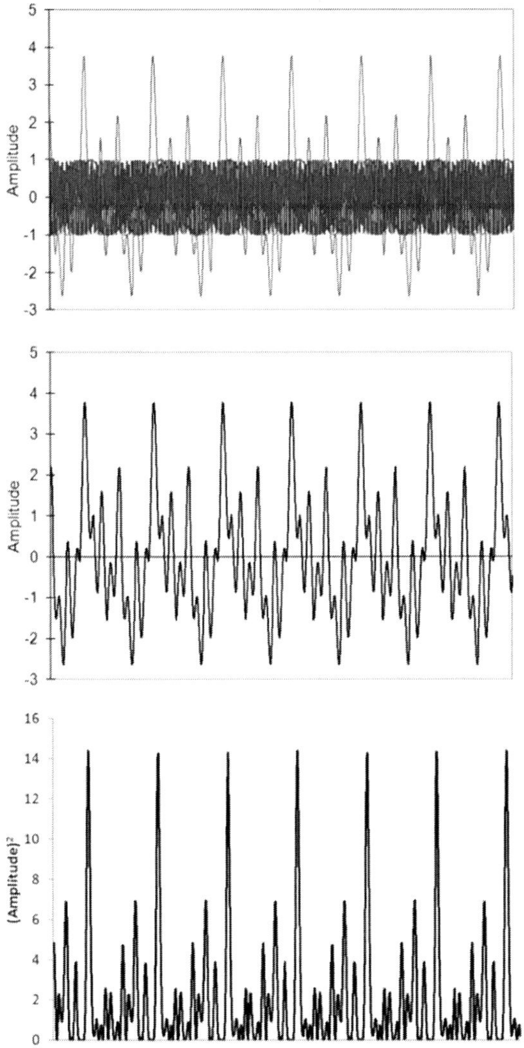

Figure 90. A power spectrogram is based on the principle that a series' variability can be understood as a superposition of cycles of different wavelengths and amplitude. In the upper pane, eight separate sine waves are shown. In the middle pane these cycles are superimposed into a single, multi-amplitude series. In the lower pane the amplitudes of the middle pane are squared, to illustrate the integrated series' spectral power. Observe that squaring removes negative values in the series.

denominator U equals window (the sum of grains). **Z** implies integer, and $\propto$ means "proportional to."

In the traditional context of power spectrograms, $1/f$ regards an interval in time or a distance in space, termed "lag" in Euclidian terms. Thus, *equal physical distance between sample units is assumed.* This is implicit from the decomposition of variability into cycles (sine waves), which is characterized by equal-sized lag between amplitude maxima and minima for respective wavelengths (Rothrock and Thorndike 1980, Voss 1989) (Figure 90). In contrast, Eq. 29 specifies indirectly that units of size u may or may not be equal-distanced in physical space (Figure 91). u/U expresses a dimensionless ratio, which means that distances like meters and intervals like hours are cancelled out. A ratio means that the expression in Eq. 29 is assumed scale-free (see also Figure 49, where this property was explained in the context of a scale-free distribution of individual step lengths). Only if window equals extent, sample units are by *necessity* equal-spaced since the sample units fill the entire extent. The spatial (or temporal) extent is the total spatial length (time period) embedding the set of sample units. This means that a given number of sample units (the window U) may cover the entire spatial extent for the study or they may be locally aggregated (U smaller than extent, owing to gaps between sample units of size u). For non-autocorrelated series, lag does not matter. If the system is complex (IC-compliant), it is the u/U ratio that according to the proposal in Eq. 29 determines the observed power and variance, not Cartesian distances in one-, two- or three-dimensional space. It is the non-denominated ratio of grain u relative to window size U that determines complex power (power spectrogram in an IC-compliant system) as a function of complex observation frequency, f_c. Like a classical power spectrogram (Eq. 28), the analogous complex spectrogram regards integrated power, P_c. However, the respective cycles do not reflect the variance of increments at different spatial or temporal lags in physical space. Instead, the lags regard magnitude of the scaling ratio u/U. While Eq. 28 is expressing a superposition of inter-independent cycles' contribution at a given frequency in terms of spatial lags, Eq. 29 regards the integrated power at respective $1/f_c$, which is postulated to be proportional to the dimensionless ratio u/U. Classical power spectrograms imply compliance with the principle of superposition (like shown in Figure 90) and thus regarding Markov-compliant processes. Complex power spectrogram also regards superposition of cycles, but along the non-denominated scaling axis quantifying u/U. Spatial and temporal lags are not reflected in Eq. 29.

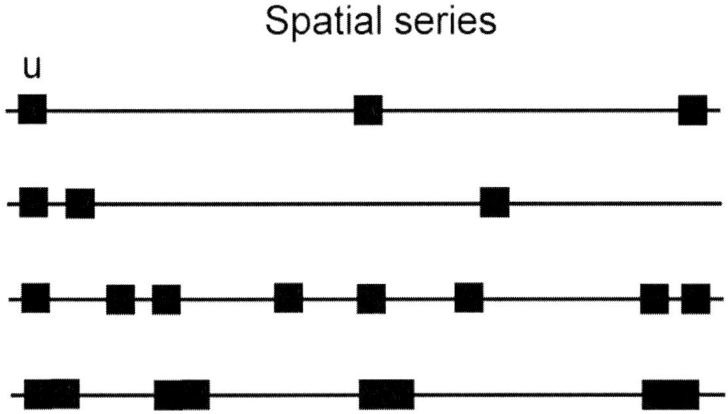

Figure 91. Sample units (u) in a spatial transect, under condition of intrinsic complexity and non-autocorrelated values, may be evenly or unevenly distributed (first and second transect line). According to Eq. 29 power is normalized by dividing by window size, which equals the sum of sample units. Thus number of units may be varied, as in row 3, and still represent same magnitude of power at the given spatial resolution (black boxes). In the lower row units are lumped into larger entities, representing power at a coarser spatial scale.

In this respect I quote from *9.5.3 The variance-mean relationship*, where this aspect was reflected upon in the context of sycamore aphids:

> *Despite relative and absolute discrepancies from uniform dispersion of sample units over space, the power law where M represents abundance at different spatial scales did not seem to be affected. Half leaves at the same leaf were closer together than half leaves at different leaves. Leaves at the respective stations had different inter-leaf distance. Larger groups of leaves represented an arc along the respective trees' perimeter, thus reducing the distance (linear scale) from first to last leaf in a group relative to the distance along the perimeter. Additionally, the distance between leaves on Station 1 and Station 2 were ca 200 m.*

In Eq. 29 I specify a conjecture for scale-free, complex population kinetics. In this frame of reference, it is a ratio that represents a scale-free demarcation of observational scale, with respect to analysing the level of power (Eq. 29, with exponent $0<\eta<2$) or variance (Eq. 12, with exponent $b>1$ over scales by varying u) over the IC-compliant scale range in space and time. In contrast, a non-complex system is expected to show power as a function of grain and extent (Eq. 28), where window size is proportional to extent (equal-distanced sample units, with or without equal-sized gaps between these units).

Small or large numbers of sample units are compensated for ("normalizing" variability) by dividing u by U. In fact, the ratio u/U is reflected in the definition both of the average and the variance of abundance. The average is μ = E[M], and the variance is the squared deviation from μ: V(M) = E[M- $\mu)^2$]. In both calculations the function E—the expectation—involves respective sums over a sample, divided by "degrees of freedom" and thus normalized by window size as in Eq. 29. Small sample sizes may produce statistical artifacts, though; this aspect is partly specified in Eq. 29 by the condition u<<U (if the ratio is small, there is not sufficient "room" to allow for a large number of sample units).

10.4.4 Complexity may require an extra Cartesian coordinate

The frame of reference described by Eq. 29 is a dramatic departure from a classical space-time-scale framework in statistical mechanics in general and in analysis of complex fluctuations (like application of power spectrograms) in particular. For the sycamore aphid data, it was the non-denominated and thus dimensionless ratio u/U that determined the level of variance, which was independent of how sample units were spatially dispersed. Linking the numerator u alone (with dimension, for example, m^2) directly to inter-unit distance in space (proportional to $1/f$, the inverse of distance) would create a paradox for the aphid results, since u is then explicitly linked to physical dimensions in an Eulerian sense. $1/f$ is explicitly linked to cycles—sine waves—with equal spatial or temporal lag between successive maxima and minima of the respective waves. Eq. 29 resolves this paradox by loosening the dependence on equal lag in physical space and time.

To summarize, Figure 91 showed the paradox of flicker noise in a nutshell. Consider all sample units congregated together (both uni-spaced and continuously distributed). Increasing the number of sample units will—if the variability is autocorrelated—show a reddened spectrum. Additional cycles with increasing amplitude will be needed to describe the variability as the series length is increased. On the other hand, if autocorrelation is absent, the spectrum is "flat," with new cycles having the same amplitude as previous cycles. So far no paradox is seen. Next, consider spacing out the sample units (whether uni-spaced or not). In the case of trivial red noise (autocorrelated series, like integrating a white noise series), new cycles in the spectrogram have larger amplitude. The 1/f paradox appears if a *non-autocorrelated* series shows a reddened spectrum whether the sample units are uni-spaced or not or continuously dispersed (without gaps) or not.

In the aphid data and the data from simulated population complexity, I verified complex 1/f noise (Figure 89) owing to non-autocorrelated series both for aphids (Figure 78) and the simulated population (Figure 87). I also showed that magnitude of power depended on the physical size of the sample unit—not on distance between sample units (Figure 83 and Figure 40a,b; respectively). This property raises the 1/f paradox.

Now the controversial and radical proposal: to resolve this paradox one may need to define an axis that stands orthogonally on the spatial axes, quantifying the u/U ratio. Recall that this ratio is not necessarily proportional with frequency or cycle lengths as in a traditional frame of reference, since the premise of equal-distanced sample units (fixed lag at respective series resolutions) is relaxed. While this added axis is dimensionless in a non-complex framework (like Cartesian axes for three-dimensional space), it requires a denomination—a currency—in the extended framework. What to call this unit? I propose to name it "teflo," which to my knowledge does not resemble the name of any other physical unit. Since u<=U and a small or large teflo is expressing a large or small scale ratio, it is convenient to define teflo as the currency for complex frequency, f_c, in Eq. 29. For example, f_c=10 teflo would then express u/U = 0.1. Teflo regards a quantification of the "scale width" between u and U and thus the number of sample units, whether u or U per se is small or large from the perspective of the classical system. The teflo unit is scaling logarithmically to reflect its scale-free nature. Log(teflo)≈0 is consequently an asymptotic limit, where the size of the numerator u is approaching the scale window U (f_c=1). Negative teflo is not possible, since it would require the forbidden magnitude u>U.

From a statistical-mechanical perspective, only by this system-geometrical extension do the empirically derived results seem to make sense for what have been defined as complex systems with non-Markovian memory capacity. In this extended frame of reference, scale is decoupled from Euclidian distances with respect to studying dispersion of population variance over a scale range and dispersion of power over a frequency range. Instead, distance and lag are replaced by a ratio, u/U, with its reverse (U/u = f_c>=1 teflo) expressing the magnitude of the quantified scale range. A complex system's power, P_c, is quantified by Q_c, which is the power unit for f_c=1 teflo at scale u=U. Thus, complex power has the currency teflo^{-1}. The distribution of power over a range of f_c>1 teflo reflects magnitude of P_c relative to power Q_c for 1 teflo. Power level Q_c is related to the variance of the series at complex frequency 1 teflo, and the distribution of variance over a larger teflo is given by the function of $P_c(f_c)$, by reducing u (increasing f_c).

At this point it is finally feasible to invoke a flashback to the hierarchical scaling axis in the scaling cube (Figure 70), which regards the three-dimensional statistical-mechanical frame of reference for complex space use at the individual level. At both levels of system abstraction—complex population kinetics and complex individual space use—a scale *ratio* is conjectured to capture the property of non-Markovian dynamics at the statistical-mechanical level. Thus, in both descriptions the scaling axis—represented by $1/f_c$ teflo^{-1} or u/U—is expressing log-linearity owing to the respective systems' hyperbolic (power law) geometry. Therefore, both the hierarchical dimension in the scaling cube and the $1/f_c$ dimension in the dispersion of power over a scale range may be termed "extended geometry for complex systems." Disregarding the respective scaling axes brings the frame of reference back to the classical system description, which is sufficient for non-complex system behaviour but invokes "issues" and paradoxes when applied to complex systems. To what extent the teflo concept may contribute to resolve the 1/f paradox in a broader context than individual space use and population kinetics (for example, systems showing self-organized criticality) remains to be seen.

10.4.5 Linking 1/f noise to the variance-mean relationship

Finding a reliable method to distinguish between intrinsically driven reddened spectra (true flicker noise) and fluctuations mirroring reddened influence from the environment (trivial red noise) has so far proven very difficult. A method involving fractional Brownian motion has been proposed (Mandelbrot 1983), but this concerns a statistical algorithm that may be applied to mimic 1/f noise statistically. It does not *explain* this phenomenon from a statistical-mechanical (physical) perspective. Eq. 29 is expressing a proposal for such an explanation (based on a postulate), by introducing $1/f_c$ noise for systems where a reddened spectrum emerges from intrinsic scaling complexity.

From another perspective of the same elephant, consider V(M) analysis in the context of rescaling (Eq. 12). If we rescale M_j into a series of fewer elements by lumping subsets of the initial elements in M to reach a more coarse-grained level as I showed in the analysis of sycamore aphid data, one should in an open environment—according to classical expectations—expect spatial variance in a homogeneous environment to increase proportionally with the change of scale (Poisson distribution): $V=aM^b$ with $b \approx 1$. The premise is that birth and death rates are not dominating the kinetics, as they typically are on very coarse spatial resolutions. At finer "local population" scales, redistribution of individuals through movement (local immigration and emigration between virtual grid cells) will be stronger than influences from birth and death. A Poisson distribution means that individuals comply with classical migration rules: Brownian motion and traditional diffusion equations. In a $1/f_c$ compliant system, this is not what happens. The variance increases non-proportionally with a change in resolution even when death and birth rates are negligible. With reference to previous chapters, $1/f$ noise is the Z-paradox in new wrapping! The magnitude of variance at coarser scales increases more than proportionally with change of spatial resolution: $V=aM^b$ with $b \approx 2$. This increase is faster than what we should expect from summing the variability in the original elements by integration. The central limit theorem is violated. This theorem, a cornerstone of classical statistics, states that the variance in a series of independent elements equals the sum of the variance of sub-sections of the series:

$$V_{tot} = V_1 + V_2 + \ldots + V_m \qquad (30)$$

In contrast, in a complex system we typically find:

$$SD_{tot} = SD_1 + SD_2 + \ldots + SD_m \qquad (31)$$

Where the SD is the series' standard deviation ($SD=\sqrt{V}$). In other words, for a complex system where $V \propto M^b$ with $b \approx 2$, $SD \propto M^{b-1} = M$. Statisticians will recognise this as a property of the so-called Cauchy distribution (Schroeder 1991). Thus, the Markov-compliant frame of reference, which is closely linked to compliance with the central limit theorem and the standard Gaussian distribution, does not apply for systems where $SD \propto M$ from intrinsic causes (complex systems). The system is Markov-breaking because the overall magnitude of variability for a given series of observations is independent of scale in absolute terms (like meters and days). Flicker noise, also called $1/f$ or pink noise, implies that the fluctuations are scale-free, or *fractal*. Eq. 29 postulates *additionally* that the power spectrum requires an independent scale axis to resolve the paradox of independence of a fixed lag at respective resolutions. This paradox was proposed to be resolved by substituting f with f_c, and make an explicit distinction between $1/f$ in Euclidian terms (a function of lag in space or time) and u as an expression of system coarse-graining (a function of relative scale, u/U teflo^{-1}).

A V(M) analysis, where M regards average abundance at a given resolution *k* and M is changed by changing *k,* is an alternative approach to study the power of the variability over scales. A good reason to apply the V(M) approach rather than calculating a power spectrogram is that a given overall variance in M at a given u may give different slopes in the spectrogram, depending on the phase for the different cycle lengths. The fluctuations in a series are a function of both power and phase, so that by changing the phase for a given cycle, one may create a somewhat different-looking trace for a given series even with the same integrated (summed) power for the cycles. This means that the degree of autocorrelation over a range of lags (i.e., at a given scale u) will vary accordingly, since the slope of the spectrum and the autocorrelation function are correlated (West and Shlesinger 1990). Autocorrelation expresses the correlation between elements in the series of M_j as a function of lag between successive M_j. Autocorrelation may be the result both of this statistical artifact and of environmental variability affecting the autocorrelation. An example of two series with the same overall V and M, but with different autocorrelation, was shown in Figure 85. V(M) is less affected by autocorrelation, and in the case of non-autocorrelated sample unit values (as for aphids), spatial coarse-graining provides a simple and easy-to-interpret variability level also of the distribution of power over a range of scales. By studying how V(M) varies as a function of u, we obtain an *alternative* measurement of the colour of the noise.

10.4.6 Variability of intrinsic and extrinsic origin

I have argued that a reddened spectrum of extrinsic causes comes from interaction dynamics that I have described in previous chapters as serial (sequential) and scale-specific (Markovian), where the dependent variable (like population abundance) follows the environmental variable(s) within a correlation range that may be described by the relaxation function. *In this case SD changes proportionally with lag, not with the size of u*. On the other hand, as I showed in the model for complex population kinetics, red noise is generated from a dynamical link between parallel processes at different scales. *In this case SD changes proportionally with the size of u, not with the lag*. Thus, complex fluctuations from serial versus parallel processing create certain differences in the observed patterns, and these differences may be used to test whether a series' red noise pattern expresses intrinsic complex dynamics or not.

The multi-scaled lattice model for population kinetics (Figure 33) also has a 1/*f* noise spectrum from intrinsic complexity as was shown by verifying the Z-paradox in the results (Figures 38-40a,b). The arena was a 32x32 lattice (1,024 grid cells at the lowest scale level). Sampling the population by taking a subset of M*j*, for example, every eighth grid cell uniformly spread over the lattice, showed that V is still proportional to M^2 (SD proportional to M) where M now is the scale-specific average for the subset of *j*. This tells us that the sample of size n=1024/8=124 is sufficient to express the true variability in the population at scale u within the 32x32 lattice. If the 128 sample units were concentrated in a continuous block in one of the corners of the lattice (*sensu* Figure 91), M varies between the four corners, but $V=aM^2$ for all corners ($a \approx 1$). Similar to the sycamore aphid results, this result from a simulated population shows that the relationship between V and M does not vary as we change the distance between the sample units. For example, if we find a block somewhere on the lattice where M for this block and M for the lattice as a whole are equal, we expect to find the same level of V.

10.5 Red noise of extrinsic origin

10.5.1 Local causality means a serial process

In this section I emphasize further one particular property of classical population dynamics, serial processing, with respect to the dynamics between the environment and the population. From this property I describe scale-specific population dynamics as a further elaboration on descriptions in previous chapters (for example, *6.2 Population kinetics: premises for the classical approach*).

Consider population fluctuations with a reddened power spectrum from perturbations and changing local and temporal conditions. The perturbations and changing conditions by definition are caused by the environment, and the population simply responds to them in a successive (serial) manner. Thus, an event in the environment triggers the population to adjust; a new extrinsic event fires a new density response, and so on. The main result is that the interaction is strictly and locally causal in space and time. In standard mechanistic terms, the adjustment represents a response to an event. There is an action, then a reaction: event, response, event, response, etc. It is a serial (sequential) process. Notice that populations of all other species are part of the environment, as are abiotic components, when we are considering a single population's dynamics.

10.5.2 A serial process has a characteristic response rate

An action-reaction process is a single-scale process, since a characteristic scale where the population interacts with the environment can be pinpointed in space as well as in time. Thus, the process is Markov-compliant and memory-less. This characteristic spatial and temporal scale is the finest scale, where a given population is "close" to dynamical equilibrium with its environment. Being close to dynamical equilibrium means that the population density at a given moment and location has adjusted itself to previous changes (time) or to local conditions in the neighbourhood (space) in the environment and also at all further distances in time or space. In a time series, for example, in absence of new environmental changes, no further change in the population is expected (beyond random fluctuations, or white noise). A local density at a given point in time is then only a function of the environment at that location at that time, and not a function of the density in the surroundings or the density at previous times. In short, the density has "forgotten" or "engulfed" the perturbation.

Observe that the equilibrium is defined as a lack of *change* of state, not the state (number of individuals) as such. Therefore, before the population fully responds to a new environmental change, the sign (negative or positive) of the change from one moment to the next will be positively correlated with the sign for the change from the previous point in time to the present. For a serial (sequential) process, the dynamical equilibrium is always approached asymptotically. In practice, the characteristic scale in space and time for the system being close to dynamical equilibrium is defined as *the correlation length and the correlation time*, respectively. For exponential decays it equals the spatial distance and time interval where only 1/e (ca 37%) of the response to a change of local conditions remains. At finer scales (shorter distances or time lags), the population change is still in the process of changing its density in response to recent or spatially close environmental changes. At these fine resolutions two adjacent measurements (in time or space) of population density under variable environmental conditions tend to be more correlated with each other than with the local environmental conditions.

At coarser scales than one correlation length or correlation time, correlated change due to intrinsic response distances and response times decays quickly towards zero. For example, already four correlation lengths after a perturbation, the change of density is less than 2% ($1/e^4$) correlated with the change four correlation lengths away or four correlation times ago.

Similarly, by coarse-graining a series (from lumping more and more successive measurements to obtain a coarser scale of observation on the same material), the effect from the response time is no longer visible. At such a coarser scale, the dependent variable (like population abundance) and the independent variable (a critical environmental variable) are expected to fluctuate more synchronously. Practically, at these lengths the distance or historic effect of the environmental change (space) or perturbation (time) would be undetectable. Since the intrinsic source of correlated change for a serial process is less than 2% beyond a lag of four correlation lengths, the actually observed correlation is in this case in the hands of the environment alone. If the spectrum of the environmental change is red (e.g., long-term fluctuations superimposed on more short-term fluctuations), the spectrum from the population variability will also be red.

In the case of scale-specific and thus Markov-compliant population kinetics, the total variability in a series can be understood as the sum of each cycle's contribution to the variability (smaller cycles superimposed on independent dynamics from larger cycles, etc.), as was shown in Figure 90. At coarser scales than a couple of correlation lengths and times, the population variability is a simple aggregation, or sum, of the influences from the environment. In other words, a proportional (linear) relationship between a change in the environment and a change in population density is expected.

For such a scale-specific process in statistical-mechanical terms, a nonlinear increase in V as we increase M(u) (M as a function of u)—in particular, $V=aM^2$—can have two explanations: (a) persistent change over time or space or (b) an environment that changes in a multi-cycle way, as described above for systems with reddened spectra. Persistent change means that the next sample unit has a substantially larger (or smaller) chance of showing a larger (or smaller) value than the foregoing unit. Examples of reddened spectra were shown in Figure 85 (Brownian motion) with its power spectrum in Figure 89. Consequently, if V changes nonlinearly with M(u) on coarser scales than four correlation lengths or so, this nonlinearity should—from *classical* premises—be a function of the summed (successively aggregated) density changes from the interactions with the environment at a characteristic scale (normally defined as the correlation length) for the dynamics.

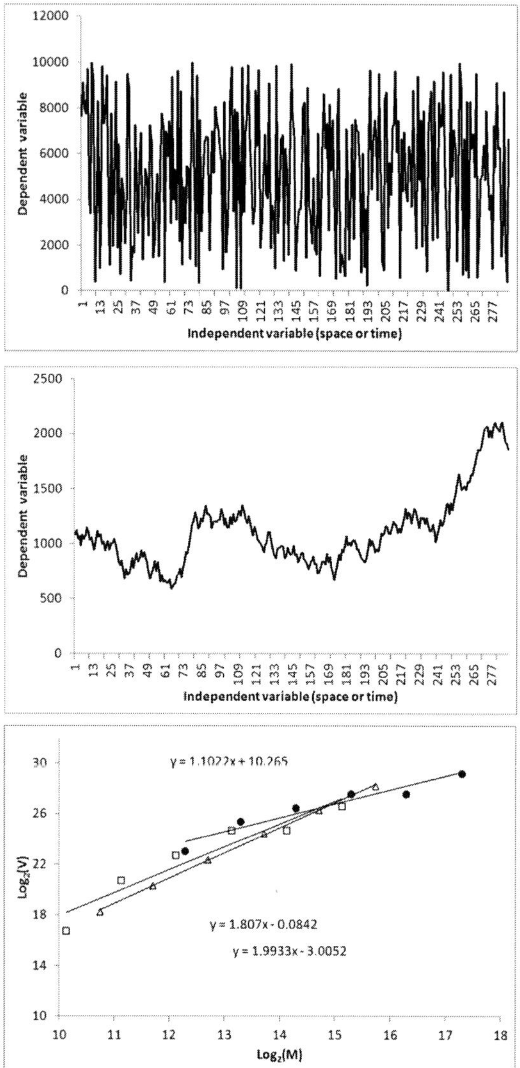

Figure 92. Relating variability of dependent variable to V-M analysis. See main text.

It is easy to see that both (a) and (b) lead to autocorrelation. Autocorrelation means that, on average, two observations of the number of individuals in a population that are close in space or time will tend to be more similar in magnitude than two observations that are more distant. Thus, when V changes nonlinearly with a change in M(u), if V is a function of *lag* size between successive measurements, this extra variance with increase in lag should be due to extrinsic factors. Figure 92 shows an example of white noise, integrated white noise (leading to Brownian motion), and V(M) analysis where M regards coarse-graining of observations. The white noise series verifies $V \propto M$ (log-slope close to 1), in compliance with the central limit theorem (adding independent sample units does not influence overall variance). The integrated series approximates the expected $V \propto M^2$, owing to the integration (the exponent increasing by one unit). Two coarse-graining methods are shown: uni-spaced sample units with no intermediate gaps in physical time or space (log-slope 1.81) and uni-spaced sample units with large gaps for small u and successively smaller gaps for larger u, as coarse-graining is approaching U (log-slope 1.99). From the premise of a serial, Markov-compliant system, the total V in a population always equals the sums of the V in its parts, if an autocorrelated series (middle pane in Figure 92) is differentiated to reveal successively independent observations (the intrinsic noise, which is assumed to be white and is shown in the upper pane in Figure 92).

By looking at a "classical" and spatially open population system at coarser spatial scales than a few correlation lengths, we are allowed to ignore the effect from movement and focus on birth and death rates. Fine-grained migration happens at a substantially larger rate and consequently determines the spatial and the temporal correlation length. Immigration to a given site (virtual grid cell) becomes close to a fixed rate: emigrating individuals from other sites at various distances and directions are entering a given site with a given average number per unit time—the immigration rate—given by standard diffusion laws from statistical mechanics. Behaviour for the individuals and the population as a whole then becomes a function of tactical (Markov-compliant), individual decisions at finer scales than the actual correlation length and time, as I described for a serial process above.

In space the upper scale limit for the tactical decisions—where statistical mechanics is a more proper modelling approach—can be simplified as the average perceptual field for an individual in the population; the idealized perimeter for the ranges of its senses. In time the upper scale limit for correlation (the relaxation time towards equilibrium) is the time an individual takes, on average, to respond to new events by movement.

10.5.3 Flicker noise, ergodicity, and environmental stochasticity

If I were to pinpoint one particular aspect I dislike about classical population dynamical modelling, it would probably be the common approach to disregard individual movement rate for the sake of simplification and focus on population growth and death rates. In spatially *implicit* models (like ordinary differential equations, where spatial structure is disregarded), the population is then assumed to be in equilibrium with its environment at all times within the spatial extent of the population. In other words, individual movement rate is sufficiently large to allow for "full mixing" of the population at the time scale for the modelled birth/death dynamics (typically the time scale equals the generation time). In this case the chosen system scales for space and time are assumed to satisfy the principle or ergodicity: the ensemble average equals the time average. Relocating a specific individual many times during a generation time should give a similar scatter of spatial fixes as a snapshot of a sample of the population (the ensemble). This aspect has been stressed many times already, for example, in *6.1 Inter-level model coherence and ergodicity*. If ergodicity is not satisfied within the actual spatial extent and Markov-compliant dynamics are assumed, ordinary differential equations should be replaced by partial differential equations or a coupled map lattice of local populations. In the latter case, ergodicity should be assumed at the scale extent of these local populations.

However—as I have stressed repeatedly—ergodicity is not the only consideration. The red thread throughout this book is the multi-faceted critique of the Markov premise, both for individual-based and population level models. This critique of course does not mean that Markov never applies! In *6. Modelling parallel processing,* I outlined and simulated a spatially structured population model which highlighted the difference between Markov and non-Markov dynamics, and in *8 Movement classes at the statistical-mechanical level,* I outlined how different movement classes could be distinguished with respect to individual space use (the scaling cube). My main critiques concern a general lack of explicit consideration of the qualitative difference between memory-influenced and classical modelling approaches, both at the population and the individual level, and whether the study regards theoretical models (including simulations) or statistical analysis of real data. That said, I have also documented how the non-Markovian system design seems to capture empirical data better than the Markovian counterparts.

10.6 Flicker noise in a time series: coffee leaf miner

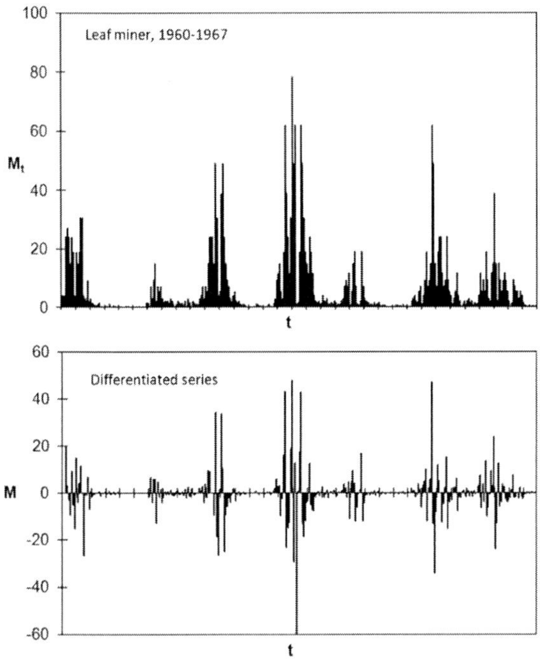

Figure 93. Temporal variability of leaf miner abundance.

In *10 Intrinsic complexity: 1/f noise and the scaling paradox,* I summarized the strange flicker noise phenomenon in the context of reddened spectra in population variability series. I also made a distinction between classical 1/*f* noise and complex $1/f_c$ noise (Eq. 29). In the former, the series length U should reveal to what extent the spectrum was reddened, and in the latter the ratio u/U (a ratio called *teflo*) should determine the degree of reddened spectra.

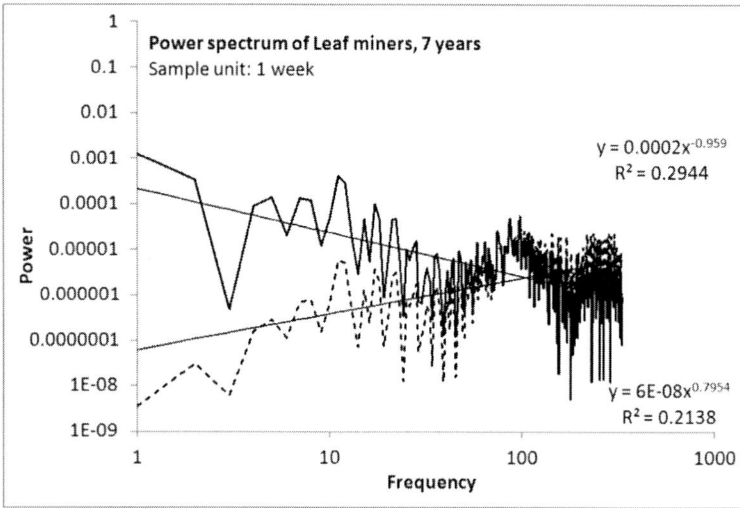

Figure 94. Power spectrum of leaf miner series.

While that chapter's overview focused on spatial variability, there is also—of course—temporal variability to consider. In particular, time series are at the centre of population dynamics, both theoretically and in empirical studies. In this section I study an eight-year time series of density fluctuations of a coffee leaf miner *Leucoptera meyricki* population on Kilimanjaro, East Africa, published by Bigger and Tapley (Bigger and Tapley). The population was sampled with suction traps on a daily basis (year-round) during 1960-67. I have estimated the average density over one-fourth month unit size (and upwards) from a direct read-off on their figure in the paper and then used these weekly units as the temporal magnitude of sample unit, u, for the power analysis (Figure 93). At finer resolutions it was difficult to measure successive counts precisely. Spatial extent of the study is difficult to interpret, since sampling regards a collection of flying individuals at a specific point in space. Clearly the temporal variability regards a local population within a narrow spatial extent, and ergodicity may be assumed to be satisfied.

Variability at coarser temporal scales was obtained by lumping weekly counts into larger units, like I described for spatial change of scale above. I also studied the effect from keeping u constant and then increasing U, for the sake of testing the degree of reddened spectrum from environmental sources (larger U means larger series). Power spectrum was calculated both for the non-differentiated and for the differentiated series. V/M regression was performed by calculating V and mean M for subsections of the series.

The power spectrum confirms flicker noise, with power increasing proportionally with the inverse of frequency (Figure 94). Yearly fluctuations ("outbreaks") cause the undulating form of the spectrum, but overall the spectrum is pink. The dotted series shows the power of the differentiated series, which confirms a change of log-log slope close to two units. The question remains, however, whether this reddened spectrum is due to environmental heterogeneity increasing over an increasing time series, or due to intrinsic complexity. To test this we should study how variability changes as a function of temporal scales, u and U.

Figure 95 confirms flicker noise of intrinsic origin, since larger u leads to larger variance, close to compliance with $V=aM^2$. The results show coherence between original (circles) and differentiated series (squares). In the latter case, absolute values of each difference were used. The upper pane illustrates the variability for small sequences of the series (with varying average M between the sequences). The lower pane shows the decisive test from M for the total series as a function of u. The lower pane also shows that the variability in larger series (larger U, same u) in overall terms is also somewhat influenced by environmental heterogeneity, since the slope is weakly positive. As expected, the original series is somewhat more "reddened" (slope 0.27) than the differentiated series (slope 0.15). For the delta-u series (differentiated series, absolute values), the variability interestingly levels off at the coarsest scales (u larger than sixty-four weeks), implying a transition scale between intrinsic complexity and environmental stochasticity. Thus, a complex frame of reference should be applied to study the variability of leaf miners at finer temporal scales than this transition, while a classical, Markov-compliant framework should be proper at coarser scales.

10.7 The next step

Hopefully the theoretical walk-through of the extended framework of memory-influenced space use has been sufficiently peppered with anecdotal evidence and logical arguments to make the reader wonder why the current paradigm is still both current and thriving well, both among ecologists and physicists. However, as shown in the foregoing chapters some recent developments that have begun to emerge in both camps may finally indicate some cracking of the paradigm's surface. In the next chapter I seek to link some of the presentations to upcoming tasks with respect to applied ecological methods.

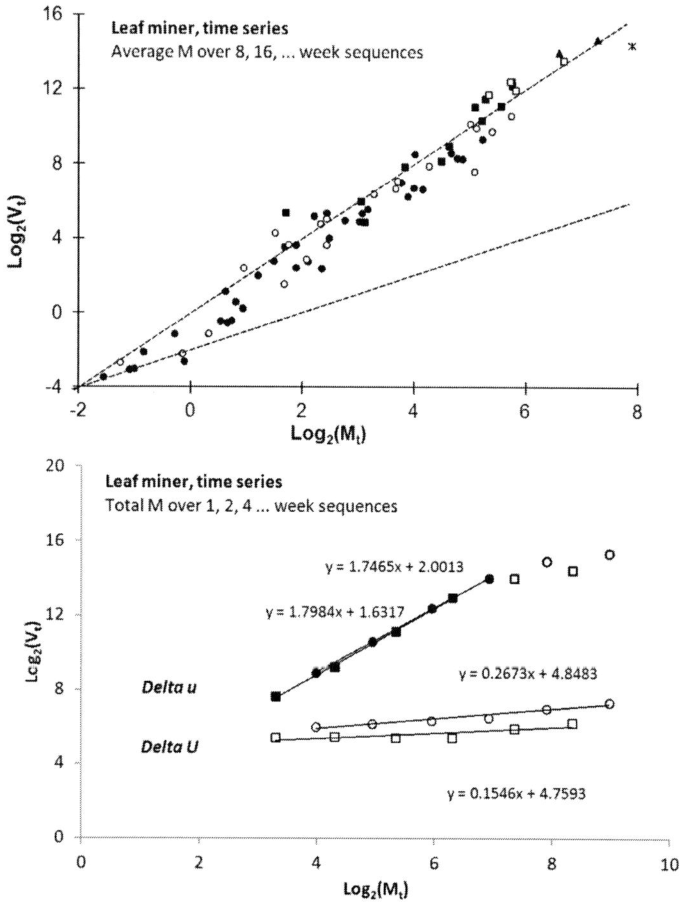

Figure 95. V-M analysis of leaf miner data.

11. IMPLICATIONS AND APPLICATIONS

11.1 What does multi-scaled movement really mean?

In the preface I stated that this book primarily covers a critical evaluation of modelling premises in movement and population ecology and presentation of biophysical model extensions in this respect. Over the ten previous chapters, I have been faithful to this outline. However, each of the many theoretical themes that have been raised with respect to memory and scale-free space use could have been accompanied by a concluding "So what?" section. For example, what are the practical consequences for wildlife management from estimating home-range sizes for moose *Alces alces* from standard protocols (the area asymptote assumption) relative to applying the "home-range ghost" formula (power law expansion, with fractal-compliant space-use intensity)? What are the expected differences with respect to inference about individual habitat selection? What is the parallel processing-based models' predictive power relative to their Markovian counterparts? In short, what may be gained—more specifically—from extending the standard statistical-mechanical toolbox where data verifies scaling patterns and spatial memory influence of a non-Markovian kind? Is it a matter of decimal adjustments of results or a potentially more profound difference? What new hypotheses may be raised and tested by the new protocols?

As stated, this book does not focus on applications, which are only fragmentarily touched upon in a more or less digressive manner throughout the chapters. Anyway, in order to improve bridge building towards actual movement ecology and ecological inference in more general terms, I propose some bricks toward the end of this chapter. However, I start with examples of implications rather than applications, despite running the risk of ditch digging rather than bridge building. Statements focusing on implications are generally more provocative than applications, which typically follow after the smoke has cleared (however, if you are sensitive to provocations, you may already have left off in previous chapters). I appreciate that you are still reading!

Multi-scaled animal perception and utilization of a landscape has been a central theme for ecologists for a long time (Senft *et al.* 1987, Wiens 1989). In this section I emphasize once again that the present book's modelling approach is compliant with the general *concept* that many animals may cognitively perceive their habitat by processing information and responding to landscape properties over a range of scales in both space and time. That said, I will—in this chapter's more sharpened context of implication—scrutinize the empirical support for how this "multi-scaled space use" concept is being mathematically formulated and dynamically modelled. The concept as such has deservingly become one of the pillars of movement ecology, in particular owing to the renewed focus on memory and scaling aspects. Thus, the multi-scale aspect presents itself as an obvious point to start rattling conventional perceptions, in particular among theoreticians. In the following sections, I dig deeper into the following questions: what is meant by multi-scaled movement? How may it be conceptualized by statistical-mechanical models?

11.2 Two variants of the standard model

In the context of stochastic modelling of scaling-compliant movement, it is at present a broad consensus that animals commonly comply with a composite (mixed) random walk model; relatively fine-grained patch utilization is occasionally disrupted by larger moves to other patches. This intra-/inter-patch movement model is typically conceptualized as a two-component correlated random walk with different characteristic diffusion rate and directional persistence strength for the two movement modes (Johnson *et al.* 2002). Using correlated random walk as building blocks implies acceptance of Markov-compliant and memory-less movement. The two-level variant may be considered the statistical-mechanical analogue to the classical ARS model (area restricted search).

A composite, scale-specific kind of movement implies that the animal at any point in time moves in compliance with either one of the movement modes at respective points in time: "either-or," not "both-and." The latter would conflict with the mechanistic and Markovian premise on which scale-specific processes firmly rest (Gautestad 2011). For example, in order to maintain a "coarse-level" movement mode in a two-mode model design, the animal needs to be assumed to be less prone to becoming interrupted by events within its trailing field of perception. If not, the switching between fine- and coarse-grained moves does not regard different movement modes but rather a single mode under which the switching—in a statistical sense—is determined by the environment alone rather than cognitive (intrinsic) change of movement state. Single-mode movement may be considered the more classical and generic variant of space use in a heterogeneous landscape: varying intensity of habitat utilization (observed, for example, as locally varying density of fixes when tracking the animal) may transcend a range of spatial scales if the environmental heterogeneity also transcends these scales. In short, the environment determines the statistical pattern of multi-scaled space use.

This is an important distinction with respect to modelling movement in heterogeneous landscapes. As far as I know, it has not previously been made explicit in the theory of movement ecology; a two-mode statistical *pattern* may be considered to appear from two qualitatively different mechanistic *processes*: extrinsic forcing or intrinsic mode switching. In the former case, a single mode is involved. When moving in a resource-rich environment, directional persistence of the movement path is on average smaller, owing to a smaller mean free path in biophysical terms (see *2.2 From mechanistic to statistical-mechanical modelling*). Alternatively, under the two-mode model condition, the animal is mode-switching to fine-grained movement with frequent turns (like in the ARS model) when a resource-rich patch is encountered. When resource density is exhausted to a critical level, the animal switches to coarse-grained movement. In simple terms, while a single-scale mode may be represented by a Gaussian distribution of response strength to the sum (superposition) of intrinsic and extrinsic input and where the mean response is the "mode," a two-mode model may be considered represented by two Gaussian distributions with sufficiently different mean to make it feasible to define two modes.

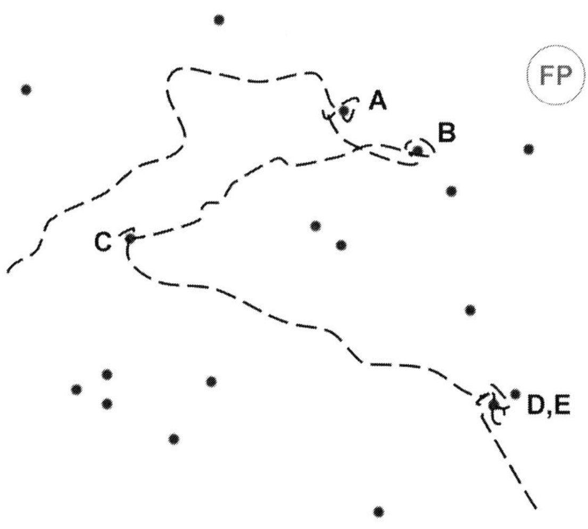

Figure 96. Hypothetical movement path. See main text.

The concept of multi-scaled single-mode movement in the context of foraging is illustrated in Figure 96. The area of the circle in the upper right corner indicates in a simplistic manner the width of the field of perception surrounding the individual at any point in time. In this scenario consider a stochastic representation of the path, owing to fix collection at a sufficiently large inter-fix interval Δt, to ensure many independent movement-influencing decisions during each of these intervals. For example, consider this interval to be one minute. Thus, we are studying the movement at a statistical-mechanical level (see *2. Basics of movement physics: the standard models*). Finally, assume that the animal is moving opportunistically and thus Markov-compliant and mechanistically; the next step is determined by current conditions (a mixture of internal state and input from the field of perception). Thus, self-crossing of the path happens by chance. Long-term, spatially explicit memory is not involved in this scenario. Many back-and-forth displacements may potentially take place during one minute, whether we consider a mouse or a moose. However, often little or nothing captures the interest of the animal during this short period, and this typically tends to maintain directional persistence during the interval and even between successive intervals. The combination of opportunistic search for food patches (marked by dots) and sufficiently large Δt means that a correlated random walk for the search is a reasonable representation of movement, given our choice of a mechanistic model framework. This stochastic search is mixed with deterministic drifts towards food items that are discovered within the trailing field of perception. In this simplistic individual-based modelling scenario, a representative path of length $T = 60\Delta t$ (one hour) is shown by the dashed line, and the animal has discovered and exploited four food patches A-E during this hour.

Next, consider Figure 97. The conditions are the same as in Figure 96, except for a higher abundance of (randomly scattered) food patches. The patches are the sum of those shown in Figure 96 and an additional set owing to stronger patch awareness (but the area of the field of perception is assumed identical with the scenario in Figure 96). Due to increased frequency of patch-targeting events, the mean free path is smaller. Consequently, the typical net displacement during an hour is smaller, and the sixty fixes are concentrated in the upper left part of the arena. Only patch C overlapped (by chance) with the utilized patches in Figure 96.

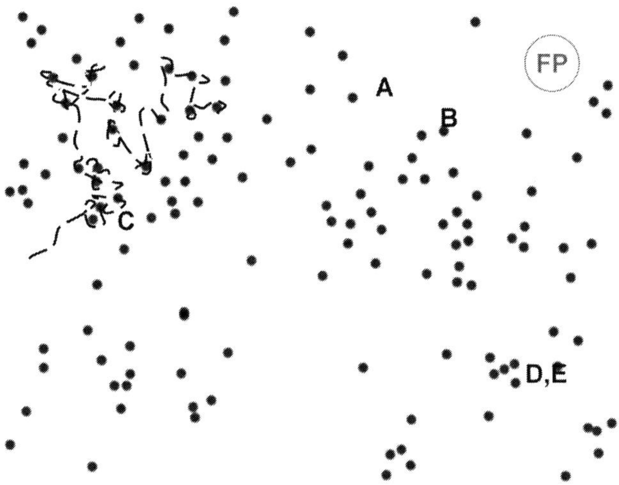

Figure 97. Hypothetical movement path in an environment with more frequent directional interrupts.

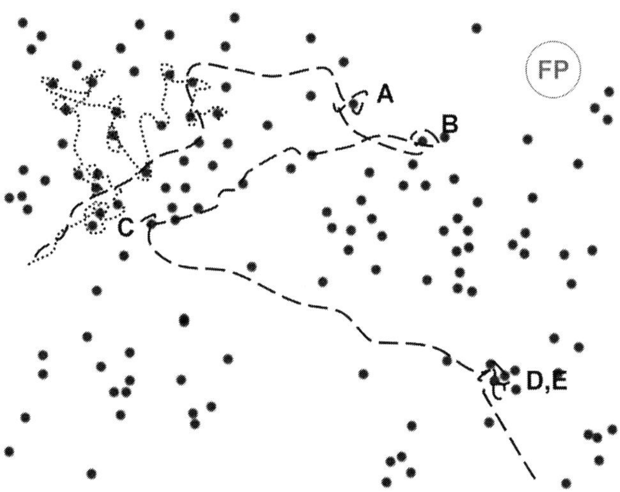

Figure 98. Two hypothetical movement paths, where the individuals differ with respect to susceptibility to environmental conditions.

In Figure 98 we compare the two paths by superimposing them. In this scenario, consider the stretched path to represent an individual moving in a "coarse-grained" mode, and the jagged path another individual moving in a finer-grained mode. Owing to a larger threshold for drifting towards a food patch when foraging in a coarse-grained mode, many patches that are targeted by the first individual are ignored by the second individual. Finally, consider a two-mode scenario where a given individual is switching between the two modes during T. It could be stochastic switching (for example, leading to 90% fine-grained mode intervals during T) or it could be based on a more or less complicated deterministic rule. For example, the animal switches from fine- to coarse-grained mode for a period after having exploited a food patch to allow for stronger displacement away from an exhausted locality, or it disregards food patches while searching for a resting place.

At this point it should be obvious that a two-modal composite random walk representation of multi-scaled movement requires the "either-or" design in terms of specifying perceptual sensitivity, not "both-and." The fine-grained mode regards relatively strong sensitivity relative to environmental attributes within the field of perception. Simultaneous execution of strong and weak sensitivity is logically prohibited under the Markovian/mechanistic framework. With reference to the scaling cube, whether the movement is two- or multi-modal, its statistical-mechanical representation resides on the Markov floor of the cube, somewhere on the scaling axis defined as "temporal memory." From the temporal perspective (resolution) of the lower-level movement mode, a coarser-level mode may be said to have temporal memory owing to stronger directional persistence of the path on average at higher levels. In short, while the process has memory at fine temporal resolutions only ("moment to moment"), a period of mechanistically strong directional persistence transcends statistically to coarser scales. Persistence prevails statistically beyond the temporal micro-scale where persistence as a process is executed.

The difference between the two modelling designs (single-mode or two-level mode) may appear subtle, but has deeper qualitative differences in process terms which are important to acknowledge. For example, under the two-mode design, one needs to accept as behaviourally realistic that the model animal is cognitively less sensitive to environmental input during inter-patch movement than during intra-patch mode. The animal is assumed to move with two strengths of input sensitivity during calculation of what to do next, and specific conditions (like local resource exhaustion) contribute to cognitively "clicking the knob" toward the other mode.

In fact, the distinction between intra- and inter-patch movements implicitly implies a high-contrast environment: strong resource aggregation (patches) that are scattered around in a landscape of otherwise non-preferred but traversable habitat (often referred to as matrix). In Figures 96-98 one could consider the white background as matrix, which then becomes subjectively defined as a function of the animal's current mode. Thus, these scenarios represent stochastic-mechanistic modelling: a mixture of correlated random walk and opportunistic directional bias towards patches appearing within the field of perception. Additionally they may regard single-scale mode walk or the intra-/inter-patch walk. The latter variant regards the two-level superposition of occasional switching between two modes, where each mode is scale-specific and characterized by either a small or a large diffusion parameter for the correlated random walk component of movement.

11.3 Standard model variants: implications for applications?

After this walk-through of the premises for Markov-compliant and memory-less multi-scaled movement, what are the implications for ecological inference whether one adheres to the one or the other variant as the premise for GPS data analysis? For example, in the context of use/availability indexing to explore habitat selection, should one expect a difference in results if the animal has executed a single- or a two-level (or more) composite walk during $T=N\Delta t$, where N equals number of fixes in the sample, and $\Delta t=2$ hours? The answer is in fact theoretically simple, since the Markov premise is also satisfied for the given space-use process (this is an implicit assumption from contrasting two model variants which both adhere to mechanistic principles).

First, consider the fact that a correlated random walk, with directional persistence between successive steps, becomes a Brownian motion with no such persistence if the path is sufficiently sub-sampled (larger Δt) (Turchin 1996). Second, adding a local bias term to the correlated random walk (response to discovering and approaching food patches) should not make any difference to this transition. Applying a sufficiently large Δt to ensure loss of directional persistence during exploratory phase is ensuring the same result both for intra-patch moves and the directional steps of magnitude up to the radius of the field of perception. At large Δt the bias effect from the latter mechanism is hidden from view; we are observing the system at a statistical-mechanical resolution (see *2.2 From mechanistic to statistical-mechanical modelling*). Third, at the statistical-mechanical path resolution, a single-mode or composite correlated random walk model should both show compliance with Brownian motion. Switching between two scale-specific modes or not, these aspects are averaged away by the coarse-graining from sufficiently large Δt (Gautestad 2013a). To conclude the question in the foregoing paragraph—the use/availability analysis should be a safe choice under the given premises of Markovian space use and a statistical-mechanical level of analysis. I stress once again: under the *Markovian* premise.

Wildlife management ecologists tend to acknowledge the two-level model conceptually but in practice apply the single-level model as a premise for ecological inference; for example, with respect to analysing habitat selection. According to the conclusion in the foregoing paragraph, this should be a safe choice, given the implicit assumptions of the actual space-use process and given sufficiently large Δt (the situation where the path is spatially autocorrelated event at the statistical-mechanical level was analysed in a previous chapter). However, what if the empirical support for these implicit assumptions is weak, or even prone to becoming falsified?

11.4 Markovian, multi-scaled movement: the evidence?

From previous chapters of this book, it should be clear that I do not buy into either variant of the mechanistic paradigm, whether composite Brownian motion is modelled as a scale-specific and single-mode process or by intrinsic mode switching between intervals of fine- and coarse-grained perception of environmental attributes. I do not question that animals seem to be capable of perceiving and utilizing the environment over a range of scales, but I'm critical of modelling this process as scale-specific by applying a Markovian/mechanistic algorithm (whether single- or multi-modal). After presenting the details of mechanistic modelling of multi-scaled movement in the foregoing sections, I now again turn towards the alternative approach: a statistical-mechanical framework that is coherent with non-mechanistic dynamics. Four questions will be addressed:

1. What is the empirical support for a mechanistic and thus Markov-compliant representation of multi-scaled movement, whether two-modal or multi-modal?
2. How could this framework be tested for, using GPS data?
3. What are the implications for applied ecology—here exemplified by the classical use/availability analysis of habitat selection—if the null hypothesis is not supported?

4. How could habitat selection be quantified under the premise of the alternative hypothesis, based on the parallel-processing postulate?

In this regard MRW has been presented in this book as the model for the alternative hypothesis. First, MRW describes a scale-free process, which deviates from a process where the individual is switching successively (i.e., mechanistically and thus Markov-compliantly) between periods of fine- and coarser-grained movement modes. Parallel processing is the key concept in this respect. Second, MRW is different from the Lévy walk counterpart of scale-free movement, by postulating a simultaneous mixture of tactical and various strategic goals involving temporal as well as *spatial* memory utilization (see *8.2 The scaling cube*).

Under the premise of parallel processing, the actual behaviour when the animal is encountering a resource patch will—according to this model postulate—depend on what coarser-grained (more strategic) goals are active for the moment and whether one of these goals is actually—and by chance—executed at this point in time. From the perspective of an actual moment in time (temporally fine-grained), a strategic goal execution will appear stochastic; only from a coarser temporal perspective will it appear deterministic (see *8.3 Parallel processing: planning for the future*). As a consequence, space-use effects from parallel processing should be expected to inflate the noise term or even support erroneous conclusions if Markov models are applied for ecological inference. I will return to this issue below, but first; what is the empirical evidence for a mechanistically based multi-scaled movement?

11.5 Scrutinizing empirical support for mode-switching

An extensive statistical analysis of movement by caribou *Rangifer tarandus* found support for a two-component movement process, called scale-specific responses to the environment (Johnson *et al.* 2002). With 141 citations (Thomson Reuters *Web of Science* as pr. August 2014), this paper has been quite influential with respect to supporting the concept of superposition of scale-specific movement modes. In this paper, Figure 1 showed a plot of the distribution frequency log(F) of movement step lengths L for a female caribou (they defined L as movement rate; displacement pr. minute). In the original semi-log histogram, which I have re-plotted in the upper pane of Figure 99 (based on their Figure 1), the distribution fitted what the authors called a "broken stick" model quite well. The semi-log linearity was proposed to be a consequence from two-level Poisson compliance and thus scale-specificity at respective levels; step length frequency in respective bin ranges fall off in proportion to a negative exponential function. This pattern is expected from a two-level space-use process, in this case with a transition between intra- and inter-patch movements at $L \approx 2.8$ m min^{-1}.

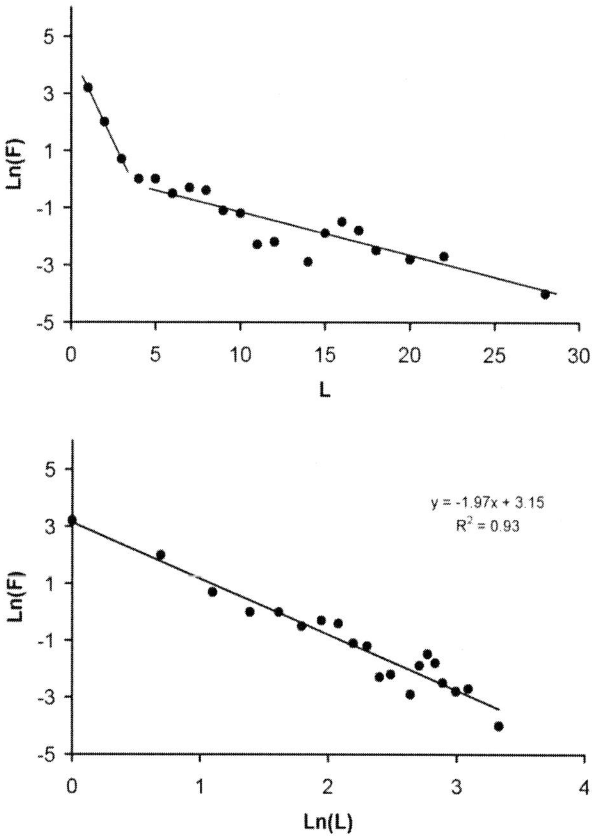

Figure 99. Distribution of step length of caribou apparently satisfies two qualitatively different movement models: "broken stick" (upper pane; observe semi-log presentation) and scale-free movement (lower pane; log-log presentation). See main text for details.

However, consider the log-log plot variant of the same data in lower pane of Figure 99, which was presented in Gautestad and Mysterud (2005). By log-transforming both axes instead of the y-axis alone, a linear fit over the total range of L implies compliance also with a scale-free movement process. Thus, we proposed that the estimated power exponent $\beta \approx -2$ indicated scale-free movement, owing to the absence of a transition at a characteristic scale. Further, the MRW model was proposed to be a more likely candidate for scale-free space use than a Lévy walk, since support for site fidelity and thus memory utilization was verified in other parts of Johnson et al.'s analyses on caribou.

After the respective papers appeared in 2002 and 2005, Benhamou (2007) illustrated by simulation that also a two-level correlated random walk model might show statistical power law compliance, and thus a pattern similar to the lower pane in Figure 99. In other words, with reference to the lower pane in Figure 99, there may be two parts of the stick, but both parts of the step length distribution have the same angle on a log-log scale (but the angle deviates in the semi-log plot). This may be achieved by fine-tuning diffusion parameters and respective statistical weight (relative frequency of occurrence) for the two components for movement mode.

The aspect of parameter tuning is exemplified by simulation results in Figure 100 where again only the y-axis is log-transformed. A broken stick is clearly visible both for λ=4 and λ=8, where λ is the ratio between the diffusion parameter for the respective levels of movement. The plots are reproduced from Gautestad (2013a) where I extended Benhamou's scale-free look-alike example for the sake of another theoretical topic (see below). For λ=8 with log-transformation of both axes, the broken stick may alternatively be interpreted as power law compliant (Figure 101).

Extending the superposition of modes to a three-level walk and accompanying parameter tuning would have improved the log-log linearity even further (this was not conducted in this study). However, more than two movement modes will influence the concept quantitatively but not qualitatively (it allows for better "curve-fitting"). With reference to the scaling cube (Figure 70 in *8.2 The scaling cube*), such a multi-mode condition is referred to as a Markov-compliant composite random walk where respective components are parametrically fine-tuned to produce a scale-free distribution of step lengths. In other words, a Lévy walk "look-alike" (Gautestad 2012).

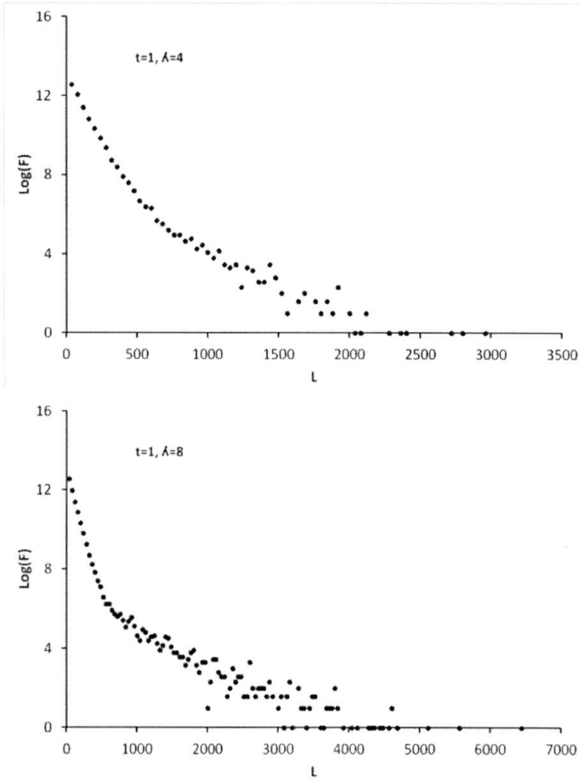

Figure 100. Distribution of step lengths from a two-level correlated random walk simulation under different conditions (see main text). At unit temporal scale the results with semi-log axes are compliant with a broken stick model.

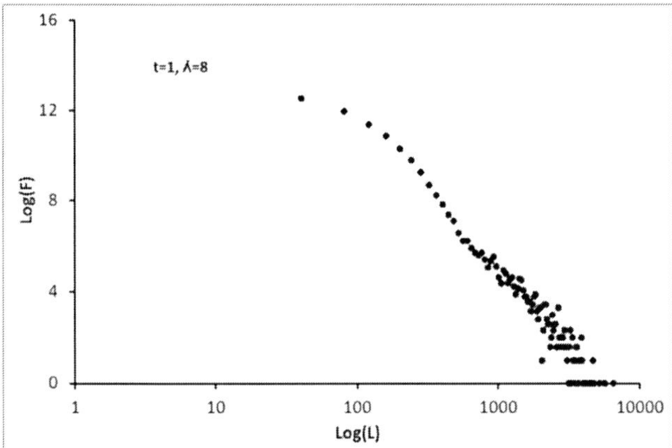

Figure 101. One of the conditions in Figure 100 presented with log-log axes.

To conclude the caribou results, in absence of supplementary testing, the power law distribution for step lengths in the lower pane of Figure 99 is—when considered in isolation—non-conclusive with respect to supporting either the two-level superposition hypothesis or one of the two hypotheses for true scale-free movement (Lévy walk and MRW). The reason was illustrated in Figures 99-101. Thus, power law compliance in step length distribution is not sufficient criterion to verify a scale-free space-use process, as pointed out by Benhamou (2007) and others. Johnson et al.'s influential "broken stick" analysis is on this basis not giving as strong support for a two-level space-use process as currently acknowledged.

There are several other examples showing apparent support for two-level movement. Killeen et al. (2014) studied migratory behaviour of elk (Cervus elaphus) in the Rocky Mountains, USA, and found a similar broken stick pattern in their GPS data. In Figure 102 I have re-plotted data from their Figure 3. In this species they found the inter-level breakpoint at $L \approx 7.1$ m min^{-1}. Owing to the semi-log linearity at both sides of the breakpoint, they interpreted the pattern as support for a statistical mixture of two movement modes; compliant with correlated random walk switching between two scales. However, Figure 102 shows that neither this support for the broken stick model is conclusive: with both axes log-transformed (and data supplementary log-binned; filled circles), the data may equally well support scale-free movement.

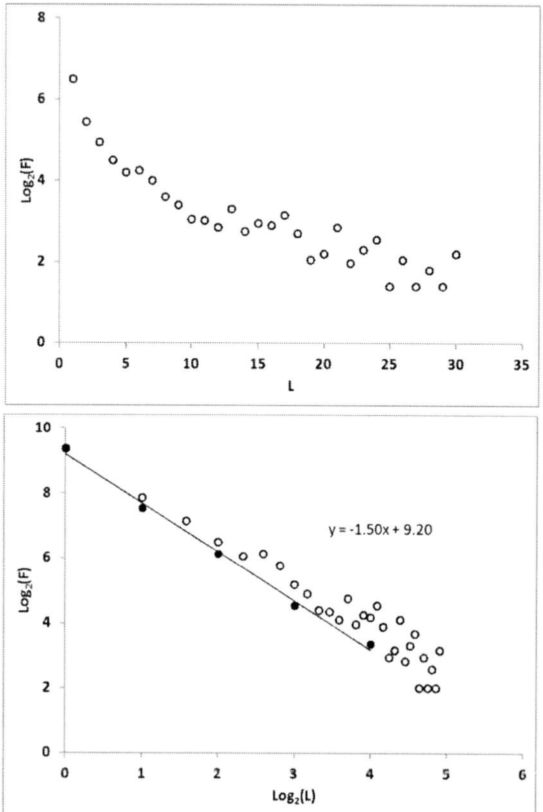

Figure 102. Three-level superposition of correlated random walk, under semi-log and log-log presentation.

In addition to the empirical data for caribou and elk, I finally want to re-point attention towards our own analysis of elk (from Norway, and thus named red deer), partly presented in Figure 57 (Gautestad *et al.* 2013, Gautestad and Mysterud 2013). Also these results showed power law compliance of the step length distribution in log-log plotting. However, in Figure 103, upper pane, I present the distribution with semi-log plotting: a broken stick pattern is apparent, similar to the caribou and elk data above.

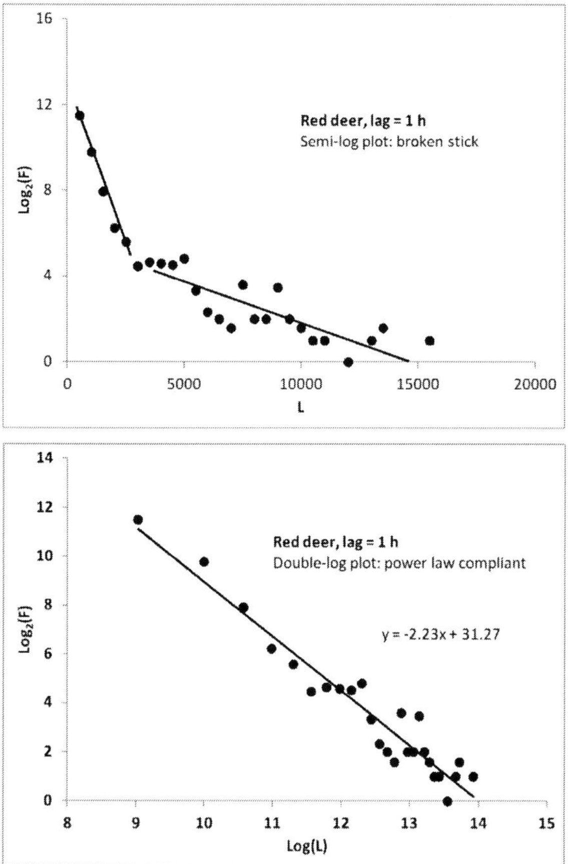

Figure 103. Apparent compliance of broken stick and scale-free movement in red deer; observe semi-log presentation in upper pane.

In summary, the three data sets on ungulate movement all show broken stick in semi-log plots but power law compliance in log-log plots. Thus, the empirical support for a two-level kind of Markov-compliant superposition of movement modes is non-conclusive. Proponents of this hypothesis are—based on Figure 99 and Figures 102-103—now confronted with an additional aspect that requires attention: what is the universal mechanism that leads to an apparent constant ratio between the two levels' respective diffusion parameter? This particular ratio is obvious from the similarity of the slope of the two step length ranges in a log-log plotting of the distribution, leading to compliance with a common power exponent in log-log plotting. In simulations, fine-tuning of model parameters is required to achieve a close match with a power law.

The support for the two-level movement hypothesis is additionally weakened by the fact that these three species have the cognitive capacity for site fidelity, as seen from the emergence of home ranges. Such memory-dependent space use basically precludes a composite correlated random walk as a realistic representation of the movement (the MemRW variant of correlated random walk, described in *3.4 Statistical-mechanical simulation of MemRW*, regarded a single-mode process with directional bias). However, an affirmative falsification of the two-level model as a realistic statistical-mechanical representation of ungulate space use will require a broader range of test methods.

Such methods have recently been proposed and already described. In *7.5 Distinguishing MRW from composite Brownian motion*, I summarized two methods—the coarse-graining test and parallel shift—which may be applied to distinguish a statistically "look-alike" scale-free process, like those presented by Benhamou (2007), from the real thing (Gautestad 2012, 2013a). With respect to the latter, we have additionally proposed methods to distinguish between Lévy walk and MRW (Gautestad and Mysterud 2010a, Gautestad 2012, Gautestad and Mysterud 2013). Thus, statistically simple methods are now available to test GPS data for compliance with composite random walk versus a true scale-free process, and whether the latter regards temporal memory scaling alone or memory scaling in both time and space (Lévy walk *vs.* MRW).

Only one test has been performed so far, owing to the short time span since the presentation of the methods: red deer GPS data from the summer season, where the MRW hypothesis for multi-scaled movement was supported and "broken stick" and other variants of Markovian models were falsified owing to a broader range of analyses than the step length distribution alone (Gautestad *et al.* 2013). In Figures 57-58, I reproduced additional results. The red deer data confirmed fractal dimension of spatial fix dispersion close to one in compliance with true scale-free space use, where two-level movement would be expected to show dimension closer to two. Incidence (number of non-empty virtual grid cells) expanded as a power law with increasing N, with power exponent in closer agreement with parallel processing ($z \approx 0.5$) than two-mode correlated random walk ($z \approx 1$) and memory-enhanced random walk of type MemRW ($z \approx 0.3$).

A single empirical falsification of the "broken stick" / two-modal model is of course inconclusive and should be regarded as a pilot test; more and broader-reaching studies are needed. In particular it would be interesting to see such tests performed on GPS data from animals other than ungulates. While waiting for additional tests, the preliminary results are at least sufficiently clear to make the following discussion not too premature. After all, it regards the thematic epicentre of this book.

11.6 Implications for theoretical ecology

The two system representations, (a) mechanistic superposition of scale-specific movement modes and (b) parallel processing, cannot be simultaneously applicable; this is an either-or kind of biophysical system validation. Either the actual animal is capable of processing previous experiences in a hierarchical and thus non-mechanistic manner (as outlined statistical-mechanically by the ceiling of the scaling cube, Figure 70), or it is switching successively between one or more modes of movement (in compliance with a mechanistic model design; the floor of the cube). Conceptually it is obviously a plurality of differences between the two frameworks in the context of multi-scaled space use, as outlined from a range of perspectives through the various chapters. If parallel processing is verified, the animal is *simultaneously* relating to various scales, in accordance to a hierarchical kind of processing transcending various process rates (see *5. Parallel processing: the concept*). Simultaneity implies that mechanistic modelling should—according to the present proposal—under these circumstances be replaced by non-mechanistic modelling, which is a novel concept. It was invoked to resolve the apparent contradiction in terms from the postulate of simultaneity, by defining two kinds of dynamics: Markov-compliant mechanistic dynamics and non-Markovian parallel processing.

Parallel processing implies that models building on Markov theory and the accompanying mechanistic principle should not be accepted a priori (as it is now), but tested for on a case-to-case basis. Figure 68 summarized some of the available protocols available at this stage. With reference to the preface and introduction, there is now gradually becoming available an alternative compartment of the statistical-mechanical toolbox. For movement data that shows parallel processing-compliant space use, one should obviously not accept the opposite premise; that the individual dynamically relates to different scales as a superposition of scale-specific moves with respective characteristic diffusion parameter.

The parallel-processing postulate may have implications for illuminating and potentially resolving long-lasting controversies and issues. From the battleground where theoretical models are confronted with empirical data, I have described three main paradoxes: (1) the home-range ghost paradox, (2) the paradox of fractal pattern-generating space use, and (3) the Z-paradox from variance increasing proportionally with the square of local fix abundance (or population abundance; see the Zoomer model). While these concepts and system properties appear paradoxical from the standard framework, I have shown over the chapters how they may find respective solutions—and thus vaporize as paradoxes—under the extended statistical-mechanical system description.

11.7 Implications for applied ecology

The non-mechanistic, parallel processing-based framework is not in opposition to central ecological aspects, like environment's influence on home-range size, expected stronger space-use intensity in resource-rich patches, and stronger weight on coarse-scale space use during the migratory phase relative to more sedentary periods. These are empirical "facts," supported by countless studies over many years. While the presently proposed framework is coherent with this body of ecological knowledge, it deviates with respect to how these aspects are actually modelled and analysed. By explicitly accounting for the qualitative difference between Markov-/ mechanistic system premises and a parallel processing/non-mechanistic approach, the latter offers alternative methods. Below I contrast these differences by three examples: home-range size, space-use intensity, and movement during sedentary versus dispersal phase.

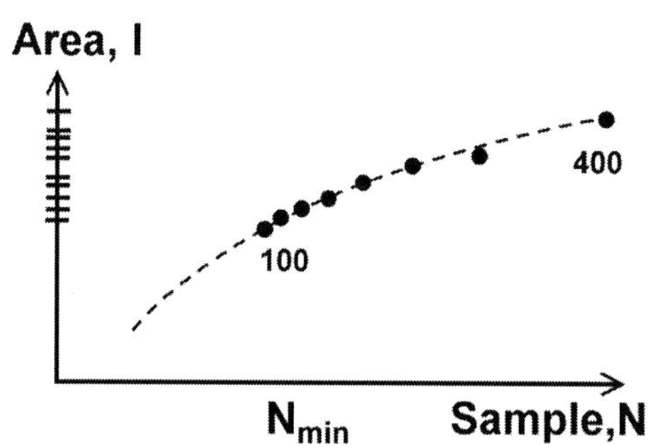

Figure 104. Outline to test for scale-free and memory-influenced space use. See main text.

The first two aspects, home-range size and selection strength, are illustrated by the hypothetical scatter plot of home-range size as a function of fix sample size in Figure 104. The animal is assumed to comply with the parallel-processing concept, and thus with the MRW model in statistical-mechanical terms. This makes it possible to contrast MRW-compliant methods to standard methods in the context of estimating home-range size. In short, we are comparing methods under the assumption that the scatter of fixes complies with properties of a statistical fractal with methods under the assumption of home range as a two-dimensional object. Let us for illustration purposes consider (a) serially non-autocorrelated GPS fixes and (b) each "dot" in Figure 104 represents a separate sample, collected independently, and fixes are spread randomly over a sampling period of a full season. Thus, the eight area estimates could represent eight researchers' independent estimates of the actual animal's home-range size for this period. Area is represented by number of virtual grid cells containing at least one fix: incidence, *I*. The same virtual grid resolution has been applied for all measurements of *I*(N). If the eight estimates alternatively were from different individuals or period, one should vary the grid resolution for the respective grids in order to "normalize" the data into a common I(N) presentation, with respective $c'=x_i c$ (x_i is here the factor for characteristic scale of space use; see *7.2.5 The characteristic scale and the home-range ghost paradox*).

While the data in Figure 104 may appear asymptotic, Figure 105 verifies that under log-log transformation, the eight plots collectively comply closely with the power law in Eq. 1b in a previous chapter: $\log(I) = \log(c) + z*\text{Log}(N)$, with power exponent z=0.5 (seen as the slope of the regression line). A minimum sample size, N_{min}, is set for the area estimates. Consider N_{min}=100, to expect to practically eliminate small-sample effect under the classical model assumption (asymptotic growth of area as a function of N). I refer to *7.2.4 Simulating space use: the generic MRW model* for further theoretical details. The reason for using incidence rather than the kernel density estimate for home-range area was explained in *1.2 Pinpointing a feasible reason for ignorance: KDE*. Empirical results from sheep, black bear, and red deer (Figures 3-4 and Figure 58) support the choice of narrow plot residuals in the linear regression in the upper pane of Figure 105.

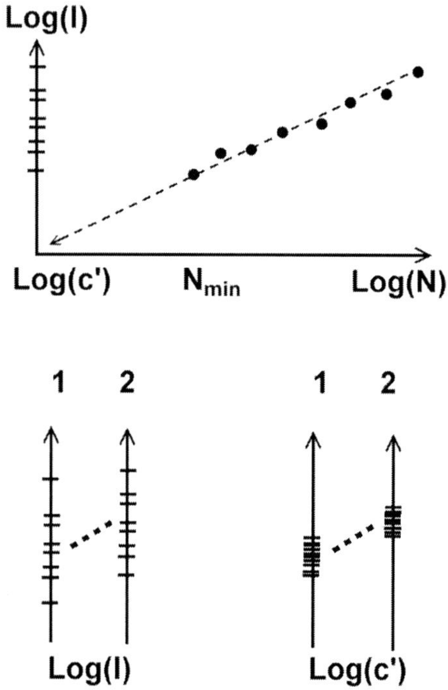

Figure 105. Outline to test for scale-free and memory-influenced space use. See main text.

The projection of respective log[*I*(N)] plots onto the y-axis provides a dispersion of home-range area representations, as illustrated under Scenario 1 in the lower pane's left part. The respective intercepts, log(c), from interpolation to log(N) = 0 for each of the eight plots show a narrower distribution than the *I*(N) distribution. In other words, the data in the hypothetical Scenario 1 comply better with the power law model (Eq. 1) than the standard model for home-range size. The standard model describes asymptotic area expansion (Eq. 2). In log-transformation Eq. 2 becomes Log(A)=log[*Q*'(N)]+ log(*K*) where *Q*'(N) estimates the small-sample effect and *K* is the asymptote representation. By using y-projections as representation of home-range area *K*, we are assuming N_{min}=100 to be sufficiently large for *Q*'(N) to be close to one. If some of the eight researcher were not satisfied with using *K* directly, they could multiply their respective estimates by Q(N)=1/Q'(N) as adjustment (see *1.5 The home-range ghost*). For N>100 such adjustment factor should typically make little difference.

The eight home-range size estimates may thus either be quantified by *c* under a premise of Eq. 1 (the MRW model) or by *K* (the standard model). When using home-range size as a quantity for ecological inference, it is obviously a goal to apply a best possible model for the estimate. In this case, *c* does obviously perform better (small standard error), since it fits the underlying space-use process (if the eight data points in the same diagram were from different individuals, places, or times, one should use $c'=x_i c$, as already described). From Eq. 1 we see that $c=I/\sqrt{N}$, or "area pr. square root of N" under the assumption that $z=0.5$. From this unit scale it is trivial to use other area standards. For example, the expected area for N=400 is twice as large as N=100 (as seen in Figure 104). In other words, since the respective eight researchers used different sample sizes for their estimate of area, they should either normalize to a common *c* or to a common N before applying this index of home-range size for further ecological analyses. On the other hand, if the data instead showed compliance with the standard expectation of area asymptote, one should of course apply—for example—the KDE method, which explicitly is developed for home-range estimate under Markov-compliant space use.

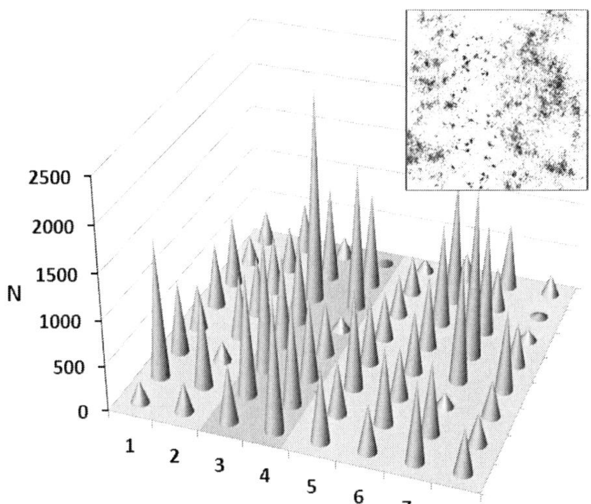

Figure 106. Simulated local dispersion of individuals, based on MRW model and a heterogeneous environment. See main text.

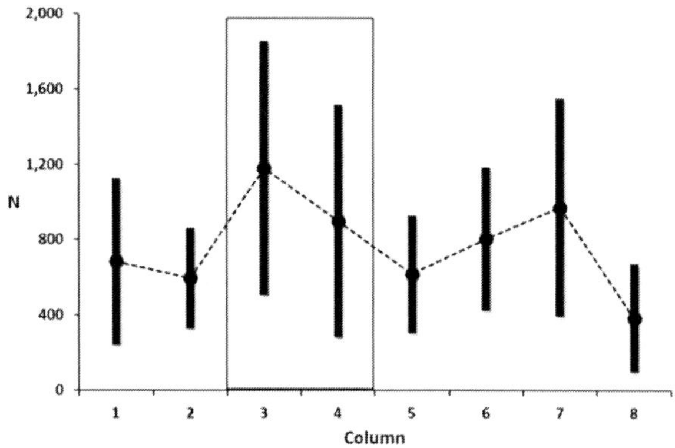

Figure 107. A classical analysis based on local density of fixes reveals the condition of local habitat heterogeneity only weakly.

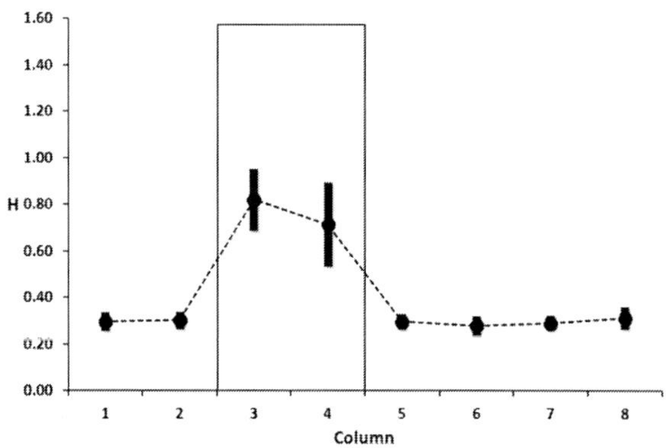

Figure 108. Alternative analysis based on MRW theory reveals the heterogeneous condition more clearly.

To conclude at this point, a pilot test should be performed (for example, based on methods summarized in Figure 68) to test whether a MRW or a standard process is the most realistic representation of the space use behaviour. Based on this test, ecological research involving home-range size should either be based on Eq. 1 or Eq. 2. However, let us now turn to the more interesting implications, in practical terms, and again assuming the same data and process as in Figures 104-105. The parameter *c* represents area pr. square root of N. Thus, as explained in a previous chapter, the inverse of c is an index of intensity of space use. A larger $1/c$ implies a smaller I(N) for a given N.

In wildlife management a smaller home range is generally found in superior habitats. Thus, fix density (N/area) is higher. Fix density is therefore proportional with the inverse of the area asymptote representation K, and the density estimate is thus the standard model's analogue to $1/c$. Given that the parallel-processing framework is found most representative for a given data set, $1/c$ should replace N/area as a proxy for intensity of use. In Gautestad and Mysterud (2010b), we defined this index as $H=1/c$. What is gained? Reflecting on Figures 104-105, the parameter H is obviously a better proxy for space-use intensity than using fix density (or its inverse, the home-range size) directly. The latter should only be applied for Markov-compliant space use. Otherwise, unnecessary noise inflation is to be expected in statistical analyses where home-range size or a fix density-based proxy is applied for space-use intensity. This is shown in the following MRW simulation scenario for 100 independently moving individuals in a given area (Gautestad and Mysterud 2010b). I refer to this paper for technical details. Figure 106 shows the locally varying abundance of fixes from these individuals' space use in superposition. Columns 3 and 4 were given superior habitat conditions. When averaging fix abundance (N) on a column-by-column basis, there is little trace of the habitat contrast (Figure 107). Five of the six columns outside numbers three to four have lower fix density, but the standard deviations are large. However, when estimating $H=1/c$ as a replacement for fix density, the difference is clear (Figure 108). In particular, observe the high precision of the estimates for each column; a result of estimating intensity of use from a parameter explicitly derived from the parallel-processing framework. To conclude, the parallel-processing framework may have substantial implication for applied ecology, given that the preliminary empirical support summarized in this book is gaining a broader verification.

As a last example, consider again the "broken stick" model above. If this two-mode model—the standard approach in this context—is replaced by the complementary power law for step length distribution (scale-free, non-mechanistic movement), it will be futile to look for transition scales between levels of space use. If this concept is applied on data from a scale-free process, the "break point" is a statistical artifact from assuming an erroneous model: superposition of two correlated random walk modes rather than a scale-free kind of movement. The slope of the log-log plot of the step length distribution provides an alternative ecological insight by studying multi-scaled use from another angle. For example, consider the caribou and the red deer data (Figure 99 and Figure 103); they both showed a power exponent close to β=2 (the negative of the slope). In *6.4.6 Dispersion of kinetic energy over a scale range,* I argued that β=2 represented the generic MRW condition, where the animal relates equally strong to all scales over the given scale range. In behavioural terms, β>2 should then mean that the animal has prioritized a more fine-grained space use. In other words, it has performed relatively fewer long displacements. The elk data in Figure 102, where $\beta\approx$1.5, indicates the opposite condition: more weight on coarser scales than finer ones. Interestingly, these data were collected by Killeen et al. during the migration period!

However, the MRW model does in fact provide a parameter expressing finer or coarser kinds of space use in average terms. For example, a wood mouse obviously has a smaller home range than red deer, just like a superior habitat is expected to allow for a smaller home range. The size concept, which is closely linked to the scale aspect, is thus an important quantity for biological and ecological inference. Above (and in previous chapters) I described how c from Eq. 1 provides a statistically much more precise description of space-use extent than the classical home-range size. In *7.2.5 The characteristic scale and the home-range ghost paradox,* I described this property in more detail. Figure 58 presented an inset of the distribution of estimates of c for various individuals of red deer. However, so far ecological hypotheses like correlation with environmental conditions have not been tested on this material under the alternative methods. Perhaps, in due course, there will appear a Volume 2 of this book; ecological inference using models from the parallel-processing framework?

12. CONCLUDING REMARKS

This book represents a personal and subjective—some may conclude one-sided—presentation of both the traditional paradigm, the scrutiny of this paradigm and the respective presentations of alternative approaches. I did in fact not intend to be objective in the present setting. The intention was to be provocative and explicit about nagging issues, and in the preface I referred to this approach as rattling the wasps' nest. This policy has probably locked some readers' minds here-and-there over the chapters, but on the other hand I hope that others have been provoked in a constructive manner. In the latter case, if you don't agree with my arguments that a home range should be regarded a fractal object (and the methodological consequences for studies on animal space use that follows from this acceptance) then I expect you to seek to falsify the present models and results based on your own data. You find several methods to do so in this book. If you don't agree on the present view that analysis of GPS fixes by default invokes a "hidden layer" of system ignorance, and thus requires a biophysical (statistical-mechanical) model representation of space use, then bring forward counter-arguments. If you don't find the scaling cube (Figure 70) helpful for organizing animal movement into eight biophysical universality classes then please reflect on why the mechanistic modelling approach (embedded on the x-y plane) is sufficient, and thus making the z-dimension and the parallel processing concept superfluous. If you believe spatial density of local accumulation of GPS fixes is better representation of "intensity" of space use than the MRW model's characteristic scale of space use (the parameter $1/c$; see Figures 107-108) then I would like to challenge you to test these qualitatively different observables against each-other. In the context of population dynamics, if you don't agree that the Zoomer model (non-mechanistic dynamics) is a feasible alternative to partial differential models and coupled map lattice models, then please consider why the classical mean field premise (the Markovian framework) may suffice as a premise under the general condition of conspecific attraction and intra-specific cohesion.

I hope that some are also genuinely positively inspired by at least some of the novel concepts, ideas and models that are presented. If you are a theoretician I hope that you grab some parts of what at this stage is a rather coarse-knitted framework, develop them further and also scrutinize the initial elements. Some ideas were given in the foregoing paragraph. If you are a field ecologist with piles of empirical data and detailed knowledge of specific animals and their ecology, I hope that you invest effort in exploring the applicability of the alternative methods and approaches that are presented here. To summarize, in this book I propose that theoreticians in movement ecology should be far more critical of their own framework of mathematical and statistical tools, and to acknowledge the qualitative difference between the biological/behavioural "micro-scale" and the biophysical "meso/macro-scale", with the hidden layer in-between. In this respect, a closer scrutiny of the universal acceptance of the Markovian paradigm (mechanistic dynamics) in movement ecology and population models is clearly needed.

In the present policy of most high-ranking scientific journals, it is unfortunately difficult to make room for explaining alternatives to problematic but still widely accepted model assumptions without making these issues the main topic of the paper. Challenging the paradigm in applications of these alternative formulations requires detailed introductions of the unfamiliar concepts at every occasion. This steals both publication space and author energy, and in practice also reduces the chances for manuscript acceptance. In the competitive publication environment where streamlined and narrow topics are favoured, too controversial directions are often left alone for a later day. This is of course unfortunate. One should not abandon one's conviction and suppress controversial ideas from the publication challenge alone.

Lack of sufficient publication space in ordinary journals has unfortunately led to several misconceptions about the present concepts and models. The MRW model has frequently been referred to as mechanistic, which it is not (owing to the parallel-processing postulate for its dynamics). It has also been claimed to unrealistically predict home-range area to increase proportionally with sample of fixes (Börger *et al.* 2006), which it does not (square root expansion is substantially slower than proportional; see Eq. 1a). It has also been referred to as unrealistic owing to lack of energy considerations when return events are targeting distant historically visited locations relative to closer ones (Powell 2000). However, since the model regards a statistical-mechanical (coarse-grained) representation of movement, distance to targets is expected to have little effect under space use that is influenced by site fidelity, and fix sampling interval consequently is sufficiently large to hide local path details.

For example, consider a thread of a given length representing a movement path during a given time span from one location of the animal to the next, and this interval on average contains one or more returns to a previously visited site. If you stretch the thread to a unidirectional line, you get a large displacement during this time interval. If you curl it, the net start-to-end displacement is small. Anyway, the energy expenditure is similar since the thread length is similar. Both return steps (expressing site fidelity) and scale-free exploratory steps—the two main components of a MRW—contribute to curling of the path, which will vary in strength from one interval to the next.

Another common critique of MRW is that it by default describes a non-asymptotic expansion of observed area use as a function of sample size of relocations. However, (a) the non-asymptotic and power law-compliant function reflects the fractal (scale-free) aspect of space use (Gautestad and Mysterud 2005), and (b) like all scale-free systems, power law compliance is bounded to a given scale range. Thus, it is expected to break down beyond these limits, which may be verified in sufficiently large data sets. Others have criticized MRW for not being specific with respect to which locations to return to (in the generic model targets are chosen randomly). However, this critique would only be valid if the model was mechanistic. It is not; it is explicitly statistical-mechanical, as repeatedly explained both in papers and throughout this book.

Merkle et al.'s recent study on bison movement (Merkle *et al.* 2014) lends explicit empirical support to the MRW's basic assumption with respect to distance and site independence for return targets. It remains to be seen how general this result is, though. For example, the bison followed another pattern during winter, when resources were generally sparser. Under this condition a modified MRW model also shows more opportunistic search (Gautestad and Mysterud 2010b).

From the onset of walking into a study on memory-influenced space use by animals, I have been strongly inspired by the scientific culture among physicists. When an odd and paradoxical observation is pinpointed in physics, peers typically flock to give it proper scrutiny. The stranger the issue, the broader the attention, it seems. Persisting problems often lead to novel research groups with accompanying economic support. Wildlife ecology generally and unfortunately seems to present a different attitude in animal space-use research. Odd observations are plenty here as well, but there seems to be a strong reluctance to step outside the prevalent paradigm to seek definitive answers to nagging questions.

In dialogue with ecologists over the years, I've experienced a lot of verbal support for my critical attitude (but fortunately also the opposite, which has been imperative for sharpening both theory and presentations of this theory). However, this engagement has—so far—not resulted in much *published* critical exploration of the ideas summarized in this book. For example, the home-range ghost—the strange observation that space use expands non-trivially as a function of observation intensity (sample size of fixes)—is challenging the fundamental view that a home range is a two-dimensional object. Instead of testing the home range's fractal properties (Eq. 1 in this book) on one's own data and publishing confirmative or opposing results, the hypothesis is ignored or rejected *a priori*, assumed to represent a temporally unstable home range (like illustrated in Figure 2a,b). In Chapter 1 I dismissed this attitude and simultaneously invited peers to consider a home range a fractal-dimensional with $0.9<D\approx<1.3$ rather than a two- or three-dimensional object. Paradox potentially solved! The consequence from such a change in perception of a central concept of animal ecology is thought-provoking. For example, two individuals may use the same space from a two- or three-dimensional perspective but different space from a fractal-dimensional perspective. In simple terms they may use different parts of a given area when this space is dissected over a range of spatial resolutions—not only different parts at different times. Overlap is possibly no longer overlap, and novel hypotheses may arise for further exploration.

The unfortunate attitude to avoid bold "out of the box" steps into novel and unfamiliar theoretical territories may be rooted in the ecologist's classical training, which is supported by courses in mathematics and statistics but not in general process theory. The latter is provided by physics. Studying animal space use based on—for example—GPS data requires a biophysical framework. My proposal for such a framework is summarized in Figure 70, the scaling cube.

All researchers will eventually recognize this statement by Schopenhauer:

> *"All truth passes through three stages before it is recognized.*
> *In the first it is ridiculed,*
> *in the second it is opposed,*
> *in the third it is regarded as self-evident."*
> Arthur Schopenhauer, philosopher

Performing active research—whether theoretical or empirical—means a risk of either becoming controversial or (worse) ignored if the proposed hypotheses are too off-pitch relative to consensus. However, scientific progress is exactly what the second stage in the quote above is all about. After a broad and often intense scrutiny by the research community, only valuable contributions should survive and reach the third stage. Choosing a direction of research far off the mainstream consensus may be interesting and potentially rewarding, but there is obviously a higher personal risk involved. Bold proposals may thus stay stuck at the first stage of being ridiculed and/or ignored, without reaching the critical second phase.

Principles from statistical mechanics have generally been considered non-relevant to ecologists. However, this attitude is now changing dramatically by the growing community of scientists with one leg in biology and one in physics. Physicists are attracted by the challenge from complex systems, which may offer fertile ground for new theory with broad implications. This aspect of biophysics is now evolving in the field of movement ecology, which includes a rapidly expanding group of engaged theoreticians and empiricists from both camps. I hope this book will contribute to further cross-field fertilization despite some rather provocative presentations and statements in some sections.

APPENDIX A

INFLUENCE FROM FIX SAMPLING INTERVAL

Figure A1 shows correlated Brownian motion (correlated random walk) that is sampled at different temporal resolutions (sampling interval). Successive sample relocations (fixes) are connected by line segments. Left pane: 1,000 steps at unit scale, with Gaussian step length distribution and directional persistence 0.8 (strong persistence) between successive steps. Middle pane: the series is expanded by a factor of 100, and sampled at 100 times larger intervals (sampling frequency 1:100). The path has now become more Brownian motion-like, owing to approximately lack of directional persistence, which is now hidden from view at finer scales. In addition, the step length distribution is no longer Gaussian, but Poisson-compliant (not shown): step length deteriorates proportionally with a negative exponential function. Right pane: sampling frequency has decreased further to 1:10,000. Except for the right-hand path traversing a larger area than the middle path (the respective images are rescaled), the paths look similar. Thus, resampling a Brownian motion to a different resolution still satisfies statistics of a Brownian motion. The two paths to the right are spatially scale-free statistically (so-called self-similar dispersion of fixes), with a similar-sized fractal dimension, despite the fact that they represent (a) a scale-specific process at fine (hidden) resolutions and (b) a scale-specific distribution of step lengths (Poisson distribution, with a specific step length scale).

A.1 Scale-specific movement process: from CorrBM to BM

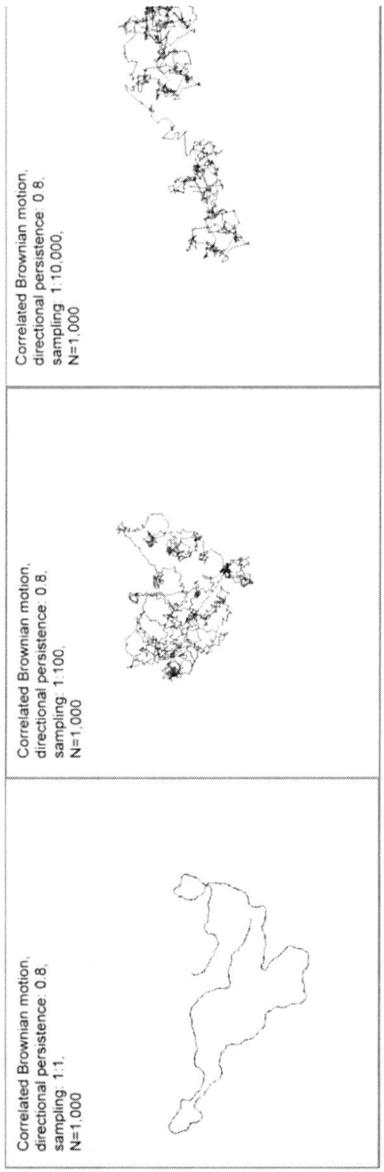

Figure A1.

A.2 Scale-free movement path: Lévy walk

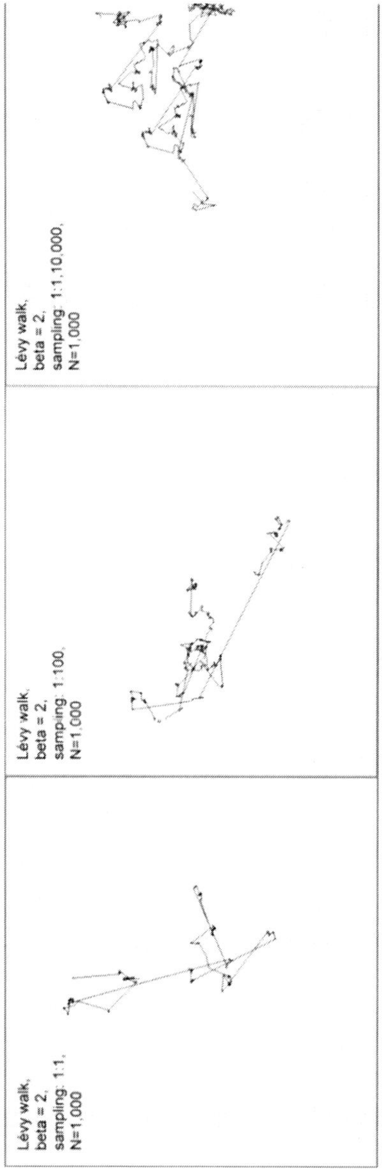

Figure A2.

The three panes show a scale-free process, Lévy walk with β=2 (see Eq. 13 in the main text). In other respects the conditions are similar to Figure A1: sampling frequency, series length, and rescaling of arena size to let the path be contained within the area shown. Successive sample relocations (fixes) are connected by line segments. In this case, with Lévy walk rather than a scale-specific process, both the step length distribution and the spatial scatter of fixes are scale-free, and the fractal dimension is smaller than in Figure A1 (stronger congregation of fixes, less space-filling).

Figure A3 shows same conditions as in Figure A2, except for introduction of a small probability p=0.0001 for a return to a previous location during next interval at unit time increment. Thus, a return, which is expressing site fidelity, takes place every 1,000th step on average (μ=1000t). If observation interval is smaller than the return interval, as in the left pane, there is little chance for site fidelity influence in a series of 1,000 observed fixes. In fact, only one return took place during an interval in this set of fixes, marked by a red line segment. In the middle and right pane, fix sampling happens at a smaller rate, and thus more return events have taken place during the sampling period for 1,000 steps. In the middle pane many of the return events are indirectly visible as line segments between two main areas of intense use. These are none coloured red, since respective returns have taken place during one or another of the intermediate unit time increments relative to the ten times larger observation interval. The series of fixes has become non-autocorrelated owing to many return events pr. fix collection (in the left pane the series was spatially autocorrelated). In the right pane some “occasional sallies” (large-range steps) are normally followed by a return to the core area of space use at next fix collection, owing to the high number of intermediate return events. Site fidelity defined in this manner leads to a self-reinforcing use of some locations (positive feedback).

A.3 Scale-free movement with site fidelity: MRW

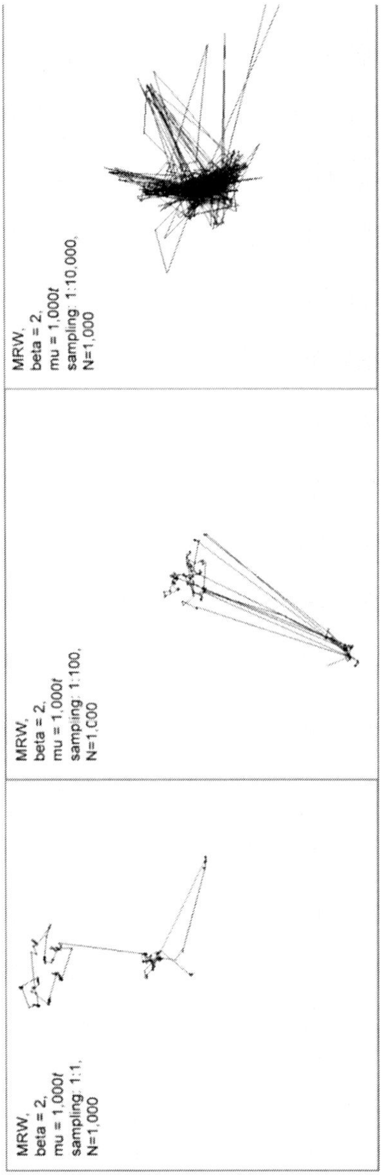

Figure A3.

REFERENCES

Allen, T. H. F. and T. B. Starr. 1982. Hierarchy: perspectives for ecological complexity. The University of Chicago Press, Chicago.

Andrewartha, H. G. and L. C. Birch. 1954. The distribution and abundance of animals. Chicago University Press, Chicago.

Auger-Méthé, M., C. C. St. Clair, M. A. Lewis, and A. E. Derocher. 2011. Sampling rate and misidentification of Lévy and non-Lévy movement paths: comment. Ecology 92:1699-1701.

Bailey, D. W. 1995. Daily selection of feeding areas by cattle in homogeneous and heterogeneous environments. Appl. Anim. Behav. Sci. 45:183-199.

Bak, P., C. Tang, and K. Wiesenfeld. 1987. Self-organized criticality: and explanation of 1/f noise. Physical Review Letters 59:381-384.

Barg, J. J., J. Jones, and R. J. Robertson. 2005. Describing breeding territories of migratory passerines: suggestions for sampling, choice of estimator, and delineation of core areas. Journal of Animal Ecology 74:139-149.

Barrett, L. and C. B. Lowen. 1998. Random walks and the gas model: spacing behaviour of grey-cheeked mangabeys. Functional Ecol. 12:857-865.

Bartumeus, F., M. G. E. da Luz, G. M. Viswanathan, and J. Catalan. 2005. Animal search strategies: a quantitative random-walk analysis. Ecology 86:3078–3087.

Bartumeus, F., L. Giuggioli, M. Louzao, V. Bretagnolle, D. Oro, and S. A. Levin. 2010. Fishery discards impact on seabird movement patterns at regional scales. Current Biology 20:215-222.

Bartumeus, F. and S. A. Levin. 2008. Fractal reorientation clocks: linking animal behavior to statistical patterns of search. Proc. Natl. Acad. Sci. USA 105:19072-19077.

Bartumeus, F., E. P. Raposo, G. M. Viswanathan, and M. G. E. da Luz. 2013. Stochastic optimal foraging theory. Pages 3-32 *in* M. A. Lewis, P. K. Maini, and S. Petrovskii, editors. Dispersal, individual movement and spatial ecology: a mathematical perspective. Springer-Verlag, Berlin.

Bédard, C., H. Kröger, and A. Destexhe. 2006. Does the 1/f scaling of brain signals reflect self-organized critical states? Physical Review Letters 97:118102 (118101-118104).

Beekman, M. and F. L. W. Ratnieks. 2000. Long-range foraging by the honey-bee, Apis mellifera L. Functional Ecol. 14:490-496.

Belant, J. L. and E. H. Follmann. 2002. Sampling considerations for American black and brown bear home range and habitat use. Ursus 13:299-315.

Beletsky, L. D. and G. H. Orians. 1989. Familiar neighbors enhance breeding success in birds. Proc. Natl. Acad. Sci. USA 86:7933-7936.

Bell, W. 1990. Searching behavior patterns in insects. Annual Review of Entomology 35:447-467.

Benhamou, S. 2007. How many animals really do the Lévy walk? Ecology 88:1962-1969.

Benhamou, S. 2008. How many animals really do the Lévy walk? Reply. Ecology 89:2351-2352.

Benhamou, S. 2014. Of scales and stationarity in animal movements. Ecology Letters 17:261-272.

Bigger, M. and R. G. Tapley. 1969. Prediction of outbreaks of coffee leaf-miners on Kilimanjaro. Bulletin of Entomological Research 58:601-618.

Bischof, R., L. E. Loe, E. L. Meisingset, B. Zimmermann, B. van Moorter, and A. Mysterud. 2012. A migratory northern ungulate in pursuit of spring: jumping or surfing the green wave? The American Naturalist 180:407-424.

Bouchaud, J.-P. and A. Georges. 1990. Anomalous diffusion in disordered media. Statistical mechanisms, models and physical applications. Physics reports 195:127-293.

Boyer, D., M. C. Crofoot, and P. D. Walsh. 2012. Non-random walks in monkeys and humans. Journal of the Royal Society Interface 9:842-847.

Boyer, D., G. Ramos-Fernándes, O. Miramontes, J. L. Mateos, G. Cocho, H. Larralde, H. Ramos, and F. Rojas. 2006. Scale-free foraging by primates emerges from their interaction with a complex environment. Proc. R. Soc. Lond. B 273:1743-1750.

Boyer, D. and J. C. R. Romo-Cruz. 2014. Solvable random-walk model with memory and its relations with Markovian models of anomalous diffusion. Physical review E. 90:1-12.

Boyer, D. and C. Solis-Salas. 2014. Random walks with preferential relocations to places visited in the past and their application to biology. arXiv 1403.6069v1:1-5.

Boyer, D. and P. D. Walsh. 2010. Modelling the mobility of living organisms in heterogeneous landscapes: does memory improve foraging success? Phil. Trans. R. Soc. A. 368:5645-5659.

Brewer, A. M. and K. J. Gaston. 2002. The geographical range structure of the holly leaf-miner. I. Population density. J. Anim. Ecol. 71:99-111.

Brown, D. and P. Rothery. 1993. Models in Biology: Mathematics, Statistics and Computing. John Wiley & Sons, Chichester.

Brown, J. H. and A. Kodric-Brown. 1977. Turnover rates in insular biogeography: effec of immigration on extinction. Ecology 58:445-449.

Brown, J. H., D. W. Mehlman, and G. C. Stephens. 1995. Spatial variation in abundance. Ecology 76:2028-2043.

Bunnefeld, N., L. Börger, B. van Moorter, C. M. Rolandsen, H. Dettki, E. J. Solberg, and G. Ericksson. 2011. A model-driven approach to quantify migration patterns: individual, regional and yearly differences. Journal of Animal Ecology 80:466-476.

Burt de Perera, T. 2004. Fish can encode order in their spatial map. Proc. R. Soc. Lond. B 271:2131-2134.

Byrne, M. E., J. C. McCoy, J. W. Hinton, M. J. Chamberlain, and B. A. Collier. 2014. Using dynamic Brownian bridge movement modelling to measure temporal patterns of habitat selection. Journal of Animal Ecology:DOI: 10.1111/1365-2656.12205.

Börger, L., B. Dalziel, and J. Fryxell. 2008. Are there general mechanisms of animal home range behaviour? A review and prospects for future research. Ecology Letters 11:637-650.

Börger, L., N. Franconi, G. de Michele, A. Gantz, F. Meschi, A. Manica, S. Lovari, and T. Coulson. 2006. Effects of sampling regime on the mean and variance of home range size estimates. Journal of Animal Ecology 75:1393-1405.

Campbell, D. 1989. Introduction to nonlinear phenomena.*in* D. L. Stein, editor. Lectures in the science of complexity. Addison Wesley Publishing Company.

Campos, F. A., M. L. Bergstrom, A. Childers, J. D. Hogan, K. M. Jack, A. D. Melin, K. N. Mosdossy, M. S. Myers, N. A. Parr, E. Sargeant, A. M. Schoof, and L. M. Fedigan. 2014. Drivers of home range characteristics across spatiotemporal scales in a Neotropical primate, *Cebus capucinus*. Animal Behaviour 91:93-109.

Charnov, E. L. 1976. Optimal foraging: the marginal value theorem. Theor. Popula. Biol. 9:129-136.

Chittka, L., N. M. Williams, H. Rasmussen, and J. D. Thomson. 1999. Navigation without vision: bumblebee orientation in complete darkness. Proc. R. Soc. Lond. B 266:45-50.

Cole, B. J. 1995. Fractal time in animal behaviour: the moment activity of *Drosophila*. Anim. Behav. 50:1317-1324.

Collett, T. S., B. A. Cartwright, and B. A. Smith. 1986. Landmark learning and visuo-spatial memories in gerbils. J Comp Physiol A 158:835-851.

Conradt, L., E. J. Bodsworth, T. J. Roper, and C. D. Thomas. 2000. Non-random dispersal in the butterfly *Manolia jurtina*: implications for metapopulation models. Proc. R. Soc. Lond. B 267:1505-1510.

Coulon, A., N. Morellet, M. Goulard, B. Cargnelutti, J.-M. Angibault, and A. J. M. Hewison. 2008. Inferring the effects of landscape structure on roe deer (*Capreolus capreolus*) movements using a step selection function. Landscape Ecology 23:603-614.

Courchamp, F., T. Clutton-Brock, and B. Grenfell. 1999. Inverse density dependence and the Allee effect. Trends Ecol. Evol. 14:405-410.

Cruse, H. and R. Wehner. 2011. No need for a cognitive map: decentralized memory for insect navigation. PLoS Computational Biology 7:1-10.

Dalziel, B. D., J. M. Morales, and J. M. Fryxell. 2008. Fitting probability distributions to animal movement trajectories: using artificial neural networks to link distance, resources and memory. American Naturalist 172:248-258.

de Roos, A. M., E. McCauley, and W. G. Wilson. 1991. Mobility versus density-limited predator-prey dynamics on different spatial scales. Proc. R. Soc. Lond. B 246:117-122.

Delestrade, A., R. H. McCleery, and C. M. Perrins. 1996. Natal dispersal in the heterogeneous environment: the case of the great Tit in Wytham. Acta Oecol. 17:519-529.

den Boer, P. J. and J. Reddingius. 1996. Regulation and stabilization paradigms in population ecology. Chapman & Hall, London.

Dobson, A. and J. Poole. 1998. Conspecific aggregation and conservation biology. Pages 193-208 *in* T. Caro, editor. Behavioral ecology and conservation biology. Oxford University Press, New York.

Doeller, C. F., C. Barry, and N. Burgess. 2010. Evidence for grid cells in a human memory network. Nature 463:657-663.

Doncaster, C. P. and D. W. Macdonald. 1991. Drifting territoriality in the red fox *Vulpes vulpes*. Journal of Animal Ecology 60:423-439.

Edwards, A. M. 2008. Using likelihood to test for Lévy flight search patterns and for general power-law distributions in nature. Journal of Animal Ecology 77:1212-1222.

Edwards, A. M. 2011. Overturning conclusions of Levy flight movement patterns by fishing boats and foraging animals. Ecology 92:1247-1257.

Edwards, A. M., M. P. Freeman, G. A. Breed, and I. D. Jonsen. 2012. Incorrect likelihood methods were used to infer scaling laws of marine predator search behaviour. PLoS One 7:e45174.

Edwards, A. M., R. A. Phillips, Watkins, N. W., M. P. Freeman, E. J. Murphy, V. Afanasyev, S. V. Buldyrev, M. G. E. da Luz, E. P. Raposo, H. E. Stanley, and G. M. Viswanathan. 2007. Revisiting Lévy flight search patterns of wandering albatrosses, bumblebees and deer. Nature 449:1044-1049.

Eisler, Z., I. Bartos, and J. Kertész. 2008. Fluctuation scaling in complex systems: Taylor's law and beyond. Adv. Phys. 57:89-142.

Fagan, W. F. and J. M. Calabrese. 2014. The correlated random walk and the rise of movement ecology. Bulletin of the Ecological Society of America 95:204-206.

Fieberg, J. 2007. Kernel density estimators of home range: smoothing and the autocorrelation red herring. Ecology 88:1059-1066.

Fieberg, J. and L. Börger. 2012. Could you please phrase "home range" a question? Journal of Mammalogy 93:890-902.

Fleming, C. H., J. M. Calabrese, T. Mueller, K. A. Olson, P. Leimgruber, and W. F. Fagan. 2014. From fine-scale foraging to home ranges: a semi-variance appraoch to identifying movement modes across spatiotemporal scales. The American Naturalist 183:E154-E167.

Fletcher, J. M., A. Revell, B. E. Reichert, W. M. Kitchens, J. D. Dixon, and J. D. Austin. 2013. Network modularity reveals critical scales for connectivity in ecology and evolution. Nature Communications 4:2572.

Fortin, D., H. L. Beyer, M. S. Boyce, D. W. Smith, T. Duchesne, and J. S. Mao. 2005. Wolves influence elk movements: behavior shapes a tropical cascade in Yellowstone national park. Ecology 86:1320-1330.

Frank, P. W. 1964. On home range of limpets. The American Naturalist 98:99-104.

Fretwell, S. D. and H. L. Lucas. 1970. On territorial behavior and other factors influencing habitat distribution in birds. I. Theoretical Development. Acta Biotheoretica 19:16-36.

Fronhofer, E. A., T. Hovestadt, and H.-J. Poethke. 2013. From random walks to informed movement. Oikos 122:857-866.

Fryxell, J. M., M. Hazell, L. Börger, B. D. Dalziel, D. T. Haydon, J. M. Morales, T. McIntosh, and R. C. Rosatte. 2008. Multiple movement modes by large herbivores at multiple spatiotemporal scales. Proceedings of the National Academy of Sciences of the USA 105:19114-19119.

Garber, P. A. 1989. The role of spatial memory in primate foraging patterns: *Saguinus mystax* and *Saguinus fuscicollis*. Am. J. Primatol. 19:203–216.

Gautestad, A. O. 2011. Memory matters: Influence from a cognitive map on animal space use. Journal of Theoretical Biology 287:26-36.

Gautestad, A. O. 2012. Brownian motion or Lévy walk? Stepping towards an extended statistical mechanics for animal locomotion. Journal of the Royal Society Interface 9:2332-2340.

Gautestad, A. O. 2013a. Animal space use: Distinguishing a two-level superposition of scale-specific walks from scale-free Lévy walk. Oikos 122:612-620.

Gautestad, A. O. 2013b. Lévy meets Poisson: a statistical artifact may lead to erroneous re-categorization of Lévy walk as Brownian motion. The American Naturalist 181:440-450.

Gautestad, A. O., L. E. Loe, and A. Mysterud. 2013. Inferring spatial memory and spatiotemporal scaling from GPS data: comparing red deer Cervus elaphus movements with simulation models. Journal of Animal Ecology 82:572-586.

Gautestad, A. O. and A. Mysterud. 2013. The Lévy flight foraging hypothesis: forgetting about memory may lead to false verification of Brownian motion. Movement Ecology 1:1-18.

Gautestad, A. O. and I. Mysterud. 1993. Physical and biological mechanisms in animal movement processes. J. Appl. Ecol. 30:523-535.

Gautestad, A. O. and I. Mysterud. 1994. Fractal analysis of population ranges: methodological problems and challenges. Oikos 69:154-157.

Gautestad, A. O. and I. Mysterud. 1995. The home range ghost. Oikos 74:195-204.

Gautestad, A. O. and I. Mysterud. 2005. Intrinsic scaling complexity in animal dispersion and abundance. The American Naturalist 165:44-55.

Gautestad, A. O. and I. Mysterud. 2006. Complex animal distribution and abundance from memory-dependent kinetics. Ecological Complexity 3:44-55.

Gautestad, A. O. and I. Mysterud. 2010a. The home range fractal: from random walk to memory dependent space use. Ecological Complexity 7:458-470.

Gautestad, A. O. and I. Mysterud. 2010b. Spatial memory, habitat auto-facilitation and the emergence of fractal home range patterns. Ecological Modelling 221:2741-2750.

Gautestad, A. O. and I. Mysterud. 2012. The Dilution Effect and the Space Fill Effect: Seeking to Offset Statistical Artifacts When Analyzing Animal Space Use from Telemetry Fixes. Ecological Complexity 9:33-42.

Gautestad, A. O., I. Mysterud, and M. R. Pelton. 1998. Complex movement and scale-free habitat use: testing the multi-scaled home range model on black bear telemetry data. Ursus 10:219-234.

Getz, W. M. and D. Saltz. 2008. A framework for generating and analyzing movement paths on ecological landscapes. Proc. Natl. Acad. Sci. USA 105:19066-19071.

Giuggioli, L., G. Abramson, V. M. Kenkre, R. R. Parmenter, and T. L. Yates. 2006. Theory of home range esitmation from displacement measurements of animal populations. J. Theor. Biol. 240:126-135.

Giuggioli, L. and F. Bartumeus. 2011. Linking animal movement to site fidelity. Journal of Mathematical Biology DOI 10.1007s00285-011-0431-7.

Gonzalez, M. C., C. A. Hidalgo, and A.-L. Barabasi. 2008. Understanding individual human mobility patterns. Nature 453:779-782.

Grimm, V. and S. F. Railsback. 2005. Individual-based Modeling and Ecology. Princeton University press, Princeton, N. J.

Gross, J. E., C. Zank, N. T. Hobbs, and D. E. Spalinger. 1995. Movement rules for herbivores in spatially heterogeneous environments: responses to small scale pattern. Landscape Ecology 10:209-217.

Grove, M. 2013. The evolution of spatial memory. Mathematical Biosciences 242:25-32.

Hagen, C. A., N. C. Kenkel, D. J. Walker, R. K. Baydack, and C. E. Braun. 2001. Fractal-based spatial analysis of radiotelemetry data. Pages 167-187 *in* J. J. Millspaugh and J. M. Marzluff, editors. Radio tracking and animal populations. Academic Press, San Diego.

Hanski, I. 1994. Spatial scale, patchiness and population dynamics on land. Phil. Trans. R. Soc. B 343:19-25.

Hanski, I. and O. E. Gaggiotti, editors. 2004. Ecology, genetics, and evolution of metapopulations. Elsevier, Amsterdam.

Hanski, I. A. and M. E. Gilpin, editors. 1997. Metapopulation biology: ecology, genetics, and evolution. Academic Press, San Diego.

Healy, S., editor. 1998. Spatial representation in animals. Oxford University Press, Oxford.

Hemson, G., P. Johnson, A. South, R. Kenward, R. Ripley, and D. W. Macdonald. 2005. Are kernels the mustard? Data from global positioning system (GPS) collars suggests problems for kernel home-range analyses with least-squares cross-validation. Journal of Animal Ecology 74:455-463.

Hiebeler, D. 1997. Stochastic spatial models: from simulations to mean field and local structure approximations. J. Theor. Biol. 187:307-319.

Hirata, M., Y. Tajiri, and K. Murakami. 2014. Cattle make two-stage discriminations in patch choice. Ecological Research DOI:10.1007/s11284-11014-11231-11285.

Hjermann, D. Ø. 2000. Analyzing habitat selection in animals without well-defined home ranges. Ecology 81:1462-1468.

Humphries, N. E., N. Queiroz, J. R. M. Dyer, N. G. Pade, M. K. Musyl, K. M. Schaefer, D. W. Fuller, J. M. Brunnschweiler, T. K. Doyle, J. D. R. Houghton, G. C. Hays, C. S. Jones, L. R. Noble, V. J. Wearmouth, E. J. Southhall, and D. W. Sims. 2010. Environmental context explains Lévy and Brownian movement patterns of marine predators. Nature 465:1066-1069.

Humphries, N. E. and D. W. Sims. 2014. Optimal foraging strategies: Lévy walks balance searching and patch exploitation under a very broad range of conditions. Journal of Theoretical Biology 358:179-193.

Humphries, N. E., H. Weimerskirch, N. Queiroz, E. J. Southhall, and D. W. Sims. 2012. Foraging success of biological Lévy flights recorded in situ. Proc. Natl. Acad. Sci. USA 109:7169-7174.

Hurst, H. E. 1951. Long-term storage capacity of reservoirs. Trans. Am. Soc. Siv. Eng. 116:770-808.

Johnson, C. J., K. L. Parker, D. C. Heard, and M. P. Gillingham. 2002. Movement parameters of ungulates and scale-specific responses to the environment. J. Anim. Ecol. 71:225-235.

Jørgensen, S. E. 1988. Fundamentals of ecological modelling. Elsevier, Amsterdam.

Kaneko, K., editor. 1993. Theory and applications of coupled map lattices. John Wiley & Sons, Chichester.

Kareiva, P. M. and N. Shigesada. 1983. Analyzing insect movement as a correlated random walk. Oecologia 56:234-238.

Kaye, B. 1993. Chaos & complexity. Discovering the surprising patterns of science an dtechnology. VCH, Weinheim.

Keeling, M. 1999. Spatial models of interacting populations. Pages 64-99 *in* J. McGlade, editor. Advanced ecological theory. Blackwell Science, Oxford.

Kehr, K. W. and R. Kutner. 1982. Random walk on random walk. Physica 110A:535-549.

Kendal, W. S. and B. Jørgensen. 2011. Taylor's power law and fluctuation scaling explained by a central-limit-like convergence. Physical review E. 83:066115 (066111-066117).

Kennedy, J. S. and L. Crawley. 1967. Spaced-out gregariousness in sycamore aphids *Drepanosiphum platanoides* (Schrank) (Hemiptera, Callaphididae). J. Anim. Ecol. 36:147-170.

Killeen, J., H. Thurfjell, S. Ciuti, D. Paton, M. Musiani, and M. S. Boyce. 2014. Habitat selection during ungulate dispersal and exploratory movement at broad and fine scale with implications for conservation management. Movement Ecology 2:1-13.

Kranstauber, B., K. Safi, and F. Bartumeus. 2014. Bivariate Gaussian bridges: directional factorization of diffusion in Brownian bridge models. Movement Ecology 2:1-10.

Lane, S. J., J. C. Alonso, and C. A. Martin. 2001. Habitat preferences of great bustard *Otis tarda* flocks in the arable steppes of central Spain: are potentially suitable areas unoccupied? J. Appl. Ecol. 38:193-203.

Levett, S. and S. Walls. 2011. Tracking the elusive life of the Emperor Dragonfly *Anax imperior* Leach. Journal of the British Dragonfly Society 27:59-68.

Lewis, M. A., P. K. Maini, and S. V. Petrovskii. 2013. Dispersal, individual movement and spatial ecology: a mathematical perspective. Springer-Verlag, Berlin.

Mandelbrot, B. B. 1983. The fractal geometry of nature. W. H. Freeman and Company, New York.

Manly, B. F. J., L. L. McDonald, D. L. Thomas, T. L. McDonald, and W. P. Erickson. 2002. Resource selection by animals: statistical design and analysis for field studies. Kluwer Academic Publishers, Dordrecht, The Netherlands.

Matthysen, E., F. Adriaensen, and A. A. Dhondt. 2001. Local recruitment of great and blue tits (*Parus major, P. caeruleus*) in relation to study plot size and degree of isolation. Ecography 24:33-42.

McArdle, B. H. 1989. Bird population densities. Nature 338:628.

Méndez, V., D. Campos, and F. Bartumeus. 2014. Stochastic foundations in movement ecology. Springer, Heidelberg.

Mendl, M., K. Laughlin, and D. Hitchcock. 1997. Pigs in space: spatial memory and its susceptibility to interference. Anim. Behav. 54:1491-1508.

Merkle, J. A., D. Fortin, and J. M. Morales. 2014. A memory-based foraging tactic reveals an adaptive mechanism for restricted space use. Ecology Letters Doi: 10.1111/ele.12294.

Merkle, T. and R. Wehner. 2009. Repeated training does not improve the path integrator in desert ants. Behav. Ecol. Sociobiol. 63:391-402.

Moorcroft, P. R. 2012. Mechanistic approaches to understanding and predicting mammalian space use: recent advances, future directions. Journal of Mammalogy 93:903-916.

Moorcroft, P. R. and A. Barnett. 2008. Mechanistic home range models and resource selection analysis: a reconciliation and unification. Ecology 89:1112-1119.

Moorcroft, P. R. and M. A. Lewis. 2006. Mechanistic home range analysis. Princeton University Press, Princeton.

Moran, P. A. P. 1953. The statistical analysis of the Canadian lynx cycle. II. Synchronization and meteorology. Australian Journal of Zoology 1:291-298.

Morellet, N., C. Bonenfant, L. Börger, F. Ossi, F. Cagnacci, M. Heurich, P. Kjellander, J. D. Linnell, S. Nicoloso, P. Sustr, F. Urbano, and A. Mysterud. 2013. Seasonality, weather and climate effect home range size in roe deer across a wide latitudinal gradient within Europe. Journal of Animal Ecology 82:1326-1339.

Morris, R. G. M. 1981. Spatial localisation does not require the presence of local cues. Learn Motivation 12:239-260.

Mueller, T. and W. F. Fagan. 2008. Search and navigation in dynamic environments - from individual behaviors to population distributions. Oikos 117:654-664.

Mysterud, A. 1999. Seasonal migration pattern and home range of roe deer (*Capreolus capreolus*) in an altitudinal gradient in southern Norway. Journal of Zoology 247:479-486.

Mysterud, A., Y. Rekdal, L. E. Loe, M. Angeloff, R. Mobæk, Ø. Holand, and G.-H. Strand. 2014. Evaluation of landscape-level grazing capacity for domestic sheep in alpine rangelands. Rangeland Ecology & Management 67:132-144.

Nabe-Nielsen, J., J. Tougaard, J. Teilmann, K. Lucke, and M. C. Forckhammer. 2013. How a simple adaptive foraging strategy can lead to emergent home ranges and increased food intake. Oikos 122:1307-1316.

Nathan, R., W. M. Getz, E. Revilla, M. Holyoak, R. Kadmon, D. Saltz, and P. E. Smouse. 2008. A movement ecology paradigm for unifying organismal movement research. Proc. Natl. Acad. Sci. USA 105:19052-19059.

Nee, S., R. M. May, and M. P. Hassel. 1997. Two-species metapopulation models. Pages 123-147 *in* I. Hanski and M. E. Gilpin, editors. Metapopulation biology: ecology, genetics and evolution. Academic Press, San Diego.

Nicholson, A. J. 1954. An outline of the dynamics of animal populations. Australian Journal of Zoology 2:9-65.

Nielsen, C. O. 1954. Studies on Enchytraeidae. 3. The micro-distribution of Enchytraeidae. Oikos 5:000-000.

Norton, D. A. and J. M. Lord. 1990. On the use of "grain size" in ecology. Functional Ecology 4:719.

O'Connor, R. J. 2002. The conceptual basis of species distribution modeling: time for a paradigm shift? Pages 25-33 *in* J. M. Scott, P. J. Heglund, M. L. Mossison, J. B. Haufler, M. G. Raphael, W. A. Wall, and F. B. Samson, editors. Predicting species occurencies. Issues of accuracy and scale. Island Press, Washington.

O'Neill, R. V., D. L. DeAngelis, J. B. Wade, and T. F. H. Allen, editors. 1986. A Hierarchical Concept of Ecosystems. Princeton University Press, Princeton.

Okubo, A. 1980. Diffusion and Ecological Problems: Mathematical models. Springer, Germany.

Ostfeld, R. S. and R. H. Manson. 1996. Long-distance homing in meadow voles, *Microtus pennsylvanicus*. J. Mammal. 77:870-873.

Ovaskainen, O. and S. J. Cornell. 2006. Space and stochasticity in population dynamics. Proceedings of the National Academy of Sciences of the USA 103:12781-12786.

Pasternak, Z., F. Bartumeus, and F. W. Grasso. 2009. Lévy-taxis: a novel search strategy for finding odor plumes in turbulent flow-dominated environments. J. Phys. A: Math. Theor. 42:1-13.

Petrovskii, S., A. Mashanova, and V. A. A. Jansen. 2011. Variation in individual walking behavior creates the impression of a Levy flight. Proc. Natl. Acad. Sci. USA 108:8704-8707.

Pimm, S. L. 1991. The balance of nature? Ecological issues in the conservation of species and communities. The University of Chicago Press, Chicago.

Pimm, S. L. and A. Redfearn. 1988. The variability of animal populations. Nature 334:613-614.

Piper, W. H. 2011. Making habitat selection more "familiar": a review. Behav. Ecol. Sociobiol. 65:1329-1351.

Plank, M. J. and E. A. Codling. 2009. Sampling rate and misidentification of Lévy and non-Lévy movement paths. Ecology 90:3546-3553.

Plank, M. J. and E. A. Codling. 2011. Sampling rate and misidentification of Lévy and non-Lévy movement paths :reply. Ecology 92:1701-1702.

Plank, M. J. and A. James. 2008. Optimal foraging: Lévy pattern or process? Journal of the Royal Society Interface 5:1077-1086.

Pokorny, T., D. Loose, G. Dyker, J. J. G. Quezada-Euán, and T. Eltz. 2014. Dispersal ability of male orchid bees and direct evidence for long-range flights. Apidologie DOI: 10.1007/s13592-014-0317-y.

Porter, J. H. and J. L. Dooley. 1993. Animal dispersal patterns: a reassessment of simple mathematical models. Ecology 74:2436-2443.

Powell, R. A. 2000. Animal home ranges and territories and home range estimators. Pages 65-110 *in* L. Boitani and D. W. Fuller, editors. Research Techniques in Animal Ecology: Controversies and Consequences. Columbia University press, New york.

Pulliam, H. R. and B. J. Danielson. 1991. Sources, sinks, and habitat selection: a landscape perspective on population dynamics. American Naturalist 137:S50-S66.

Pyke, G. H. 2014. Understandingmovements of organisms: it's time to abandon the Lévy foraging hypothesis. Methods in Ecology and Evolution 6:1-16.

Pärt, T. 1995. The importance of local familiarity and search costs for age- and sex-biased philopatry in the collared flycatcher. Animal Behaviour 49:1029-1038.

Pöysä, H., J. Elmberg, K. Sjöberg, and P. Nummi. 1998. Habitat selection rules in breeding mallards (*Anas platyrhynchos*): a test of two competing hypotheses. Oecologia 114:283-287.

Quaglietta, L., V. C. Fonseca, A. Mira, and L. Boitani. 2014. Sociospatial organization of a solitary carnivore, the Eurasian otter (Lutra lutra). Journal of Mammalogy 95:140-150.

Ranta, E., P. Lundberg, and V. Kaitala. 2006. Ecology of populations. Cambridge University Press.

Revilla, E. and T. Wiegand. 2008. Individual movement behavior, matrix heterogeneity, and the dynamics of spatially structured populations. Proceedings of the National Academy of Sciences of the USA 105:19120-19125.

Reynolds, A. 2008. How many animals really do the Lévy walk? Comment. Ecology 89:2347-2351.

Reynolds, A. 2012. Olfactory search behaviour in the wandering albatross is predicted to give rise to Lévy flight movement patterns. Anim. Behav. 83:1225-1229.

Reynolds, A. M. 2010. Animals That Randomly Reorient at Cues Left by Correlated Random Walkers Do the Lévy Walk. The American Naturalist 175:607-613.

Reynolds, A. M. and C. J. Rhodes. 2009. The Lévy flight paradigm: random search patterns and mechanisms. Ecology 90:877-887.

Rivrud, I. M., L. E. Loe, and A. Mysterud. 2010. How does local weather predict red deer home range size at different temporal scales? Journal of Animal Ecology 79:1280-1295.

Rodriguez, F., E. Duran, J. P. Vargas, B. Torres, and C. Salas. 1994. Performance of goldfish trained in allocentric and egocentric maze procedures suggests the presence of a cognitive mapping system in fishes. Animal Learning & Behavior 22:409-420.

Rothrock, D. A. and A. S. Thorndike. 1980. Geometric properties of the underside of sea ice. Journal of Geophysical Research 85:3955-3963.

Salt, G. and F. S. J. Hollick. 1946. Studies of wireworm populations. II. Spatial distribution. J. Exp. Biol. 23:1-46.

Schick, R. S., S. R. Loarie, F. Colchero, D. B. Best, A. Baustany, D. A. Conde, P. N. Halpin, L. N. Joppa, C. M. McClellan, and J. S. Clark. 2008. Understanding movement data and movement processes: current and emerging directions. Ecology Letters 11:1338-1350.

Schieck, J. O. and S. J. Hannon. 1989. Breeding site fidelity in willow ptarmigan: the influence of previous reproductive success and familiarity with partner and territory. Oecologia 81:465-472.

Schoener, T. W. 1971. Theory of feeding strategies. Ann. Rev. Ecol. Syst. 2:369-403.

Schroeder, M. 1991. Fractals, Chaos, Power Laws - Minutes from an Infinite Paradise. W. H. Freedman and Company, New York.

Schuler, K. L., G. M. Schroeder, J. A. jenks, and J. G. Kie. 2014. Ad hoc smoothing parameter performance in kernel estimates of GPS-derived home ranges. Wildlife Biology 20:259-266.

Senft, R. L., M. B. Coughenour, D. W. Bailey, L. R. Rittenhouse, O. E. Sala, and D. M. Swift. 1987. Large herbivore foraging and ecological hierarchies. Bioscience 37:789-799.

Seuront, L. 2010. Fractals and multi-fractals in ecology and aquatic science. CRC Press, Boca Raton.

Sherry, D. F. 1996. Middle-scale navigation: the vertebrate case. J. Exp. Biol. 199:163-164.

Sims, D. W. and N. E. Humphries. 2012. Lévy flight search patterns of marine predators not questioned: a reply to Edwards et al. arXiv 1210.2288:[q-bio.PE].

Sims, D. W., N. E. Humphries, R. W. Bradford, and B. D. Bruce. 2012. Lévy flight and Brownian search patterns of a free-ranging predator reflect different prey field characteristics Journal of Animal Ecology 81:432-442.

Sims, D. W., E. J. Southall, N. E. Humphries, G. C. Hays, C. J. A. Bradshaw, J. W. Pitchford, A. James, M. Z. Ahmed, A. S. Brierley, M.

A. Hindell, D. Morritt, M. K. Musyl, D. Righton, E. L. C. Shepard, V. J. Wearmouth, R. P. Wilson, M. J. Witt, and J. D. Metcalfe. 2008. Scaling laws of marine predator search behaviour. Nature 451:1098-1103.

Sims, D. W., M. J. Witt, A. J. Richardson, E. J. Southhall, and J. D. Metcalfe. 2006. Encounter success of free-ranging marine predator movements across a dynamic prey landscape. Proc. R. Soc. B 273:1195-1201.

Skellam, J. G. 1951. Random dispersal in theoretical populations. Biometrika 38:196-218.

Sklar, L. 1993. Physics and chance. Philosophical issues in the foundations of statistical mechanics. Cambridge University Press, Cambridge.

Smith, A. T. and M. M. Peacock. 1990. Conspecific attraction and the determination of metapopulation colonization rates. Conservation Biology 4:320-323.

Smouse, P. E., S. Focardi, P. R. Moorcroft, J. K. Kie, J. D. Forester, and J. M. Morales. 2010. Stochastic modelling of animal movement. Phil. Trans. R. Soc. B 365:2201-2211.

Solé, R. W. and J. Bascompte. 2006. Self-organization in complex ecosystems. Princeton University Press, Princeton.

Solstad, T., N. Boccara, E. Kropff, M. Moser, and E. I. Moser. 2008. Representation of geometric borders in the entorhinal cortex. Science 322:1865-1868.

Song, C., T. Koren, P. Wang, and A.-L. Barabási. 2010. Modelling the scaling properties of human mobility. Nature Physics 6:818-823.

Spencer, W. D. 2012. Home ranges and the value of spatial information. Journal of Mammalogy 93:929-947.

Stamps, J. 1995. Motor learning and the value of familiar space. American Naturalist 146:41-58.

Stamps, J. A. 1988. Conspecific attraction and aggregation in territorial species. American Naturalist 131:329-347.

Stamps, J. A. 1991. The effect of conspecifics on habitat selection in territorial species. Behav. Ecol. Sociobiol. 28:29-36.

Stephens, D. W. and J. R. Krebs. 1986. Foaging theory. Princeton University Press, princeton.

Stephens, P. A. and W. L. Sutherland. 1999. Consequences of the Allee effect for behaviour, ecology and conservation. Trends Ecol. Evol. 14:401-405.

Sugihara, G. and R. M. May. 1990. Applications of fractals in ecology. Trends Ecol. Evol. 5:79-86.

Swihart, R. K. and N. A. Slade. 1985. Influence of sampling interval on estimates of home-range size. Journal of Range Management 49:1019-1025.

Taylor, L. R. 1961. Aggregation, variance and the mean. Nature 189:732-735.

Taylor, L. R. 1986. Synoptic dynamics, migration and the Rothamsted insect survey. J. Anim. Ecol. 55:1-38.

Tilles, P. F. C. and S. Petrovskii. 2015. Statistical mechanics of animal movement: Animals's decision-making can result in superdiffusive spread. Ecological Complexity 22:86-92.

Toffoli, T. 1990. How cheap can mechanics' first principles be? Pages 301-308 *in* W. H. Zurek, editor. Complexity, entropy and the physics of information. Addison-Wesley Publishing Company, Redwood City.

Travis, J. M. J. and D. R. French. 2000. Dispersal functions and spatial models: expanding our dispersal toolbox. Ecology Letters 3:163-165.

Turchin, P. 1991. Translating foraging movements in heterogeneous environments into the spatial distribution of foragers. Ecology 72:1253-1266.

Turchin, P. 1996. Fractal analyses of animal movement: a critique. Ecology 77:2086-2090.

Turchin, P. 1998. Quantitative analysis of animal movement: measuring and modeling population redistribution in animals and plants. Sinauer Associates, Sunderland, Massachusetts.

Turchin, P. 2003. Complex Population Dynamics: A Theoretical/Empirical Synthesis. Princeton University Press, Princeton.

van Moorter, B., D. Visscher, S. Benhamou, L. Börger, M. S. Boyce, and J.-M. Gaillard. 2009. Memory keeps you at home: a mechanistic model for home range emergence. Oikos 118:641-652.

Vergara, P. M., S. Saura, C. G. Pérez-Hernández, and G. E. Soto. 2015. Hierarchical spatial decisions in fragmented landscapes: Modeling theforaging movements of woodpeckers. Ecological Modelling 300:114-122.

Viswanathan, G. M. 2010. Fish in Lévy-flight foraging. Nature 465:1018-1019.

Viswanathan, G. M., V. Afanasyev, S. V. Buldyrev, S. Havlin, M. G. E. da Luz, E. P. Raposo, and H. E. Stanley. 2000. Lévy flights in random searches. Physica A 282:1-12.

Viswanathan, G. M., S. V. Buldyrev, S. Havlin, M. G. E. d. Luz, E. P. Raposo, and H. E. Stanley. 1999. Optimizing the success of random searches. Nature 401:911-914.

Viswanathan, G. M., M. G. E. da Luz, E. P. Raposo, and H. E. Stanley. 2011. The Physics of Foraging: An Introduction to Random Searches and Biological Encounters. Cambridge University Press, Cambridge.
Voss, R. F. 1989. Random fractals: self-affinity in noise, music, mountains, and clouds. Physica D: Nonlinear phenomena 38:362-371.
Välimäki, P. and J. Itämies. 2003. Migration of the clouded Apollo butterfly *Parnassius mnemosyne* in a network of suitable habitats - effects of patch characteristics. Ecography 26:679-691.
Walters, C. 1992. Trends in applied ecological modelling. Pages 117-122 *in* D. R. McCullough and R. H. Barrett, editors. Wildlife 2001: Populations. Elsevier Applied Science, London.
Ward, D. and D. Saltz. 1994. Foraging at different spatial scales: dorcas gazelles foraging for lillies in the Negev desert. Ecology 75:48-58.
Webb, S. L., S. K. Riffell, K. L. Gee, and S. Demarais. 2009. Using fractal analyses to characterize movement paths of white-tailed deer and response to spatial scale. Journal of Mammalogy 90:1210-1217.
West, B. J. and M. Shlesinger. 1990. The noise in natual phenomena. American Scientist 78:40-45.
White, K. A. J. and R. A. Garrot. 1990. Analysis of wildlife radiotracking data. Academic Press, San Diego.
Wiegert, R. G. 1993. Compartment models.*in* J. C. Fry, editor. Biological data analysis: a practiacal approach. IRL Press, Oxford.
Wiens, J. A. 1989. Spatial scaling in ecology. Functional Ecology 3:385-397.
Wiens, J. A. 1990. On the use of "grain" and "grain size" in ecology. Functional Ecology 4:720.
Wiens, J. A. 1995. Landscape mosaics and ecological theory. Pages 1-26 *in* L. Hansson, L. Fahrig, and G. Merriam, editors. Mosaic landscapes and ecological processes. Chapman & Hall, London.
Wiens, J. A. 1997. Metapopulation dynamics and landscape ecology.*in* I. Hanski and M. E. Gilpin, editors. Metapopulation biology. Ecology, genetics, and evolution. Academic Press, San Diego.
With, K. A. and A. W. King. 1997. The use and misuse of neutral landscape models in ecology. Oikos 79:219-229.
Wohlgemuth, S., B. Ronacher, and R. Wehner. 2001. Ant odometry in the third dimesion. Nature 411:795-798.
Wolf, M., J. L. Frair, E. Merrill, and P. Turchin. 2009. The attraction of the known: the importance of spatial familiarity in habitat selection in wapiti *Cervus elaphus*. Ecography 32:401-410.
Worton, B. J. 1989. Kernel methods for estimating the utilization distribution in home-range studies. Ecology 70:164-168.

Yasukawa, K. and W. A. Searcy. 1985. Song repetoires and density assessment in red-winged blackbirds: further tests of the Beau Geste hypothesis. Behav. Ecol. Sociobiol. 16:171-175.

Yodzis, P. 1989. Introduction to theoretical ecology. Harper & Row, Publishers, New York.

Zaburdaev, V., S. Denisov, and J. Klafter. 2014. Lévy walks. arXiv 1410.5100v1:1-50.

SUBJECT INDEX